Semiconductors (Second Edition)

Bonds and bands

Semiconductors (Second Edition)

Bonds and bands

David K Ferry

School of Electrical, Computer, and Energy Engineering, Arizona State University, USA

IOP Publishing, Bristol, UK

Media content for this book is available from https://iopscience.iop.org/book/978-0-7503-2480-9.

ISBN   978-0-7503-2480-9 (ebook)
ISBN   978-0-7503-2478-6 (print)
ISBN   978-0-7503-2479-3 (mobi)

DOI   10.1088/978-0-7503-2480-9

Version: 20191101

IOP ebooks

British Library Cataloguing-in-Publication Data: A catalogue record for this book is available from the British Library.

Published by IOP Publishing, wholly owned by The Institute of Physics, London

IOP Publishing, Temple Circus, Temple Way, Bristol, BS1 6HG, UK

US Office: IOP Publishing, Inc., 190 North Independence Mall West, Suite 601, Philadelphia, PA 19106, USA

Contents

Preface to the second edition

Semiconductors are not metals, nor are they insulators. Most semiconductors are described by their covalent bonds and tetrahedral coordination. The covalent bond is, of course, shared with organic compounds. But, it is a massively different type of bonding than found in metals. And, they have become important as the basis of the integrated circuit and the micro- and nano-electronic revolution of the past 70+ years.

The rationale for the use of semiconductors lies entirely in the basic property that, at low temperatures and in their pure form, they are insulators, since the nature of the crystal structure and the covalent bond means that the bonding valence band will be completely occupied (all the available states are full) while the anti-bonding conduction band is completely empty. This means that the semiconductor is an empty canvas upon which industry can paint whatever properties are desired with the controlled introduction of various impurities that have an extra electron, or a lack of an electron, as defined by the tetrahedral coordination. Hence, the semiconductor is a base upon which an enormous number of different electronic devices can be created.

The first edition of this book was relatively successful, in that the publisher says that it was one of the dozen or so top downloaded books in their portfolio. With that as an underlying endorsement, it seems timely to undertake a second edition. There are many reasons for this, one of which is that there is substantial material related to the chapters of the book, and that was included in the 1991 book, that has been omitted. This omission was more for the purpose of a compact book that could be used in a one semester graduate course on the basic properties of semiconductors. In this revision, a great deal of this material, related strongly to the material actually included in the first edition, has been re-introduced, and of course updated. While giving additional chapters and a longer product, most readers will find this material helpful and a better, and more complete, result will also help them. It is clear that the book can benefit from adding the material that was omitted, and this has been done, particularly in the individual chapters, but also with the additional chapter on statistics, as it is the carrier statistics that largely dominate the behavior of the final canvas developed by industry. The ability to embed transistors within the base material is, of course, the entire idea of the integrated circuit, and understanding these statistics is at the heart of the design process for such circuits. From the first transistor, the continued drive to add more and more transistors to each individual chip followed an economic trend known in the industry as Moore's law. Many believe that this 'law' is a pattern for continued down-scaling of transistor size. In reality, however, it is a cost curve which implies that the more functions (necessary for, e.g., a complete computer) that are placed upon an individual integrated circuit 'chip,' the cheaper each function will be. Thus, the power of a 1970 Cray supercomputer is now only a small fraction of that in the latest edition micro-processor in a modern smart phone. Needless to say, this smart phone cost a miniscule fraction of what that Cray cost.

Then, the revisions have also addressed the problem that the original chapters gave the theory at a relatively meaningful level, but left out sufficient examples to provide the insight to understand what the theory was telling us. Also, some advanced topics could have given more insight, and these have been included in this version. Hence, the individual chapters have been augmented with addition material to help provide that understanding. Hopefully, the user will appreciate this additional material and its usefulness to their own application world.

Finally, I have to thank the people who contributed their time to reading the new chapters and providing some feedback to me. Thanks goes especially to Steve Goodnick, Josef Weinbub, Jon Bird and Dragica Vasileska for their help in this.

Preface to the first edition

This book is an outgrowth of a portion of the 1991 book *Semiconductors*, which I wrote many years earlier, and is now totally out of print. Nevertheless, we have continued to use the material as a textbook for a graduate course on the electronic properties of semiconductors. It is important to note that semiconductors are quite different from either metals or insulators, and their importance lies in the base they provide to a massive microelectronics and optics community and industry. In this book, the electronic band structure, the lattice dynamics, and the electron–phonon interactions, which provide the basis for the electronic transport that is important, particularly for semiconductor devices. As noted, this material covers the topics that we teach in a first year graduate course.

David K Ferry

Author biography

David K Ferry

David K Ferry is Regents' Professor Emeritus in the School of Electrical, Computer, and Energy Engineering at Arizona State University. He was also graduate faculty in the Department of Physics and the Materials Science and Engineering program at ASU, as well as Visiting Professor at Chiba University in Japan. He came to ASU in 1983 following shorter stints at Texas Tech University, the Office of Naval Research, and Colorado State University. In the distant past, he received his doctorate from the University of Texas, Austin, and spent a postdoctoral period at the University of Vienna, Austria. He enjoyed teaching (which he refers to as 'warping young minds') and continues active research. The latter is focused on semiconductors, particularly as they apply to nanotechnology and integrated circuits, as well as quantum effects in devices. In 1999, he received the Cledo Brunetti Award from the Institute of Electrical and Electronics Engineers, and is a Fellow of this group as well as the American Physical Society and the Institute of Physics (UK). He has been a Tennessee Squire since 1971 and an Admiral in the Texas Navy since 1973. He is the author, co-author, or editor of some 40 books and about 900 refereed scientific contributions. More about him can be found on his home pages http://ferry.faculty.asu.edu/ and http://dferry.net/.

IOP Publishing

Semiconductors (Second Edition)
Bonds and bands
David K Ferry

Chapter 1

Introduction

As we settle into the third decade of the twenty-first century, it is generally clear to us in the science and technology community that the advances that micro-electronics has allowed have truly revolutionized our normal day to day lifestyle. This process began in the middle of the last century with what blossomed into the *information revolution*, or *The Micro Millennium* as it also has been called [1], but it rapidly expanded to impact every aspect of our life today. There is no obvious end to this growth or to the impact which it continues to make in our everyday life.

While this process began with the arrival of the transistor, followed soon by the so-called microprocessor, the continued growth of first microelectronics, and then nano-electronics, has fueled this information revolution. From the microprocessor came the exponential increase in computing power which truly drives the modern lifestyle. While most people remain unaware of much of the actual computer power which enables their lifestyle, they nevertheless rely upon it even for such a simple act as turning on the morning source of news, or checking the current time. (As I stand in the kitchen on a typical morning, there are almost a dozen digital clocks within reach, many of which are kept current by the Wi-fi network. These range from mobile phones and tablets to stoves, microwave ovens, and even coffee brewers, all of which have embedded microprocessors.) When the first Cray supercomputer arrived (in the last century), it had an amazing amount of computing power, which is, of course, why it was called a supercomputer. Yet, today, the typical cell phone has far more computing power than that Cray.

This growth in computing power has been driven, and in turn is calibrated, by the growth in the density of transistors on a single integrated circuit, and this has come to be described by Moore's Law. This growth is shown in figure 1.1 for several manufacturers (the single point for MOS is included as this was the prominent 6502 microprocessor that powered the early Apple personal computers). Moore's Law is not a physical law, but an empirical law coming from economics rather than physical science. Transistors and computer components tend to be laid out on the

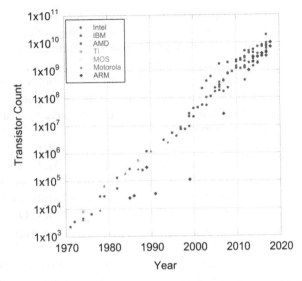

Figure 1.1. The increase in transistors on a given chip by year for several manufacturers.

microchip in a planar, rectilinear fashion much like the houses in any block of a modern southwestern city. This allows their length scale to be reduced quadratically (linearly with area) with any reduction in a critical dimension. The main technology utilizes silicon as the semiconductor of choice. This is important, as one finds that it is the cost of silicon area on the chip that drives Moore's Law. Originally, the rule was composed of three parts: (i) reducing the transistor size, (ii) increasing the size of the microchip itself (no longer followed due to heat dissipation problems), and (iii) circuit cleverness, by which the number of transistors needed to perform a function is reduced, thus saving area [2]. Originally, the microcomputers contained a great many peripheral chips and these were laid out on a large circuit board. As the transistors were shrunk, more of these peripheral chips could be included on the main processor chip. Since the basic cost of the main chip has not changed much over the intervening decades, this means that the cost per functional block has been greatly reduced. Thus, the cost of computing power has exponentially gone down. It is this economic argument that drives Moore's Law. And, as we near the limiting dimensions for silicon transistors, it remains the economics that determines what will, or will not survive, although it is absolutely clear that nano-electronics is not going away any time soon, regardless of any arguments about the end of Moore's Law. Such arguments are pointless, as economics will always be a driver for the high-tech industry, and thus Moore's Law *in some form* will survive as well.

Considering that the first transistor appeared only at the middle of the last century, it is remarkable that today tens of billions of transistors appear on a single chip, whose size is roughly 1–2 cm^2, and this single chip may contain a great many actual processors, or *cores*. As mentioned, the cornerstone of this technology is silicon, a simple semiconductor material whose properties can be modified almost at will by proper materials processing, and which has a stable insulating oxide, SiO_2.

But, Si has come to be supplemented by many important new materials for specialized applications, particularly in infrared imaging, microwave communications, and optical technology. The ability to grow one material upon another has led to artificial superlattices and heterostructures, which mix disparate semiconducting compounds to produce structures in which the primary property, the band gap, has been engineered for special values suitable to the particular application. What makes this all possible is that semiconductors quite generally, have very similar properties across a wide range of possible materials. This follows from the fact that nearly all of the useful materials mentioned here have a single-crystal structure, the zinc-blende lattice, or its more common diamond simplification. And, the atoms of these semiconductors are predominantly tetrahedrally coordinated in the structure, so that each atom has four nearest neighbors. It is this tetrahedral, covalent bonding of the semiconductors that produces the comparable properties. Of course, there are unique other materials, such as the single layered compound graphene, which is not even a three-dimensional material. Nevertheless, it remains true that the wide range of properties found in semiconductors arise from very small changes in the basic positions and properties of the individual atoms, yet the over-riding observation is that these materials are dominated by their similarities.

Semiconductors were discovered by Michael Faraday in 1833 [3], but most people suggest that they became usable when the first metal-semiconductor junction device was created [4]. The behavior of these latter devices was not explained until several decades later, and many suggestions for actual transistors and field-effect devices came rapidly after the actual discovery of the first (junction) transistor at Bell labs [5]. Until a few years ago, the study of transport in semiconductors and the operation of the semiconductor devices made from them could be covered in reasonable detail with simple quasi-one-dimensional device models and simple transport based upon just the mobility and diffusion coefficients in the materials. This is no longer the case, and a great deal of effort has been expended in attempting to understand just when these simple models fail and what must be done to replace them. Today, we find full-band, ensemble Monte Carlo transport being used in both commercial and research simulation tools. Here, by full-band, we mean that the entire band structure for the electrons and holes is simulated throughout the Brillouin zone, as the carriers can sample extensive regions of this Brillouin zone under the high-electric fields that appear in nanoscaled devices. The ensemble Monte Carlo technique addresses the exact solution of transport by a particle-based representation of the Boltzmann transport equation. Needless to say, the success of these simulation packages relies upon a full understanding of the electronic band structure, the vibrational nature of the lattice dynamics—the phonons, and the manner in which the interactions between the electrons and the phonons vary with momentum and energy throughout the Brillouin zone. Hence, we arrive at the purpose of this book, which is to address these topics that are relevant and needed to carry out the creation of the simulation packages mentioned above. (Now, it is worth noting that it is assumed that the reader has a basic knowledge of crystal structure and the Brillouin zone, and is familiar with quantum mechanics.) But, even this is not sufficient these days to do the simulation of a semiconductor device, as one must start at the beginning and do a lot of work before we ever get to the device and transport simulation. We discuss this in the next section, as it is a massive, multi-scale problem of which the device model or simulation is only a small part.

1.1 Multi-scale modeling in semiconductors

For a great many years, semiconductor devices were modeled with simple approaches based upon the gradual channel approximation, and with the use of simple drift mobility and diffusion constants to treat the transport. In fact, this is still the basis upon which the basic theory of semiconductor devices is taught in undergraduate classrooms. Indeed, with proper short-channel corrections and the inclusion of velocity saturation, relatively good results still can be obtained. However, the small size of common devices such as the MOSFET that are found in modern circuits are difficult to handle in this manner. There are many reasons for this, most of which have to do with the small physical size of the small device. In today's such devices, a critical dimension may be only a very few nanometers. In Si, the distance between two neighboring atoms is 0.235 nanometer, so a distance of a very few nanometers does not contain a whole lot of atoms! As a consequence, we are not talking about bulk materials when we begin to describe a modern semi-conductor device. In fact, this device is primarily a number of interfaces between dissimilar materials, although there remains a core of Si. Today, it is recognized that modeling begins with atomistic modeling applied to parts of the device, if not the entire device. The atomistic approach begins with the full band structure mentioned above, but must extend to the multi-material structures that appear within the device.

Moreover, the atomistic modeling actually begins before the device. Devices are quite like specialized modern buildings, only much smaller. When the building is designed, the architect produces the design drawings. However, in creating the building itself, changes occur, and, at the end, one hopes to have an accurate set of 'as built' drawings of the building. In our semiconductor system, there similarly is a design for the transistors, but what is important is exactly what is actually built. It is the 'as built' transistors that must be analyzed and modeled to be able to then extract the parameters that are necessary to simulate the circuit (and the entire chip) and its performance. The device model is at a different level of detail than the atomistic model. And, no one would ever (at the present time) simulate the entire chip with an atomistic model. So, the circuit performance and the chip performance must be modeled at an even higher level in the hierarchy of simulation (see figure 1.2). We will describe some of these levels of simulation in the next section, where we will discuss the fabrication of a reasonably small planar MOSFET.

It is important to note that these various levels of simulation do not apply to just a single part of the problem, such as the transport of carriers in the device. They point to more systematic modeling, that includes solution of the actual electrostatics via the Poisson equation *and* the transport of the carriers via e.g. a Monte Carlo simulation. Even so, in some cases one can obtain very good agreement with modeling tools that couple the Poisson equation to relatively simple transport

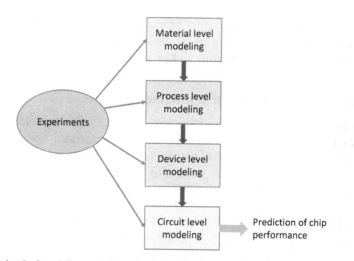

Figure 1.2. The level of modeling and simulation necessary in today's semiconductor world encompasses a multiple level of detail. This multi-scale modeling addresses everything from the basic material, such as silicon, to the final chip containing more than 10^{10} individual transistors.

models for the delay time (switching speed) and the product of energy dissipation and delay time. But, for careful study of the details of the device physics, such as the result of strain on effective mass and mobility and the effect of tunneling through gate oxides, more complicated approaches are required.

Numerical simulations are generally regarded as part of any device physicist's tools and are routinely used in several typical cases: (1) when the device transport is nonlinear and the appropriate differential equations cannot be solved in an exact closed form, (2) as surrogates for laboratory experiments that are either too costly and/or not feasible for initial investigations, to explore the exact physics of new processing approaches, such as the introduction of strained Si in today's CMOS technology, and (3) in computer-aided design, especially at the circuit and chip level. Interestingly, point 2 brings the study of complicated transport into the general realm of *computational science*, which has been termed a third paradigm of scientific investigation, adding to the earlier ones of experiment and theory [6]. Originally, it was thought that this new (at the time) approach amounted to theoretical experimentation, or experimental theory. We now know that it can go beyond just the extension of one or the other original concepts, and is relatively essential to modern semiconductor device design.

Simulation and modeling of semiconductor devices entails a number of factors. To begin, the field has taken on the more formal name of Technology for Computer-Aided Design (TCAD). And it begins really with making the device, where it is necessary to predict the results of the individual fabrication processes in order to know what the 'as built' device looks like. (Again, these processes will be shown by example in the next section.) At the next level, we must take the as-built design and produce a proper model for the devices themselves. And, this model/simulation must be sufficiently good that a variety of circuit level parameters can be extracted from it.

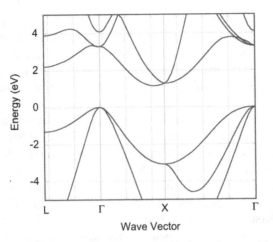

Figure 1.3. The band structure of Si, computed with an empirical pseudo-potential method. The band gap exists in the region from 0 to 1 eV, where no wave states exist.

These parameters must themselves describe the operation of the device in sufficient detail that the circuit performance can be described. At the end of the ladder, we want to know with some confidence that the chip performs properly as the computational engine for which it was designed.

For example, it has long been known that Si devices, at high bias levels, can emit light whose photon energy extends from about 0.4 eV upwards. But, this lower energy was a problem, as the band gap is just over 1.0 eV, which means that these 0.4 eV photons are not conduction band to valence band transitions. While several exotic explanations have appeared, the answer is simpler but also more complicated. In figure 1.3, the lower conduction bands (at the top of the figure) and the upper valence bands are shown. The band gap extends from the top of the valence band at the point Γ to the bottom of the conduction band near the point X. Within the gap, no propagating, wave-like states exist, so that optical transitions from conduction to valence band must have an energy greater than the band gap. Similarly, optical absorption occurs when the photon energy is larger than the band gap. Now, we notice that, at the point labeled X, the lowest conduction connects to a second conduction band. From detailed transport simulations, it is now thought that these low energy photons ($\geqslant 0.4$ eV) are coming from optical transitions from the second conduction band to the first conduction band near X, which is a totally unexpected result. Thus, it is clear that the carriers are getting distributed through large regions of the Brillouin zone, and exist in a great many bands, rather than merely staying around the minima of the conduction band. Thus, the transport itself becomes even more complicated in some sense, if we are to clearly study the detailed physics of scattering in semiconductors and in semiconductor devices. In using the full band structure, e.g. the coupling strength for the electron–phonon interaction can be determined throughout the Brillouin zone, and varies with the momentum state **k** (that is, where the electron actually sits in the Brillouin zone). Approaches, such as the cellular Monte Carlo [7] utilize a scattering formulation based upon the initial and final momentum states and can thus take into account this momentum dependent coupling strength to improve the Monte Carlo approach.

The above example illustrates the fact that a fuller consideration of the entire band structure is needed in modern device simulations. Consideration of the full conduction band in an ensemble Monte Carlo simulation was first done by Hess and Shichijo [8], who dealt with impact ionization in silicon. The approach was adapted then by Fischetti and Laux [9] in developing the *Damocles* simulation package at IBM. Today, such full-band Monte Carlo simulation approaches are available in many universities, as well as from a number of commercial vendors. However, one must still be somewhat careful, as not all full-band approaches are equal and not all Monte Carlo approaches are equivalent. If a rational simulation of the performance and detailed physics of a semiconductor device is to be set up, then it is essential that the user fully understand what is incorporated into the code, and what has been left out. This extends to the band structure, the nature of the lattice vibrations, the details of the electron–phonon interactions, and the details of the transport physics and the methodology by which this physics is incorporated within the code. One cannot simply acquire a code and use it to get meaningful results without understanding its assumptions and its limitations.

1.2 Building a planar MOSFET

In order to fully comprehend the broad range of simulations that have to be carried out in a multi-scale simulation effort, it is useful to actually go through the levels of detail that are encountered in creating a MOSFET, although this is only one type of device. Most of the necessary processing is quite similar in any type of semiconductor device, and this should be kept in mind while reviewing the topic. To do this, we follow a detailed process flow for a MOSFET with an effective 25 nm gate length. Although this process is about a quarter of a century old, much of it is still used in detail today, so it remains surprisingly relevant [10]. As we remarked in the previous section, before we can model the actual device, we have to know what its local structure actually is. The device that has been built may not be the one that was either desired or designed. One would hope that it is quite close, especially if we want to say that it we understand the various processes from beginning to end. The global level of the process involves isolation of the device (from other devices), channel doping, gate oxidation, gate electrode formation, source/drain formation, and metallization, but each of these global steps involves many other steps.

The first step is isolation of the individual device, which begins with oxidizing the bare silicon wafer, particularly in the region where the device is to be located, as shown in figure 1.4. Shown in panel (a) is the oxidized wafer. After this, the sample is coated with a positive photoresist. With a positive photoresist, the molecules are usually long-chain molecules for which the light breaks up these chains. Then the developer dissolves these short chains leaving a hole where the light was projected. Then the oxide can be etched away leaving a hole, as in panel (b). The remaining oxide isolates this device from others. To create the *p*-type layer that will be the channel region, boron is implanted through the hole at an energy of 30 keV and a

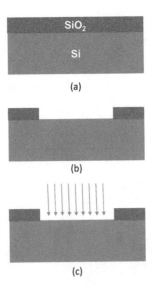

Figure 1.4. Isolation and forming the channel. (a) The oxidized silicon waver. (b) A hole is opened by lithographic techniques (see text). (c) Dopants are implanted into the waver at the desired location.

dose of 3.6×10^{13} cm^{-2}. The energy determines the depth of the implant while the dose will determine the final doping density in the p-type region. Now, we have to correct a problem. When the Si is implanted the energetic ions severely damage the crystal structure of the Si. To repair this damage, the structure must be annealed. But, the annealing proceeds via a regrowth process. In Si, the fast growth direction is the (001) direction, so the wafer we use has to be one in which the surface normal is the (001) direction. If it is any other direction, the fast growth direction will not be normal to the interface, and we will not be able to anneal out all of the damage. This follows as the growth proceeds in the (001) direction, which if it is not the surface normal causes many growth fronts to interfere, leading to dislocations and grain boundaries. In the case under discussion, the annealing is carried out at 1000 C for 2 h. Of course, the implanted atoms will diffuse during the anneal procedure, and this has to be accounted for during the 'design' of the device. The oxide is then regrown over the implanted region and the surface planarized for the next step.

The next step is to form the central gate oxide. Once more a positive photoresist is deposited on the oxidized wafer, and exposed and developed. This time, however, the new hole is smaller than the previous one and must be located in the center of the p-type layer formed with the previous hole. This new hole is shown in figure 1.5(a), and is expanded in panel (b) (red dashed lines indicate the expansion). Once the hole is opened in the resist, the oxide is etched away so that the surface of the p-region is exposed. Now, the gate oxide is grown at 800 °C for 8 min, producing a 3 nm thickness. Over the top of this is deposited/grown polysilicon which will serve as the actual gate material (blue in figure 1.5(b)). This polysilicon is heavily phosphorous dope to make it n-type and to have a low resistivity.

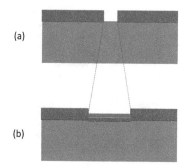

Figure 1.5. (a) The gate region opening is produced in the oxide. (b) The thermal oxide gate is then produced followed by deposition of the heavily doped polysilicon gate.

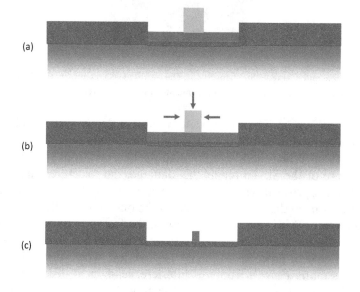

Figure 1.6. (a) The negative photoresist after exposure and development. (b) This resist is then thinned in an oxygen plasma. (c) The resist is used to protect the desired gate material during reactive ion etching.

Now that the gate material is deposited, it is time to pattern it to the actual short gate length desired. This time, a negative photoresist is deposited on the gate material. With a negative photoresist, we desire the exposing light to cross-link the material which is dominantly short molecules. Hence, when developed, the exposed pattern is all that remains. This exposure is typically carried out with an excimer laser, but even this has a wavelength that is too large to make the small gate desired. So, the photoresist that remains, shown in figure 1.6(a) is thinned in an oxygen plasma (figure 1.6(b)). The resist is thinned until it is 40 nm in length in the source-drain direction. Hence this is called a 40 nm *drawn* (lithographically defined) gate. This pattern is then used to protect the desired gate while the rest of the polysilicon is etched away with a reactive ion etching process leaving the desired gate in figure 1.6(c).

It should be remarked that the gate is oriented so that the carriers (in this case, electrons) will move in the (110) direction in the surface layer. From figure 1.3, it may be observed that the minimum of the conduction band lies along the Γ to X line, which is the (100) direction, and lies about 85% of the way to X. Because of the crystal symmetry, there will be six such minima, all of which are equivalent. A constant energy surface near this energy will be an ellipsoid of revolution (a cigar-like shape). The long axis of the ellipsoid is parallel to one of the (100) directions. Hence, in our device, four of these valleys lie in the surface plane, and two of them are oriented with the long axis normal to the plane. Now, the (110) direction makes an equal angle with all four of the ellipsoids that lie in the plane, so that the transport in the device does not see differences in the four ellipsoids. Hence, the transistors are almost always oriented in the (110) direction. This will turn out to be important in the next few section. Further discussion of these ellipsoids, and their masses, will be presented in chapters 2, 4, and 6.

To proceed, we have to face another problem. The natural approach is to now implant the source and drain dopants, in order to convert the p-type layer to a heavily doped n-type layer (creating the n-p-n doping profile down the channel). If we do this directly at this point, using the gate as a mask to keep the implanted ions from the region under the gate, these dopants will diffuse into this masked region during the annealing process after the implant. This will do away with the p-layer entirely and ruin the transistor. We could place a spacer layer on either side of the gate to account for the diffusion but this makes the size of this space hyper-critical to the entire process. However, this is the approach we shall follow, but the spacer will be made much larger than the anticipated diffusion along the surface, and it will be made of a special material. In this case, the spacer is a phospho-silicate glass (PSG). It is deposited, and again patterned by a lithographic process to result in a structure such as that of figure 1.7, where the PSG is shown in green. This will serve two purposes. First, it protects the channel area from diffusing of the implanted atoms. Secondly, it will provide a source of phosphorous atoms that will diffuse out of the PSG during the anneal and these will form a shallow doped layer that extends from the source/drain regions to the channel. These regions are known as source(drain)-extensions. To create the source and drain regions, As atoms are implanted at 30 keV with a dose of 5×10^{15} cm^{-2}. The PSG has a phosphorous concentration of about 3×10^{21} cm^{-3}, and provides a relatively constant source of this dopant during

Figure 1.7. The addition of sidewall spacers, formed from PSG (in green), helps in the transistor fabrication and also helps keep the implanted source/drain atoms from diffusing into the channel regions. The polysilicon gate is the top blue region, while the narrow grey region is the gate oxide.

the anneal. The anneal itself is done by rapid thermal annealing (RTA) at 1000 °C for 5–10 s. The implantation and anneal of the As gives source and drain regions that are doped at approximately 5×10^{19} cm^{-3}, and extend some 70 nm below the surface. The extensions created by the phosphorous out-diffusion from the PSG, are doped much higher and extend some few nm from the surface, as shown in figure 1.8, which displays what is essentially the completed transistor, lacking only the metallization layers. One should note that even the shallow phosphorous diffusion lets the atoms penetrate into the region under the gate. This gives the distance between the two (red) extensions as about 25 nm, which is called the *effective* gate length.

Now, it is clear that before we even consider trying to simulate the actual transistor, we have to simulate a large number of processes that are used to create the transistor. We have used deposition, exposure, and stripping of two types of photoresist, oxidation, etching, ion implantation, annealing, material deposition, and so on. If we do not understand these processes in great detail, and cannot simulate their action, we stand no chance to determine the final device structure. Hence, we have already gone through two levels of the multi-scale simulation (figure 1.2), and we haven't even begun with the device. While our focus in this book is on the transport and physics of the device, it is clear we first have to understand the physical processes involved in creating the device.

1.3 Modern modifications

The above prescription for the creation of the short-gate MOSFET involves many processes and design parameters. Yet, almost everything about the device has been changed in recent years to overcome problems that arise from the small physical size. In this section, we will discuss a number of those problems and describe how they have been minimized or overcome. This will show that the planar MOSFET has evolved greatly over the past quarter of a century, even though most of the actual processing techniques are still utilized, but others have been added.

1.3.1 Random dopants

In the previous section, it was noted that the p-layer forming the channel was implanted with a B dose of 3.6×10^{13} cm^{-2}. The effective channel length is about 25 nm, and if we take this dimension as a number for the width (into the page in the figures), we find that the number of B atoms in this width of channel is only 225. In Moore's Law, the reduction of channel area goes directly with the length scaling.

Figure 1.8. The finished transistor. The red regions represent the heavily doped n-type areas that form the source and drain regions of the transistor.

Hence, if we want to reduce the effective gate length by a factor of 2, the area decreases as a factor of 4, and this results in only 56 atoms in the selected area. The problem with this is that the fluctuation in this number is ±7–8 atoms, and this is intolerable for a modern device. The potential under the gate is no longer a smooth potential [11]. Rather, it has peaks near each dopant and valleys in between, and it becomes important to know exactly where each dopant is located. This is because current is controlled in the MOSFET by the potential at the source end of the channel, and this is sensitively affected by the nearby dopants. The fluctuation in the dopants produces a fluctuation in the threshold (turn-on) voltage of the device, and this variation in the threshold is strongly correlated to the distance the dopant is from the source [12]. Indeed, in a 100 micron gate length MOSFET, with a channel doping of 8×10^{17} cm^{-3}, it was found that the threshold voltage for 20 different implementations of the discrete dopants (all giving the same doping level) varied over 100 mV, clearly unacceptable [13].

It is not just the random dopant effect upon the threshold voltage that is a problem. Due to the peaks in the potential near the dopants, the current through the device is no longer homogeneous. Rather, the carriers will follow the paths between the peaks, which leads to the formation of current filaments [14]. The random potential landscape and carrier motion can be seen in figure 1.9 for a similar structure. Transport of carriers with random dopants will be discussed further in chapter 7.

The approach to minimize the role of the random nature of the dopants in the channel has followed a couple of paths. First, there was an attempt to minimize the effect by implanting additional B atoms quite near the gate with a local implantation using a focused ion beam [15], the effect of which has been simulated as well [16]. This approach, of course, requires additional processing, which is undesirable. What was finally adopted was to just ignore any doping in the substrate itself; e.g. no p-type layer, only an intrinsic region under the gate. This had the result of essentially cutting the potential difference between the source and the channel in half. But, since the voltages were being scaled downward as device size was reduced, this was deemed acceptable at the time and the move was being made to smaller devices. Yet, in time, this would lead to a more concerning problem, discussed below.

1.3.2 Roughness at the interface and mass changes

There is an unavoidable interface in the MOSFET, and that is the interface between the gate oxide and the channel. This interface is created during the process of forming the gate oxide. With SiO$_2$, the oxide is thermally grown by oxidizing the bare surface in figure 1.4(a). While this forms an almost atomically abrupt interface, 'almost' isn't good enough. There remains about a 1–2 atom variation along the interface [17–19]. Although small, this is enough to affect the performance of the transistor through the introduction of a scattering process. We will discuss the details of this scattering process in chapter 5. Here, we note that the gate electric field pulls the carriers closer to the surface. As the devices get smaller, they naturally are pulled even closer to the surface, until the scattering is dominated by this surface roughness

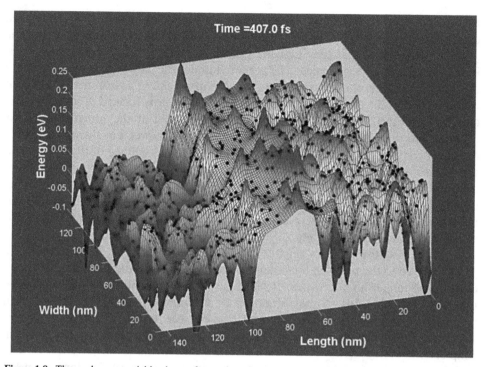

Figure 1.9. The random potential landscape for random dopant in the MOSFET. The drain is to the left, which lies at a lower potential energy corresponding to the drain bias of approximately 150 mV. The small black dots are the carriers, in this case electrons, which move from right to left. The video was produced by Stephen Ramey and is used with his permission. Video available at https://iopscience.iop.org/book/978-0-7503-2480-9.

[17]. As a result of the increased scattering, the mobility of the carriers is reduced which leads to a degradation in performance which is outside the theory of scaling the size of the transistor. As a counter to this effect, the industry introduced strain into the device as a method to increase the mobility [20], through effectively reducing the effective mass in the transport direction.

As we discussed above, there are six equivalent ellipsoids in the conduction band of Si. If uniaxial tensile strain is introduced along the channel, these six ellipsoids respond differently. The four lying in the surface plane are raised in energy relative to the two that are normal to the surface. The individual ellipsoids have an anisotropic mass with a longitudinal (heavy) mass along the major axis of the ellipse and a smaller transverse mass in the other two directions. This means that the two ellipsoids normal to the surface will exhibit this smaller mass, and they will also be preferentially populated as the other four lie at a higher energy. Thus, this tensile strain serves the purpose of changing the effective nature of the conduction band with the result of a smaller conductivity mass, thus improving the mobility. This tensile strain is usually accomplished by putting a silicon nitride layer over the gate stack, so that as the device cools after processing, strain appears along the channel [20].

However, for the valence band transport in the p-channel devices, we require a different approach. One can note in figure 1.3 that the top of the valence band is triply degenerate. One of these bands is split-off by the spin–orbit interaction, and the remaining two are termed the heavy hole and light hole bands (this and the mass are discussed in the next chapter). These bands are warped, which means that they are anisotropic which gives a very complex mass (discussed in chapter 2). However, if the channel is uniaxially compressively strained along the channel, this warpage is changed and results in a reduced effective mass for the holes in the device [21]. To obtain compressive strain in these p-channel devices, Ge is introduced into the source and drain regions. Germanium has a larger lattice constant, so having SiGe in the source drain requires these regions to expand, which produces the desired compressive strain. This is accomplished by etching away the Si p-layer in the source and drain regions to produce a recessed region. A SiGe compound is then deposited by chemical vapor deposition.

The need for tensile strain on one device of the CMOS pair and compressive strain on the other means that different processing, and more masking steps, are required for the two types of transistors. Nevertheless, the need meant that it had to be made manufacturable, and Intel (followed immediately by others) apparently first introduced the strain technology with the 45 nm gate length technology [20].

1.3.3 New oxides

In figure 1.5(b), it was pointed out that the gate oxide was 3 nm thick. As the device sizes are reduced, this oxide has to get thinner by the same factor in order to preserve the electrostatics of the scaled device. However, when the oxide gets much thinner than 2 nm, one has to begin worrying about tunneling between the channel and gate through this oxide. This is a leakage mechanism that greatly reduces the performance of the device. So for many years, there was a search for a replacement oxide which would have a higher dielectric constant (called a *high-K* material, where K stands for the dielectric constant). The rationale for a new oxide lies in the nature of the gate capacitance

$$C_{\text{gate}} \sim \frac{K}{t_{\text{oxide}}}, \tag{1.1}$$

where t_{oxide} is the thickness of the gate oxide. Thus, if we increase the dielectric constant K, we can increase the thickness of the oxide. The increased thickness will reduce the tunneling likelihood. There are a number of oxides that have these higher dielectric constants (SiO_2 is about 3.7–3.9 depending upon how dense the amorphous oxide is). But, the requirements are that it have the higher K, but also needs to be reliable, stable, and manufacturable with dimensional control over the large area of the wafer being processed. Nevertheless, this was achieved and the technology was developed at Intel (and elsewhere) [22], using HfO_2.

The process was introduced in the 45 nm 'node', where the effective gate length is about 25 nm, in 2007–08. In the process flow, the deposition of the polysilicon gate is considered to be a dummy, in that it will be removed later and replaced with the new oxide. After the source and drains and stressor components have all been introduced, the polysilicon gates and the gate oxides are removed. Then, the HfO_2 is deposited by atomic layer epitaxy, which basically is a set of chemical reactions that leave the desired material layer in place [22, 23]. One problem with this material is that the oxide has a number of polar vibrational modes which cause additional scattering of the channel carriers. Hence, the physical gate is a metal, such as TiN or a properly doped metal, which will screen these modes from the carriers. The use of the metal increases the mobility by almost a factor of 2 over the polysilicon gate, and gives performance comparable to the thin SiO_2 oxide with a polysilicon gate.

1.3.4 New structures

In section 1.3.1, we commented that the reduction of dopant atoms in the channel would decrease the potential barrier between the source and the channel when the transistor was switched off, with the view that the leakage current was still manageable. This was perhaps over-optimistic, as it was not too many generations after this that the leakage current became unreasonable. That is, the problem with the transistors was not the 'on' current, which is affected by the mobility, but the 'off' current, which is attributable to leakage between the source and drain. With the 45 nm gate length generation, the saturation current for a low threshold logic transistor was about 2.5 mA μm^{-1} (gate width, which is a standard normalization). The saturation current was slightly less for a high threshold device. The standard scaling theory says that the saturation current scales as [24]

$$I_{\text{sat}} \sim \frac{V_D^2}{L_G},$$

(1.2)

where V_D is the drain voltage and L_G is the gate length. So, as the size of the transistor is reduced, both the gate length and the drain voltage are comparably reduced, while the saturation current is increased. By 2010, we were approaching the 22 nm node, although the transistor gate length was no longer one-half the node size any more. So, for this node, we take a conservative estimate of 5 mA/μm and a gate width of 0.4 μm, so that we get a saturation current of 2 mA. The on/off ratio was about 3.5×10^3 for that low threshold 45 nm device and perhaps an order of magnitude higher for a high threshold device. Let us consider a value of 10^4 for this ratio, which produces a leakage current of 0.2 nA. In 2019, a typical chip has about 10^{10} transistors/cm^2, and the chip size is typically around 1 cm^2. So, the leakage current for such a chip is about 2 A. If the voltage is 1 V, this is 2 W of dissipation when the chip is idling (doing nothing). However, if the on/off ratio dropped to 10^3, the dissipation would be 20 W, which is clearly intolerable. Hence, at the 22 nm node, a new structure was introduced, in which the transistor was turned vertically so that gates could be applied to each side of the vertical 'fin', as shown in figure 1.10. Such a structure typically has a variety of names, but the most common are FinFET [25] and tri-gate FET [26], where the former usually does not have the gating from the top that is shown in the figure. All of the previous changes in the

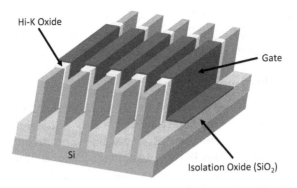

Figure 1.10. The tri-gate MOSFET, where the transistor is now turned vertical. The channel exists on the two sides and top of each fin. The gate runs over the fins and has to be clearly isolated from the main Si body. The source and gate are at the front and back of the figure.

MOSFET are still included here. As a single fin usually carries much less current than a normal logic transistor, multiple fins are connected together to get the desired width for the transistor. While 5 such fins are shown in the figure, more are typically used. The advantage of this structure is that there are gates on both sides of the fin, and this leads to a more secure shut-off of the transistor and thus much lower leakage current. Intel introduced the three dimensional tri-gate transistor in late 2011 with the 22 nm node technology. Other manufacturers quickly followed, as this new technology turned out to offer more than just low leakage current. It also provided a better transistor, with the opportunity to move to higher frequencies as well.

1.3.5 Through the crystal ball

Progress in the MOSFET has continued even since the introduction of the tri-gate technology, but the changes have not been so dramatic as those of the previous decade or so. Nevertheless, there is considerable interest in the next major changes. There is significant speculation that Intel and others might move to new materials for the channel itself. One suggestion is that the n-channel devices might be replaced with a strained $In_{0.7}Ga_{0.3}As$ quantum well channel, while the p-channel devices might be replaced with a Ge channel device. In both cases the newer material offers higher mobilities than the current strained Si, but at the expense of added manufacturing difficulty. In comparisons with strained-Si devices, the quantum well channel has offered similar to slightly better current and transconductance than a strained-Si device of half the gate length [27]. This would mean that a larger transistor can be used for comparable performance. But, one can ask if a factor of two is enough gain to warrant the change, and the expected introduction of the technology continues to be pushed back (as of 2019).

Other changes have been made in manufacturing technology throughout this time period, which has made the cost of making the chips increase at the same time. Many companies have invested in EUV lithography, a form using 13.5 nm light. This revolutionizes the nature of the photoresist, light source, and stepper technology, and is a major transition point. Consequently, it also changes the simulation problems for the exposure and development processes that must be included in the multi-scale modeling that extends from wafer to system.

1.4 What is in this book?

In the preceding sections, we discussed the need for detailed understanding of the physics in the simulation of semiconductors and semiconductor devices and demonstrated the intricate number of levels that are involved in making the transistors. The purpose of this book is to provide the important concepts that go into the simulations, particularly for the semiconductor material itself. These include the electronic band theory, the lattice dynamics, carrier statistics, the understanding of the electron–phonon interaction, and the methods for transport simulation. We can perhaps see how this fits together if we examine the total Hamiltonian for the entire semiconductor crystal:

$$H = H_{el} + H_L + H_{el-L}, \tag{1.3}$$

where the electronic portion is

$$H_{el} = \sum_i \frac{p_i^2}{2m_0} - \sum_{i,j \neq i} \frac{q_i q_j}{4\pi\epsilon_0 x_{ij}}, \tag{1.4}$$

where x_{ij} is the distance between the two charges. In this equation, the first term represents the kinetic energy of the electrons, while the second term represents the Coulomb interaction between the electrons. We have used lower case letters to indicate the electronic variables. Similarly, we can write the lattice portion of the Hamiltonian as

$$H_L = \sum_j \frac{P_j^2}{2M_j} - \sum_{r,s \neq r} \frac{Q_r Q_s}{4\pi\epsilon_0 X_{rs}}, \tag{1.5}$$

where, once again, the first term represents the kinetic energy of the atoms and the second term represents the Coulombic interaction between the atoms. In this equation, we have used capital letters to indicate the coordinates of the atoms. Now, this is kind of a pictorial view, because in semiconductors the net bonding forces between the atoms arises from the covalent bond sharing between the valence electrons. The above equations represent all of the electrons of the atoms and, of course, all of the atoms. We will take a rather different formulation when we treat the electronic structure in the next chapter and the lattice vibrations in the third chapter.

Finally, the interaction term between the electrons and the atoms can be expressed as

$$H_{el-L} = \sum_{i,s} \frac{q_i Q_s}{4\pi\epsilon_0 |x_i - X_s|}. \tag{1.6}$$

As before, we can really expand this into two terms, which can be seen if we rewrite this equation as

$$H_{el-L} = \sum_i q_i V(x_i),$$ (1.7)

where

$$V(x_i) = \sum_s \frac{Q_s}{4\pi\varepsilon_0 |x_i - X_s|}$$ (1.8)

is the potential seen by an electron due to the presence of the atoms. We can get two terms from this by separating the potential as (1) that part due to the exact position of the atoms residing precisely on the central positions, which define the crystal structure (their average positions in that sense), and (2) the motion of the atoms about this position which can perturb the electronic properties of the electrons.

To understand the above separation, it is important to understand that we are just not capable of solving the entire problem. Instead, we invoke the adiabatic approximation, which arises from the recognition that the electrons and the atoms move on different time scales. Thus, when we investigate the electronic motion, we admit that the atoms are moving too slowly to consider, so we treat them as if they are frozen rigidly to the lattice sites defined by the crystal structure. So, when we calculate the energy bands in the next chapter, we ignore the atomic motion and treat their presence as only a rigid shift in the energy. Hence, we can compute the energy bands in a rigid, periodic potential provided by these atoms are stuck in their places. On the other hand, when we examine the interactions of the slowly moving atoms, we consider that the electrons are so fast that they instantaneously follow the atomic motion. Hence, the electrons appropriate to an atom are frozen to it—they adiabatically adjust to the atomic motion. Thus, we can ignore the electronic motion when we study the lattice dynamics in chapter 3. Finally, what is still important to us is the small vibration of the atoms about their average positions that the electrons can actually see. This is a small effect, and therefore is treated by perturbation theory, and this is the electron–phonon interaction that gives us the scattering properties. This is the subject of chapter 5.

After, we compute the energy bands, we need to know how many electrons are in the conduction band at a given temperature, as the bands are normally computed at $T = 0$. This provides the statistics for the carriers, electrons or holes. We also need to know how these statistics are perturbed by the presence of impurity atoms, either donors or acceptors, as these change the number of electrons in the conduction band or the number of holes in the valence band. This will be treated in chapter 4.

Finally, in chapter 6, we discuss both simple transport theory for electrons which remain near the band edges, and can be described by the relaxation time approximation. This allows us to discuss mobility, conductivity, the Hall effect, and other transport concepts. In chapter 7, we move to truly non-equilibrium transport theory and the role of high electric fields which can actually exist in semiconductor devices. Consider, for example, the case of a transistor with a 25 nm effective gate length. If the drain bias is 0.5 V, then the average electric field in the channel is 200 kV cm^{-1}, which is an astonishingly high electric field. In fact, this field is approaching the breakdown field for Si. Thus, while the mobility governs the behavior of the carriers at the source end, the extremely non-equilibrium behavior at these high fields determines the transport at the drain end of the device.

References

[1] Evans C 1979 *The Micro Millennium* (New York: Washington Square Press)
[2] Moore G 1965 *Electronics* **5** 114
[3] Faraday M 1833 *Experimental Researches in Electricity ser.* **IV** 433–39
[4] Braun F 1874 *Ann. Phys. Pogg.* **153** 556
[5] Bardeen J and Brattain W 1948 *Phys. Rev.* **74** 23
    Shockley W 1949 *Bell Syst. Tech. J.* **28** 435
[6] Wilson K G 1984 *High Speed Computation* ed J S Kowalik (Berlin: Springer)
    Wilson K G 1984 *Proc. IEEE* **72** 6
[7] Saraniti M, Zandler G, Formicone G, Wigger S and Goodnick S 1988 *Semicond. Sci. Technol.* **13** A177
[8] Shichijo H and Hess K 1981 *Phys. Rev.* B **23** 4197
[9] Fischetti M V and Laux S E 1988 *Phys. Rev.* B **38** 9721
[10] Ono M, Saito M, Yoshitomi T, Fiegna C, Ohguro T and Iwai H 1995 *IEEE Trans. Electron. Dev.* **42** 1822
[11] Wong H S and Taur Y 1993 *Intern. Electron Dev. Mtg. Proc.* (New York: IEEE), 705
[12] Vasileska D, Gross W J and Ferry D K 2000 *Superlatt. Microstruc.* **27** 147
[13] Vasileska D, Gross W J and Ferry D K 1998 *Intern. Workshop Comp. Electron.* (New York: IEEE), 259
[14] Ramey S M and Ferry D K 2004 *Semicond. Sci. Technol.* **19** S238
[15] Vignaud D, Etchin S, Liao K S, Musil C R, Antoniadis D A and Melngalis J 1992 *Appl. Phys. Lett.* **60** 2267
[16] Vasileska D, Knezevic I, Akis R, Ahmed S and Ferry D K 2002 *J. Comp. Electron.* **1** 453
[17] Goodnick S M, Ferry D K, Wilmsen C W, Lilienthal Z, Fathy D and Krivanek O L 1985 *Phys. Rev.* **B32** 8171
[18] Yoshinobu T, Iwamoto A and Iwasaki H 1993 *Proc. 3rd Int. Conf. Sol. State Dev. Mater.* (Japan: Makuhari)
[19] Feenstra R M 1994 *Phys. Rev. Lett.* **72** 2749
[20] Thompson S E *et al* 2004 *IEEE Trans. Electron. Dev.* **51** 1790
[21] Wang E X, Matagne P, Shifren L, Obradovic B, Kotylar R, Cea S, Stettler M and Giles M D 2006 *IEEE Trans. Electron. Dev.* **53** 1840
[22] Chau R, Datta S, Doczy M, Doyle B, Kavalieros J and Metz M 2004 *IEEE Electron. Dev. Lett.* **25** 408
[23] Natarajan S *et al* 2008 *IEEE Intern. Electron Dev. Mtg.* (New York: IEEE)
[24] Baccarani G, Wordeman M R and Dennard R H 1984 *IEEE Trans. Electron. Dev.* **31** 482
[25] Hisamoto D *et al* 2000 *IEEE Trans. Electron. Dev.* **47** 2320
[26] Doyle B S *et al* 2003 *IEEE Electron. Dev. Lett.* **24** 263
[27] Dewey G *et al* 2008 *IEEE Electron. Dev. Lett.* **29** 1094

Chapter 2

Electronic structure

An electron moving through a crystal, in which there is a large number of atomic potentials, will experience a transport behavior significantly different from an electron in free space. Indeed, in the crystal, the electron is subject to a great many quantum mechanical forces and potentials. The point of developing an understanding of the electronic structure is to try to simplify the multitude of forces and potentials into a more condensed form, in which the electron is replaced by a *quasi-particle* with many of the properties of the electron, but with significant differences in these properties. Important among these is the introduction of an *effective* mass, which is representative of the totality of the quantum forces. To understand how this transition is made, we need to first understand the electronic structure of the semiconductor, and that is the task of this chapter.

The calculation of the electronic structure of an atom is quite old, older even than quantum mechanics. The first model of an atom, both planar and non-planar appears to have been done by Nagaoka in 1904 [1]. As this work was known to Rutherford at the time Bohr came to him as a postdoc, it must also have been known to the latter. In fact, it may have affected Rutherford's own work on α particle scattering and his planetary model of the atom [2]. Bohr finished his first paper on his own model of the atom shortly after this [3], but he never formalized a proper three dimensional version. It was left to Sommerfeld to provide the three dimensional mathematical model [4]. Following this time, the study of atomic structure exploded with many groups around the world contributing to its development. Many methods were developed to determine increasingly accurate simulations of the atomic structure, as well as extending the work to condensed matter systems. An excellent introduction to the world as it stood in the 1960s is found in Slater's first volume [5]. Generally, these approaches, as well as all modern approaches, relied upon a central potential as well as Schrödinger's equation. Poisson's equation for the central potential and the wave functions must be solved self-consistently to produce the energies for the electrons. However, problems arise as this potential is not spherically symmetry due to the many-body interactions between the electrons and the nucleus and between the electrons themselves, and therein lies the difficulty even in atomic structure. Ignoring the interaction between the electrons gives rise to the Hartree model, in which the potential is merely the solution to Poisson's equation [6], although his main point is that the potential is non-Coulombic in nature due to the electrons. Fock added the role of the exchange interaction slightly later [7], primarily as the Hartree approach did not contain the proper symmetry of the wave functions when spin is included.

Modern approaches, especially for condensed matter systems, began in 1964 with the Hohenberg–Kohn theorems [8]. The first of the theorems demonstrated that the ground state of a many-electron system was uniquely determined by the three dimensional electron density. The second theorem proved that an energy functional could be written in terms of the density and that the correct ground state electron density minimized this functional. In later work, Kohn and Sham showed that the many-electron problem could be reduced to non-interacting electrons moving in an effective potential [9]. Thus, if one could find the correct effective potential, the electronic structure could be determined by solving the one electron Schrödinger equation. This is the basis upon which all modern band structure theory is based. The density-functional theory thus relies upon two potentials: (1) the sum of all external potentials which is constructed from the crystal structure and the atomic compositions of the materials (e.g., the atomic potentials), and (2) the effective potential that arises from the inter-electronic interactions. At this point, it must be pointed out that the 'electrons' include all of the electrons, bonding electrons and inner shell electrons which are strongly bound to the individual nuclei. Since these strongly bound inner shell electron wave functions do not change much in the computation, it was apparently first suggested by Hellman that they be removed from the potential in a manner similar to equations (1.6) and (1.7) [10]. This inner shell potential is subtracted from the external potential to produce a pseudo-potential; a modified potential that allows the simulation to proceed without need to consider the inner electrons.

But, there are more than one way to determine the electronic structure in condensed matter. One can do *ab initio* calculations, such as those discussed above. These so-called first principles calculations begin with the atomic wave functions and the two potentials and solve everything by self-consistent procedures which yield the 'correct' wave functions, potentials, and energy levels for the electrons. Beginning with the actual atomic wave functions means that we are working in real space—the three dimensional world we live in. And, in many cases, such as for atoms and molecules, this is the proper approach to take. But, in periodic structures, such as crystal lattices, there is another way, and that is to Fourier transform everything into momentum space. This Fourier approach is enabled by the very repetitive nature of the crystal structure. When the crystal is Fourier transformed, the relevant region of momentum space is the Brillouin zone (the unit cell of the reciprocal lattice). Now, the momentum wave functions are mainly the plane waves arising from the various reciprocal lattice vectors. We deal with these planes waves and the Fourier transform of the potentials, again in an *ab initio* self-consistent manner. In semiconductors, however, these *ab initio* approaches have had problems, the most important of which is called the band-gap problem. That is, they give a band gap that is much too small. This problem arises from the fact that the conduction band corresponds to excited states, not the ground state, so that the H-K theorems are not guaranteed to give the excitation energies that correspond to the band gap. In recent times, corrections have been developed, and these will be discussed later in the chapter.

The mere presence of the band gap problem led to an older approach, which is termed the empirical approach. With most semiconductors, the values for the principal points of the energy structure have been measured by a variety of techniques. Hence, the argument goes 'why should I spend a lot of time and effort to find the full spectrum by *ab initio* methods, when I can fit the various values for parameters to the experimental values?' These empirical approaches are used fairly extensively in semiconductor simulations. Again, however, there are both real space and momentum space approaches. In a real space approach, we begin with the nature of the outer shell orbitals that will provide the bonding; hence typically the atomic *s*- and *p*-orbitals in semiconductors. In chemistry, this is termed the linear combination of atomic orbitals (LCAO), while in physics this is usually termed the semi-empirical tight-binding method (SETBM) [11, 12]. Here, the values of the atomic energies for these orbitals and the values of overlap integrals between them (on neighboring atoms) are treated as adjustable parameters used to match the computed bands to the experimental data. The corresponding momentum space method is termed the empirical pseudopotential method (EPM) [13]. Now, the main adjustable parameters are the Fourier coefficients for the transform of the pseudo-potential itself. We will discuss each of these approaches later in this chapter.

First, however, it is necessary to discuss how the presence of the atomic lattice and its periodicity affect the nature of the electronic structure. Then we discuss the manner in which the Bloch functions for the crystal arise from the atomic functions and the bonding in the crystal. This leads us to discuss how the directional hybrid states are formed and these then lead to the bands when the periodicity is invoked. Following that, we will discuss a variety of real-space and momentum-space situations that illustrate the various methods for computing the actual energy bands in the semiconductor. We will then turn to the perturbative spin–orbit interaction to see how spin affects the bands, and then discuss the effective mass approximation. Then, we will discuss how scaling between various semiconductors can be used. We finish the chapter with discussion of heterostructures and alloys between different semiconductors.

2.1 Periodic potentials

In most crystals, the interaction with the nuclei, or lattice atoms, is not negligible. But, the lattice has certain symmetries which the energy structure must also possess. The most important one is periodicity, which is represented in the potential that will be seen by a nearly free electron. Suppose we consider a one-dimensional crystal, which will suffice to illustrate the point, then for any vector L, which is a vector on the lattice, we will have

$$V(x + L) = V(x). \tag{2.1}$$

We say that L is a vector in the lattice. This means that it may be written as $L = na$, where n is an integer and a is the (uniform) spacing of the atoms on the lattice. Thus, L can take only certain values and is not a continuous variable. L and a then represents the periodicity of the lattice. The important point is that this L periodicity must be imposed upon the wave functions which arise from the Schrödinger equation

$$-\frac{\hbar^2}{2m_0}\frac{\partial^2 \psi(x)}{\partial x^2} + V(x)\psi(x) = E\psi(x). \tag{2.2}$$

Here, and throughout, we take m_0 as the free electron mass. If the potential is weak, the solutions will be close to those of the free electrons, which we will address shortly. The important point here is that, if the potential has the periodicity of (2.1), the solutions for the wave functions $\psi(x)$ must exhibit behavior that is consistent with this periodicity. The wave function itself is complex, but the real probability that arises from this wave function must have the periodicity. That is, we cannot really identify one atom from all the others, so the probability relating to the presence of the electron must be the same at each and every atom. This means that

$$|\psi(x + L)|^2 = |\psi(x)|^2, \tag{2.3}$$

and this must hold for each and every value of L. This must also hold for two adjacent atoms, so that we can say that the wave function itself can differ by at most a phase factor, or

$$\psi(x + a) = e^{i\varphi}\psi(x). \tag{2.4}$$

Generally, at this point one realizes that the line of atoms is not infinite, but has a finite length. In order to assure that the results are not dependent upon the ends of this chain of atoms, we invoke periodic boundary conditions. If there are N atoms in the chain, then

$$e^{iN\varphi} = 1, \quad \varphi = \frac{2n\pi}{N}, \tag{2.5}$$

where n is again an integer. We note that the smallest value of φ (other than 0) is $2/L = 2/Na$, while the largest value is $2n\pi/L = 2\pi/a$. The invocation of periodicity means that the Nth atom is actually also the 0th atom.

2.1.1 Bloch functions

The value $2\pi/a$ has an important connotation, as we recognize it as a basic part of the Brillouin zone—the Fourier transform of the lattice. To see this, let us write the wave function in terms of its Fourier transform through the definition

$$\psi(x) = \sum_k C(k)e^{ikx}. \tag{2.6}$$

At the same time, let us introduce the Fourier transform of the potential in terms of the basic lattice constant over which it is periodic, as

$$V(x) = \sum_G U_G e^{iGx}, \quad G = n\frac{2\pi}{a}, \tag{2.7}$$

where n is an arbitrary integer. Hence, we see that the quantities G are the harmonics of the basic spatial frequency of the potential $2\pi/a$. If we put these two Fourier transforms into the Schrödinger equation (2.2), we obtain

$$\sum_k \left[\frac{\hbar^2 k^2}{2m_0} C(k) + \sum_G U_G C(k)e^{iGx} - EC(k) \right] e^{ikx} = 0. \tag{2.8}$$

In the Fourier transform space, the analogy to (2.4) is that there is a displacement operator in momentum space by which

$$C(k + \lambda) = e^{i\lambda x}C(k), \tag{2.9}$$

so that we recognize the shift inherent in the second term of the square brackets of (2.8). It is important to recognize, both here and later in this chapter, that the exponential term is (2.9) is an operator, in that (quantum mechanically) x is a differential operator in momentum space [14, 15]. The role of this displacement operator is specifically to shift the position (in momentum space) of the wave function-like quantity $C(k)$ to $C(k - G)$. A sufficient condition for (2.8) to be satisfied is that the quantity in the square brackets vanishes and, with the shift indicated above, this leads to

$$\left(\frac{\hbar^2 k^2}{2m_0} - E \right) C(k) + \sum_G U_G C(k - G) = 0. \tag{2.10}$$

This result represents an entire set of equations, one for each value of k, that must be solved to find the Fourier coefficients $C(k)$. The second term represents a convolution summation of these coefficients with the Fourier coefficients of the potential. Throughout this chapter, we will continually see this equation in a variety of slightly different forms, but it is the basis of the determination of the band structure. Hence, the results depend upon the Fourier transform of the potentials, or pseudo-potentials, through the coefficients U_G. If we are doing first principles, then we must also self-consistently determine the Fourier coefficients of the wave functions, represented by the $C(k)$'s. This is the heart of the momentum space approaches.

From the above discussion, it is apparent that a continuous spectrum of Fourier coefficients is not present. In fact, only a discrete number of values of the vector k are allowed by the discretization introduced by the periodic boundary conditions. This number is N, which is the number of unit cells (each of length a) in the crystal. This is often thought to be the number of atoms, but this is true only for systems with a single atom per unit cell. We note that the values of k are selected by the values of G. These latter values form the *reciprocal lattice* in momentum space, and the set of values k, formed via (2.5), span one unit cell of this reciprocal lattice. This cell is called the (first) Brillouin zone of the reciprocal lattice. (As a side note, we will usually use a value of k which runs through $-\pi/a < k < \pi/a$ to provide a centered cell.) Now, let us return to (2.6), and write it in terms of the shifted vector in the second term of (2.10) as

$$\psi(x) = \sum_G C(k - G)e^{i(k-G)x} = \left[\sum_G C(k - G)e^{-iGx}\right]e^{ikx}. \tag{2.11}$$

The term in the square brackets is a function that is periodic in the lattice, and in the reciprocal lattice. Normally, we can rewrite (2.11) as the *Bloch function*

$$\psi(x) = e^{ikx}u_k(x). \tag{2.12}$$

The term in the square brackets of (2.11) is just the Fourier representation of the cell periodic expression $u_k(x)$. Thus, it is clear that the general solutions of the Schrödinger equation in a periodic potential are the Bloch functions (2.12). These functions are general properties of a wave in a periodic structure and are not unique to quantum mechanics [16].

2.1.2 Periodicity and gaps in energy

We have reached an interesting point. The wave functions for our crystal are now Bloch functions which represent the presence of the periodic potential, and this changes dramatically the nature of the propagating waves characteristic of the electrons. If we turn off the crystal potential, while retaining the periodicity of this potential (essentially just letting the amplitude become extremely small), then (2.10) reduces to just the free particle energy

$$E = \frac{\hbar^2 k^2}{2m_0}. \tag{2.13}$$

The Bloch wave function, however, is not unique, as it has been sufficient to define k only in the first Brillouin zone. When we use a value of k which runs through $-\pi/a < k < \pi/a$ to provide this first Brillouin zone, we are using what is called a Wigner–Seitz cell, which are those values of k closer to the Γ point ($k = 0$) than to any point shifted from this one by a reciprocal lattice vector $G = n \cdot 2\pi/a$, where n is any integer. This means that the momentum vector k is only defined up to a reciprocal lattice vector G, so that (2.13) must also be satisfied for any value of the shifted momentum vector, as

$$E = \frac{\hbar^2(k - G)^2}{2m_0}. \tag{2.14}$$

We show this in figure 2.1 for three such parabolas. The red curve represents (2.13), while the blue and green curves represent (2.14) for $G = 2\pi/a$ and $G = -2\pi/a$,

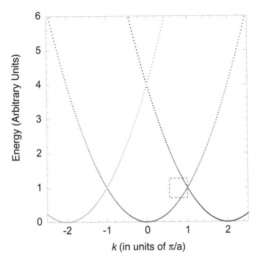

Figure 2.1. The periodicity of the free energy requires that multiple parabolas overlap. The dashed box will be examined further in figure 2.2.

respectively. The energy is degenerate at ±1 as well as at 0 in this limited plot. It must be noted that parabolas arise from all values of G, not just those shown.

If only the values of k that lie in the first Brillouin zone (the Wigner–Seitz cell) are taken, the energy is a multi-valued function of k, and different branches are characterized by different lattice periodic parts of the Bloch function. Each branch, indeed each energy value, in this first Brillouin zone is now labeled with both a momentum index k and a band index n. As mentioned, the bands are degenerate at a few special points, which are limited in this one-dimensional discussion (they will be more complicated and numerous in multiple dimensions). It is at these degeneracies that the crystal potential is expected to modify the basic nearly-free electron picture by opening gaps at the crossing points. These gaps will replace the degenerate crossings. Let us shift the momentum k in (2.10) an amount G', so that it becomes

$$(E_{k-G} - E_k)C(k - G') + \sum_{G'} U_{G'}C(k - G - G') = 0. \tag{2.15}$$

Here, the E in (2.10) is taken as E_k for the red curve in figure 2.1, while the rest of the equation is the blue curve in the figure. Just as in the case for (2.10), this equation is true for the entire family of reciprocal lattice vectors. However, let us focus on the two parabolas that cross at $k = \pi/a$, indicated by the dashed box in figure 2.1. At this point, we have

$$E_{k-G} = E_k, \tag{2.16}$$

or $k = \pm G'/2 = \pm G/2$. Thus, we select just these two terms from (2.10) and (2.15) as [17]

$$(E_k - E)C(k) + U_G C(k - G) = 0$$
$$(E_{k-G} - E)C(k - G) + U_G C(k) = 0. \tag{2.17}$$

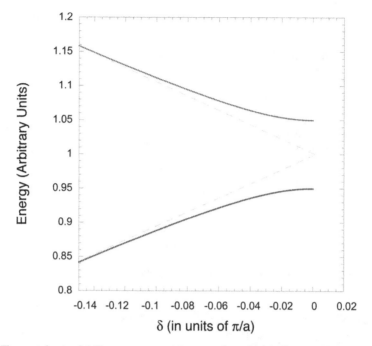

Figure 2.2. The crystal potential U_G opens a gap at the zone edge, which is illustrated here. The dashed lines would be the normal free electron parabolic behavior without the gap. The abscissa scale is discussed in the text.

Obviously, the determinant of the coefficient matrix must vanish if solutions are to be found, and this leads to

$$E = \frac{E_k + E_{k-G}}{2} \pm \sqrt{\left(\frac{E_k - E_{k-G}}{2}\right)^2 + U_G^2} = E_{\pi/a} \pm U_G, \qquad (2.18)$$

where the last form is that given precisely at the crossing point. Hence, the gap that opens is $2U_G$ and is exactly proportional to the potential interaction between these two bands. The lower energy state is a cooperative interaction (thus lowering the energy) which is termed the bonding band, while the upper state arises from the competition between the two parabolas (thus raising the energy) and is termed the anti-bonding band. Later, we will call these the valence and conduction bands, respectively. The new, split energy bands are shown in close-up detail in figure 2.2.

The argument can be carried a bit further, however. Suppose, we consider a small deviation from the zone edge crossing point, and ask what the bands look like in this region. To see this, we take $k = (G/2) - \delta = (\pi/a) - \delta$. Then, each of the energies may be expanded as

$$E_k = \frac{\hbar^2}{2m_0}\left(\frac{G^2}{4} - \delta G + \delta^2\right)$$

$$E_{k-G} = \frac{\hbar^2}{2m_0}\left(\frac{G^2}{4} + \delta G + \delta^2\right). \qquad (2.19)$$

Using these values of the energy in the first line of (2.18) gives us the two energies

$$E = E_{G/2} + \frac{\hbar^2\delta^2}{2m_0} \pm \sqrt{4E_{G/2}\frac{\hbar^2\delta^2}{2m_0} + U_G^2}. \qquad (2.20)$$

Here, $E_{G/2} = \hbar^2 G^2/8m_0 = \hbar^2\pi^2/8m_0 a^2$ is the nearly free electron energy at the zone edge and is the energy at the center of the gap. We can write $E_+ = E_{G/2} + U_G$ and $E_- = E_{G/2} - U_G$, and the variation of the bands becomes

$$E_C = E_+ + \frac{\hbar^2\delta^2}{2m_0}\left(\frac{2E_{G/2}}{U_G} + 1\right)$$

$$E_V = E_- - \frac{\hbar^2\delta^2}{2m_0}\left(\frac{2E_{G/2}}{U_G} - 1\right), \qquad (2.21)$$

for small values of δ. Here, the C and V subscripts refer to the upper (conduction) and lower (valence) bands of figure 2.2. The crystal potential has produced a gap in the energy spectrum and the resulting bands curve much more than the normal parabolic bands, as may be seen in figure 2.2. Equation (2.21) also serves to introduce an effective mass, in the spirit that the band variation away from the minimum should be nearly parabolic, similar to the normal free electron parabolas. Thus, for small δ, the bonding (valence) and anti-bonding (conduction) band effective masses are defined from (2.21) as

$$\frac{1}{m_V^*} = \frac{1}{m_0}\left(1 - \frac{2E_{G/2}}{U_G}\right), \quad \frac{1}{m_C^*} = \frac{1}{m_0}\left(1 + \frac{2E_{G/2}}{U_G}\right). \qquad (2.22)$$

One may notice that, since the second term in the parentheses is large, the bonding mass is negative as the energy decreases as one moves away from the zone edge. By using these effective masses, we are introducing our quasi-particles, or quasi-electrons and quasi-holes, which have a characteristic mass different from the free electrons. In the bonding case, the quasi-hole is a positive hole, or empty state, and this sign change of the charge compensates for the negative value of the mass—hence we normally talk about the holes having a positive mass due to the opposite charge of the hole. It may also be noted that the two bands are not quite mirror images of one another as the masses are slightly different in value due to the sign change between the two terms of (2.22). This also makes the anti-bonding mass the slightly smaller of the two in magnitude.

For larger values of δ (just how large cannot be specified right now), it is not fair to expand the square roots that are in the first line of (2.18). The more general case is given by

$$E(\delta) = E_{G/2} \pm \frac{E_{\text{gap}}}{2}\sqrt{1 + \frac{2\hbar^2\delta^2}{m^*E_{\text{gap}}}}. \qquad (2.23)$$

In this form, we have ignored the free electron term (the second term in (2.20)), as it is negligible for small effective masses. We also have introduced the gap $E_{\text{gap}} = 2U_G$ apparent in the last line of (2.18). It is clear that the bands become very non-parabolic as one moves away from the zone edge and this will also lead to a momentum dependent effective mass, a point we return to later in the chapter, where we will encounter it in the discussion of the $k \cdot p$ and spin–orbit interactions. In general, any time there is an interaction between the wave functions of the bonding, or valence, band and the anti-bonding, or conduction, band, this interaction will lead to a splitting which gives rise to non-parabolic band. In many cases, especially in three dimensions, the presence of the spin–orbit interaction greatly complicates the solutions, so that it is normally treated as a perturbation. On the other hand, we will see in the momentum space solutions that it can be incorporated rather easily without too much increase in complexity of the system (the Hamiltonian matrix is already fairly large, and isn't increased by the additional terms in the energy).

2.2 Potentials and pseudopotentials

In the development of the last chapter, as well as the first equations of this chapter, the summations within the Hamiltonian were carried out over all of the electrons for all of the atoms. In the tetrahedrally coordinated semiconductors, as well as for the lower dimensional layered semiconductors, the bonds are actually formed by only the outer shell electrons. These are the so-called valence electrons. In general, the inner core electrons play very little role in this bonding process which actually determines the crystal structure. For example, the Si bonds are composed of $3s$ and $3p$ levels, while GaAs and Ge have bonds composed of $4s$ and $4p$ electrons. This reliance on the outer shell electrons is not completely true, as the occupied inner d levels (where they exist) often lie quite close to the lowest energies of these bonding s and p orbitals. This can lead to a slight modification of the bonding energies in those materials composed of atoms lying lower in the periodic table. Although this correction is usually small, it can be important in a number of cases, and will be mentioned at times in that context. Normally, however, we treat only the 4 outer shell electrons (or 8 from the two atoms in the zinc-blende and diamond structures per face-centered cell), although the actual primitive unit cell in the tetrahedrally bonded structure holds only 2 atoms.

As mentioned in chapter 1, we can simplify the Hamiltonian when we are treating only the outer shell bonding electrons. In this case, we can rearrange (1.4) as

$$H_{el} = \sum_{i,val}\left\{\frac{p_i^2}{2m_0} - q_i\left[V(x_i) - \sum_{j,core}\frac{q_j}{4\pi\varepsilon_0 x_{ij}}\right] - \sum_{j,val}\frac{q_j}{4\pi\varepsilon_0 x_{ij}}\right\} + E_{core}. \quad (2.24)$$

Here, we have split the Coulomb interaction of (1.4) into 3 terms. The first term represents the interaction between the outer shell (subscript *val* for valence) electrons and the inner core (subscript *core*) electrons. This term appears in the square bracket. The second term represents the interactions between just the outer shell electrons, and this is the last term in the curly brackets. Then there are the interactions between just the inner shell electrons, and this is represented by the last term in (2.24). The interpretation is that the first term represents a potential through Coulomb's law, and this potential modifies the total potential, which is the first term in the square brackets. The second term creates an interaction that leads to the Hartree and higher order corrections to the one electron approximation of the H-K theorems. The third term can be treated as a rigid energy shift of the band structure. This shift is important for many applications, such as photoemission, but is not particularly important in our discussion of the electronic structure since we will usually reference our energies to either the bottom of the conduction band or the top of the valence band. Because the second sum in the square brackets represents another potential, we can rewrite the terms in the square brackets as a new potential, which is termed a *pseudo-potential*. This can be written as

$$V_P(x_i) = V(x_i) - \sum_{j,core} \frac{q_j}{4\pi\varepsilon_0 x_{ij}}. \tag{2.25}$$

We illustrate how this conversion to the pseudo-potential looks like in space in figure 2.3.

Now, the problem is to find these pseudo-potentials. In these problems, the first-principles approach is to solve for the pseudo-potentials and the bonding wave functions in a self-consistent approach [18]. The effect of including the core electron contributions in the potential is to remove the deep Coulombic core of the atomic potential and give a smoother overall interaction potential, which should ease the complications in the self-consistent solution.

Still, one needs to address the remaining interaction between the bonding electrons, the last term in the curly brackets of (2.24). This leads to a nonlinear behavior in the Schrödinger equation. Various approximations to this term have been pursued through the years. The easiest is to simply assume that the bonding electrons lead to a smooth general potential, and this this interaction arises from the

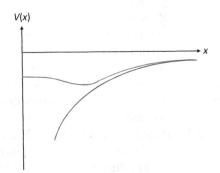

Figure 2.3. When the potential of the inner shell electrons is removed, the external potential (magneta) is simplified to the pseudo-potential (blue).

role this potential imposes upon the individual electron. This quasi-single electron approach is known as the Hartree approximation, which treats the interaction of the single electron with a mean-field background potential, and gives the normal electronic contribution to the dielectric function. The next approximation is to explicitly include the exchange terms—the energy correction that arises from interchanging any two electrons (on average), which leads to the Hartree–Fock approximation. The more general approach, which is widely followed, is to follow the H-K theorems and adopt an energy functional term, in which the energy correction is a function of the local density. This energy functional is then included within the self-consistent solution for the wave functions and the energies. This last approach is known as the local-density approximation (LDA) within density-functional theory (DFT) [19]. In spite of this range of approximations, the first-principles calculations all have a problem with getting the band gaps correct in semiconductors. Generally, they find values for the energy gaps that are roughly as much as an order of magnitude too small. Even though there are a number of corrections that have been suggested for LDA, none of these has solved the band gap problem. Humorously, this approach has led to the use of the 'scissors operator,' in which a plot of the band structure is cut through the gap with scissors and the two parts moved to give the right band gap!

To my knowledge, only two approaches have come close to solving this problem, and these are the GW approximation [20] and exact exchange [21]. In the former approach, one computes the total self-energy of the bonding electrons and uses the single particle Green's function to give a new self-energy which lowers the energies of the valence band and corrects the gap in that manner. This is usually very effective, but computationally expensive. In the second case, one uses an effective potential based upon Kohn–Sham single particle states and a calculation of the interaction energy through what is called an effective potential. Neither of these will be discussed further here, as they go beyond the level of our discussion, although advances to both methods have been made [18, 19]. One would think that the wave functions and pseudo-potentials for e.g. the Ga atoms would be the same in GaAs as in GaP. But, this has not generally been the case. However, there has been a significant effort to find such so-called *transferable* wave functions and potentials. This is true regardless of the approach and approximations that have been utilized. There has been some significant progress on this front, and several sets of wave functions and pseudo-potentials can be found in the literature and on the web that are said to be transferable between different compounds. It is my experience that one must check these out carefully on your own before accepting the claim for transferability, whether you are using real space or momentum space approaches.

The above discussion focused upon the self-consistent first-principles approaches to electronic structure. There is another approach, which is termed *empirical*. Especially because of the band gap problem, it is often found that rather than doing the full self-consistent calculation, one could replace the overlap integrals involving different wave functions and the pseudo-potential with a set of constants, one for each different integral, and then adjusting the constants for a best fit to measured experimental data for the band structure in a real space method [12]. In the momentum space approach, we will see that it is the Fourier coefficients of pseudo-potential that are adjusted. The positions of many of the critical points in the band structure are known from a variety of experiments, and they are well-known enough to use in such a procedure. To be sure, in the first-principles approach, one does try to get agreement with some experimental data. In the empirical approach, however, one sheds the need for self-consistency by adopting the experimental results as the 'right' answer, and just adjusts the constants to fit this data. The argument is that such a fit already accounts for all of the details of the inter-electron interactions, because they are included exactly in whatever material is measured experimentally. The attraction for such an approach is that the electronic structure is obtained quickly, but the drawback is that, once again, the set of constants so obtained is not assured of being transferable between materials, in spite of many claims to the contrary.

2.3 Real-space methods

In real-space methods, we compute the electronic structure using the Hamiltonian and the wave functions written in real space, just as the name implies, and as we discussed in the opening of this chapter. The complement of this, momentum-space approaches, will be discussed in the next section. In the real-space approach, we want to solve the pseudo-potential version of the Schrödinger equation, which may be written as

$$H(x)\psi(x) = H_0(x)\psi(x) + V_P(x)\psi(x), \qquad (2.26)$$

where H_0 includes the kinetic energy of the electron and V_P describes the pseudo-potential at the particular site. Any multi-electron effects are included through a term added to H_0 that is described through H-K theorems as an effective potential that depend upon the electron density. The pseudo-potential V_P is just (2.25), depicted by the blue curve in figure 2.3. As discussed above, whether or not we include this effective potential for the interactions depends upon the exact approach we take to find the bands, such as *ab initio* or empirical for example.

To proceed, we need to specify a lattice, and the basis set of the wave functions. Of course, since we are interested in a real-space approach, the basis set will be an orbital (or more than one) localized on a particular lattice site, and the basis at each site is assumed to satisfy the atomic orthonormality that is imposed on each different lattice site. We will illustrate this further in the treatment below. We will go through this first in one spatial dimension, and for both one and two atoms per unit cell of the lattice. Then, we will proceed to two spatial dimensions. We will treat graphene as a specific example of a two-dimensional lattice that actually occurs in nature, but will discuss some other two-dimensional materials. Finally, we will move to the three-dimensional crystal with four orbitals per atom, the sp^3 basis set common to the tetrahedral semiconductors. One very important point in this is that, throughout this discussion, we will only consider two-point integrals and interactions. That is, we will ignore integrals in which the wave functions are on atoms 1 and 2, while the potential may be coming from atom 3. While these may be important is some cases of first-principles calculations, they are not usually considered necessary for empirical approaches.

2.3.1 Bands in one dimension

As shown in figure 2.4, we assume a linear chain of atoms uniformly spaced by the lattice constant a. As above in section 2.1, we will use periodic boundary conditions, although they do not appear specifically except in our use of the Brillouin zone and its properties. With the periodic boundary conditions, atom N in the figure folds back onto atom 0, so that these are the same atom. We adopt an index j, which designates with which atom in the chain we are dealing. As we discussed above, the basis set for our expansion is one in which each wave function is localized upon a single atom so that orthonormality for basis functions r and s on the atom at point j appears as

$$\langle r|s\rangle_j = \delta_{rs}, \tag{2.27}$$

where we have utilized the Dirac notation for our basis set and the right-hand side is the Kronecker delta function which vanishes unless $r = s$. Generally, the use of Dirac notation simplifies the equations, and reduces confusion and clutter, and we will follow it here as much as possible. We further assume that these are energy eigenfunctions, and that the diagonal energies are the same on each atom as we cannot distinguish one atom from the next, so that

$$H_0|r\rangle = E_r|r\rangle = E_1|r\rangle. \tag{2.28}$$

In this last step, we have assumed for simplicity that only a single atomic wave function is involved and labeled the energy accordingly.

A vital assumption which we will follow throughout this section is that the nearest-neighbor interaction dominates the electronic structure. Hence, we will not go beyond nearest-neighbor interactions other than to discuss degeneracies at critical points where one might use longer-range interactions to some advantage. Let us now apply (2.26) to a wave function at some point j, where $0 \leqslant j \leqslant N$, that is to one of the atoms in the chain of figure 2.4. However, we must include the interaction between the atom and its neighbors that arises from the pseudo-potential between the atoms. Hence, we may write (2.26) as

$$H_0|j\rangle + V_P|j+1\rangle + V_P|j-1\rangle = E|j\rangle. \tag{2.29}$$

If we pre-multiply this equation with the complex conjugate of the wave function at this site, we obtain

$$E_1 + \langle j|V_P|j+1\rangle + \langle j|V_P|j-1\rangle = E. \tag{2.30}$$

a

0   1   2   N-1 N

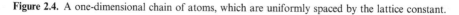

Figure 2.4. A one-dimensional chain of atoms, which are uniformly spaced by the lattice constant.

The second and third terms are the connections that we need to evaluate. To do this, we utilize the properties of the displacement operator to set

$$|j \pm 1\rangle = e^{\pm ika}|j\rangle, \qquad (2.31)$$

where the exponential is exactly the real-space displacement operator in quantum mechanics and shifts the wave function by one atomic site [22]. Similarly, the third term produces the complex conjugate of the exponential. We are then left with the integration of the onsite pseudo-potential

$$\langle j|V_P|j\rangle = -A, \qquad (2.32)$$

where A is an arbitrary constant. One could actually use exact orbitals and the pseudo-potential to evaluate this integral, or one could fit it to experimental data. The difference is the difference between first-principles approaches and empirical approaches, which we discuss above and will return to later. Here, we just assume that the value is found by one technique or another.

Using the above expansions and evaluations of the various overlap integrals that appear in the equations, we may write (2.30) as

$$E = E_1 - A(e^{ika} + e^{-ika}) = E_1 - 2A\cos(ka). \qquad (2.33)$$

This energy structure is plotted in figure 2.5. The band is $4A$ wide (from lowest to highest energy), and is centered about the single site energy E_1, which says that the band forms by spreading around this single atom energy level. The band contains N values of k, as there are N atoms in the chain, and this is the level of quantization of the momentum variable, as was shown earlier. That is, there is a single value of k for each unit cell in the crystal. If we incorporate the spin variable, then the band can hold $2N$ electrons, as each state holds one up spin and one down spin electron. But, we have only N electrons from the N atomic sites. Hence, this band would be half full, with the Fermi energy lying at mid-band.

Suppose we now add a second atom per unit cell, so that we have a diatomic basis. This is illustrated in figure 2.6. Here, each unit cell contains the two unequal atoms (one blue and one green in this example). The lattice can be defined either

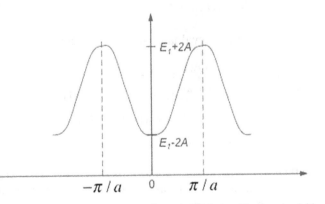

Figure 2.5. The resulting bandstructure for a one dimensional chain with nearest neighbor interaction.

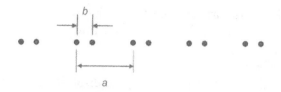

Figure 2.6. A diatomic lattice. Each unit cell contains one blue atom and one green atom, and this will change the electronic structure even though it remains a one-dimensional lattice.

with the blue atoms or with the green atoms, but each lattice site contains a *basis* of two atoms. This will significantly change the electronic structure. We will have two Bloch functions, one for each of the two atoms in the basis.

This, in turn, leads us to have to write two equations to account for the two atoms. And, there are two distances involved in the *Blue–Green* atom interaction. There is one interaction in which the two atoms are separated by b and one interaction in which the two atoms are separated by a-b. The lattice constant, however, remains a. To proceed, we need to adopt a slightly more complicated notation. We will assume that the blue atoms will be indexed by the site i, when i is an even number (including 0). Similarly, we will assume that the green atoms will be indexed by the site i, when i is an odd number. We have to write two equations, one for when the central site is a blue atom and one for when the central site is a green atom. It really doesn't matter which sites we pick, but we take the adjacent sites so that these equations become

$$E_i|i\rangle + V_P|i+1\rangle + V_P|i-1\rangle = E|i\rangle$$
$$E_i|i+1\rangle + V_P|i+2\rangle + V_P|i\rangle = E|i+1\rangle. \tag{2.34}$$

We will assume here that i is an even integer (blue atom) in order to evaluate the integrals. Once again we premultiply the first of these equations with the complex value of the central site i, and the second by the complex value of the central site $i+1$, and integrate, we then obtain

$$E_1 + \langle i|V_P|i+1\rangle + \langle i|V_P|i-1\rangle = E$$
$$E_1 + \langle i+1|V_P|i+2\rangle + \langle i+1|V_P|i\rangle = E \tag{2.35}$$

Now, we have four integrals to evaluate, and these become

$$\langle i|V_P|i+1\rangle = e^{ikb}\langle i|V_P|i\rangle \equiv e^{ikb}A_1$$
$$\langle i|V_P|i-1\rangle = e^{ik(b-a)}\langle i|V_P|i\rangle \equiv -e^{ik(b-a)}A_2$$
$$\langle i+1|V_P|i+2\rangle = e^{-ik(b-a)}\langle i+1|V_P|i+1\rangle \equiv -e^{ik(a-b)}A_2$$
$$\langle i+1|V_P|i\rangle = e^{-ikb}\langle i|V_P|i\rangle \equiv e^{-ikb}A_1. \tag{2.36}$$

Here, we have taken the overlap integrals to be different for the two different atoms. The choice of the sign on the A_2 terms assures that the gap will occur at $k = 0$. The choice of in which direction to translate the wave functions is made to insure that the

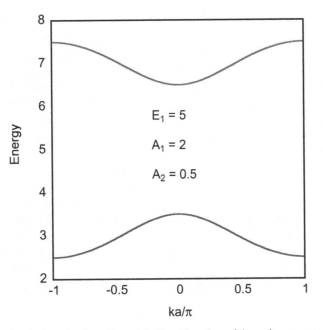

Figure 2.7. The two bands that arise for a diatomic lattice. The values of the various parameters are shown on the figure.

Hamiltonian is Hermitian. The two equations now give us a secular determinant which must be solved, as

$$\begin{vmatrix} (E_1 - E) & (A_1 e^{ikb} - A_2 e^{ik(b-a)}) \\ (A_1 e^{-ikb} - A_2 e^{-ik(b-a)}) & (E_1 - E) \end{vmatrix} = 0. \tag{2.37}$$

It is clear that the two off-diagonal elements are complex conjugates of each other, and this assures that the Hamiltonian is Hermitian and the energy solutions are real. This basic requirement must be satisfied, no matter how many dimensions we have in the lattice, and is a basic property of quantum mechanics—in order to have real measurable eigen-values, the Hamiltonian must be Hermitian. We now find that two bands are formed from this diatomic lattice, and these are mirror images around an energy midway between the lowest energy of the upper band and the highest energy of the lower band. The energy is given by

$$E = E_1 \pm \sqrt{A_1^2 + A_2^2 - 2A_1 A_2 \cos(ka)}. \tag{2.38}$$

The two bands are shown in figure 2.7 for the values $E_1 = 5$, $A_1 = 2$, $A_2 = 0.5$. The two bands are mirrored around the value 5, and each has a band width of 1.

As in the monatomic lattice, each band contains N states of momentum, as there are N unit cells in the lattice. As before, each band can thus hold $2N$ electrons when we take into account the spin degeneracy of the states. Now, however, the two atoms per unit cell provide exactly $2N$ electrons, and these will fill the lower band of figure 2.7. Hence, this diatomic lattice represents a semiconductor, or insulator, depending upon how wide the band gap is. Here, it is 3 (presumably eV) wide, which would normally be construed as a wide band gap semiconductor.

While it is easy to consider this discussion of the one dimensional material a mainly tutorial of limited interest, one dimensional (or quasi-one dimensional) materials have appeared in recent years. Foremost among these are the molecules bonded to metals or semiconductors for studies of their conduction properties [23]. But, one can also have pseudo-one dimensional conduction in biological ion channels which translate potassium and sodium into, and out of, individual cells [24]. But, there have also been quasi-one dimensional structures created by putting single phosphorous atoms in a chain embedded in undoped silicon to produce an n^+ silicon nanowire [25]. Then, one can make atomic chains which create nanowires in e.g. gold [26]. But, the primary approach of interest to us is the growth of nanowires on a silicon substrate [27]. While these latter nanowires are not atomically thin, they remain quasi-one dimensional in nature due to the strong quantization arising from their small lateral dimensions [28]. The one-dimensional band structure provides a first estimate of the nanowire bands, while the transverse quantization provides more detail.

2.3.2 Two-dimensional lattices

The natural progression from one dimension to two dimensions is of course a reasonable step. However, the world is full of real two-dimensional semiconductors, which are called layered compounds. The best known one is graphite, in which layers of atomically thin carbon are weakly bound together to produce the bulk. The single layer of carbon atoms is called graphene, and it has been isolated only recently [29]. In this single layer graphene, the carbon atoms are arranged in a hexagonal lattice, which has two atoms per unit cell. There are other layered compounds, which have similar structure, and a well-known group are the transition metal di-chalcogenides (TMDC), such as MoS_2 and WSe_2 [30]. In the TMDC, the layer is actually an atomic tri-layer, with the metal atoms forming the central layer, and the chalcogenide atoms forming the top and bottom layers. Each metal atom has a triangle of three chalcogenide atoms above it and below it. Yet the basic structure still consists of hexagonal coordination with two so-called atoms per unit cell—these consist of a metal atom and an up and down pair of chalcogenide atoms, which sit above one another to form a pseudo-atom. As a result, both graphene and the main TMDCs have a quite similar band structure. In the graphene case, the conduction and valence bands are composed of the p_z orbitals of carbon [31]. In the TMDC case, the conduction and valence bands are determined by the metal d orbitals [30], as these bands lie in the gap between the bonding and anti-bonding sp hybrids. As our example for the two-dimensional case, we will take the case of graphene.

The single layer of graphene is exceedingly strong, and has been suggested for a great many applications. Here, we wish to discuss the energy structure. To begin, we refer to the crystal structure and reciprocal lattice shown in figure 2.8. As mentioned,

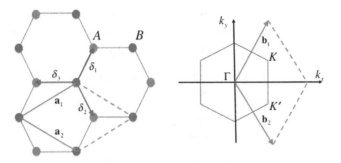

Figure 2.8. (left) The crystal structure of graphene. The unit cell contains one red C atom and one blue C atom. The red arrows are the primitive vectors for the unit cell, while the blue vectors are the nearest neighbor vectors. The unit cell is closed by the dashed red lines. (right) The reciprocal lattice also has a (rotated) hexagonal shape, although the primitive cell is also a diamond defined by the two red momentum vectors. The primitive cell is closed by the dashed red lines. The major points in the Brillouin zone are defined by Γ, K, and K'.

graphene is a single layer of C atoms, which are arranged in a hexagonal lattice. The unit cell contains two C atoms, which are non-equivalent. Thus the basic unit cell is a diamond which has a basis of 2 atoms. In figure 2.8, the unit vectors of the diamond cell are designated as the red arrows $\mathbf{a}_1$ and $\mathbf{a}_2$, and the cell is closed by the red dashed lines. The two inequivalent atoms are shown as the A (red) and B (blue) atoms. The three nearest neighbor vectors (blue) are also shown pointing from a B atom to the three closest A atoms. The reciprocal lattice is also a diamond, rotated by 90 degrees from that of the real space lattice, but the hexagon shown is usually used. There are two inequivalent points at two distinct corners of the hexagon, which are marked as K and K'. As we will see, the conduction and valence bands touch at these two points, so that they represent two valleys in either band. The two unit vectors of the reciprocal lattice are $\mathbf{b}_1$ and $\mathbf{b}_2$, with $\mathbf{b}_1$ normal to $\mathbf{a}_2$ and $\mathbf{b}_2$ normal to $\mathbf{a}_1$. The nearest neighbor distance is $a = 0.142$ nm, and from this, one can write the lattice vectors and reciprocal lattice vectors as

$$a_1 = \frac{a}{2}(3a_x + \sqrt{3}\,a_y)\,, \quad a_2 = \frac{a}{2}(3a_x - \sqrt{3}\,a_y)$$
$$b_1 = \frac{2\pi}{3a}(a_x + \sqrt{3}\,a_y)\,, \quad b_2 = \frac{2\pi}{3a}(a_x - \sqrt{3}\,a_y),$$

(2.39)

and the K and K' points are located at

$$K = \frac{2\pi}{3a}\left(1, \frac{1}{\sqrt{3}}\right),\quad K' = \frac{2\pi}{3a}\left(1, -\frac{1}{\sqrt{3}}\right).$$

(2.40)

(In this last equation, the coordinates are in the k_x, k_y frame.) From these parameters, we can construct the energy bands with a nearest-neighbor interaction. This was first done apparently by Wallace [31], and we basically follow his approach.

Just as in our diatomic one-dimensional lattice, we will assume that the wave function has two basic components, one for the A atoms and one for the B atoms. Thus, we write the wave function in the following form

$$\psi(x, y) = \varphi_A(x, y) + \lambda\varphi_B(x, y)$$

$$\varphi_A(x, y) = \sum_A e^{ik \cdot r_A}\chi(r - r_A) \tag{2.41}$$

$$\varphi_B(x, y) = \sum_B e^{ik \cdot r_B}\chi(r - r_B)$$

Here, we have written both the position and momentum as two-dimensional vectors. Each of the two component wave functions is a sum over the Bloch functions for each type of atom. Without fully specifying the Hamiltonian, we can write Schrödinger's equation as

$$H(\varphi_A + \lambda\varphi_B) = E(\varphi_A + \lambda\varphi_B). \tag{2.42}$$

At this point, we pre-multiply (2.42) first with the complex conjugate of the first component of the wave function and integrate, and then with the complex conjugate of the second component of the wave function. This leads to two equations which can be written as (the indices are now switched from A,B to 1,2 for normal matrix notation)

$$H_{11} + \lambda H_{12} = E$$
$$H_{21} + \lambda H_{22} = E, \tag{2.43}$$

with

$$H_{11} = \int \varphi_A^* H \varphi_A dr$$

$$H_{22} = \int \varphi_B^* H \varphi_B dr \tag{2.44}$$

$$H_{21} = \int \varphi_B^* H \varphi_A dr = H_{12}^*.$$

As mentioned, we are only going to use nearest neighbor interactions, so the diagonal terms become

$$H_{11} = \int \chi^*(r - r_A)H\chi(r - r_A)dr \equiv E_0$$

$$H_{22} = \int \chi^*(r - r_B)H\chi(r - r_B)dr \equiv E_0. \tag{2.45}$$

In graphene, the in-plane bonds that hold the atoms together are sp^2 hybrids, while the transport is provided by the p_z orbitals normal to the plane. For this reason the local integral at the A atoms and at the B atoms should be exactly the same, and this is symbolized in (2.45) by assigning them the same net energy. By the same process, the off-diagonal terms become

$$H_{21} = H_{12}^* = \int \varphi_B^* H \varphi_A dr = \sum_{A,B} e^{ik \cdot (r_B - r_A)} \int \chi_B^* H \chi_B dr$$
$$\equiv \gamma_0 \sum_{A,B} e^{ik \cdot (r_B - r_A)} = \gamma_0 (e^{ik \cdot \delta_1} + e^{ik \cdot \delta_2} + e^{ik \cdot \delta_3}). \tag{2.46}$$

The three position vectors in the last term are the nearest neighbor vectors shown in figure 2.8. The sum of the three exponentials shown in the parentheses is known as a *Bloch sum*. Each term is a displacement vector that moves the A atom basis function to the B atom where the integral is performed. We can write down the coordinates of the nearest neighbor vectors, relative to a B atom as shown in the figure 2.8, as

$$\delta_1 = \frac{a}{2}(1, \sqrt{3})$$
$$\delta_2 = \frac{a}{2}(1, -\sqrt{3}) \tag{2.47}$$
$$\delta_3 = -a(1,0).$$

With a little algebra, the off-diagonal element can now be written down as

$$H_{12} = \gamma_0 \left[2e^{ik_x a/2} \cos\left(\frac{\sqrt{3}k_y a}{2}\right) + e^{-ik_x a} \right]. \tag{2.48}$$

Now that the various matrix elements have been evaluated, the Hamiltonian matrix can be written down from these values. This leads to the determinant

$$\begin{vmatrix} (E_0 - E) & \lambda H_{12} \\ H_{21} & \lambda(E_0 - E) \end{vmatrix} = 0 \tag{2.49}$$
$$E = E_0 \pm \sqrt{|H_{21}|^2}.$$

This leads us to the result

$$E = E_0 \pm \gamma_0 \sqrt{1 + 4\cos^2\left(\frac{\sqrt{3}k_y a}{2}\right) + 4\cos\left(\frac{\sqrt{3}k_y a}{2}\right)\cos\left(\frac{3k_x a}{2}\right)}. \tag{2.50}$$

The two energy surfaces are shown in figure 2.9. The most obvious fact about these bands is that the conduction and valence bands touch at the six K and K' points (where the cosine terms add to -1) around the hexagon reciprocal cell of figure 2.8. This means that there is no band gap. Indeed, expansion of (2.50) for small values of momentum away from these six points shows that the bands are linear. These have come to be known as massless Dirac bands, in that they give similar results to solutions of the Dirac equation with a zero rest mass. If this small momentum is written as ξ (which is zero at K and K'), then the energy structure has the form (with $E_0 = 0$)

$$E = \pm \frac{3\gamma_0 a}{2} \xi. \tag{2.51}$$

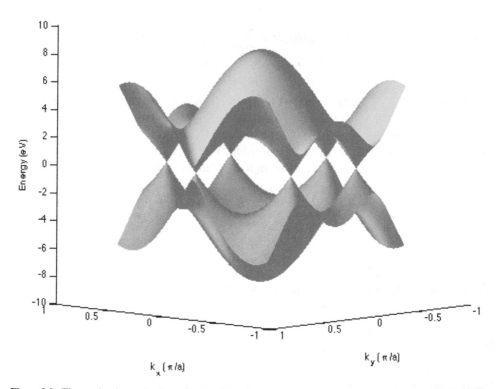

Figure 2.9. The conduction and valence bands of graphene according to (2.50). The bands touch at the K and K' points around the hexagonal cell.

The maximum value of the square root expression in (2.50) arises when $k = 0$, and the square root takes the value 3. Up until now, we have expressed explicitly whether the approach is first principles or empirical, although we have not talked about the full Hamiltonian that goes into E_0. At this point, we shall proceed down the empirical path for the evaluations of the two parameters. Experiments show that the width of the valence band is about 9 eV, so that $\gamma_0 \sim 3$ eV. Using this value, and the energy structure of (2.50), we arrive at an effective Fermi velocity (determined from the slope of (2.51) to be $3\gamma_0/\hbar a$, as we will explain in section 2.6) for these linear bands of about 9.7×10^7 cm s^{-1}. We should point out that, while the rest mass is zero, the dynamic mass of the electrons and holes is not zero, but increases linearly with ξ. This variation has also been seen experimentally shown, using cyclotron resonance to measure the mass [32]. The above approach to the band structure of graphene works quite well. At higher energies, the nominally circular nature of the energy 'surface' near the Dirac points becomes trigonal, and this can have a big effect upon transport. For a more advanced approach, which also takes into account the sp^2 bands arising from the in-plane bonds, one needs to go to the first-principles approaches described in a later section [33].

As with the diatomic lattice, there are N states in each two-dimensional band, where N is the number of unit cells in the crystal. With spin, the valence band can thus hold $2N$ electrons, which is just the number available from the two atoms per

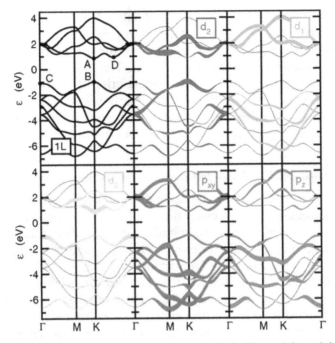

Figure 2.10. Band structure and orbital character of single-layer MoS$_2$. The top left panel shows the full band structure while, in the other panels, the thickness of the bands represents the orbital weight of the various orbitals (discussed in the text). Reprinted with permission from Cappelluti *et al* [35], copyright 2013 by the American Physical Society.

unit cell. Hence, the Fermi energy in pure graphene resides at the zero point where the bands meet, which is termed the Dirac point.

In other two-dimensional systems, such as the TMDCs, there is some preference to use real space methods such as the linear combination of atomic orbitals, as done here for graphene. This is because of the highly localized orbitals of the transition metals (which would require an enormous number of plane waves in the momentum space approach). In general, however, it is believed that these approaches are not sufficiently accurate to give details such as surface morphology that can be measured with scanning tunneling microscopy [34]. For that reason, most of the explorations of the TMDCs has been with plane wave DFT approaches. Indeed the properties of MoS$_2$ have been studied with tight-binding for the TMDCs suitable for both the bulk and single-layer cases [35]. Using the empirical approach of Slater and Koster [12], the latter authors have the orbital character of the states at various points in the zone and found that the transition from an indirect band gap to a direct band gap occurs as the structure is reduced to a single layer of the material. The band structure for the single layer system is shown in figure 2.10. Here the orbital character arises from the d states of the Mo atom's $4d$ orbitals ($d_2 = d_{x^2-y^2}$, d_{xy}, $d_1 = d_{xz}$, d_{yz}, $d_0 = d_{3z^2-r^2}$) and the p character arises from $2p$ orbitals of S ($p_{xy} = p_x$, p_y). From this type of figure, it is easy to see that, while there is some p contribution to the top of the valence band and bottom of the conduction band, the main contribution to the top of the valence band comes from the d_2 orbital and the bottom of the conduction band comes mainly from the d_0 orbital. A major point that appears here is that, in multi-atom materials, the nature of the various atomic contributions at different points in the Brillouin zone are quite different. This is different from the case of graphene above, where the entire band structure arises from the pure p_z orbitals of the C atoms, the only atoms in the structure. As we will see later, it is necessary to identify these orbitals, as they will be used in the perturbation theory that gives rise to the various scattering processes of the electrons by the lattice (chapter 5).

2.3.3 Three-dimensional lattices—tetrahedral coordination

We now turn our attention to the three-dimensional lattices. As most of the tetrahedral semiconductors have either the zinc-blende or diamond lattice, we focus on these. First, however, we have to consider the major difference, as these atoms have, on average, four outer shell electrons. These can be characterized as a single s state and three p states. Thus, the four orbitals will hybridize into four directional bonds, each of which points toward the four nearest neighbors. As these four neighbors correspond to the vertices of a regular tetrahedron, we see why these materials are referred to as tetrahedral materials. The group 4 materials (C, Si, Ge, and grey Sn) all have four outer shell electrons. On the other hand, the III–V and II–VI materials have only an average of four electrons. These latter materials have the zinc-blende lattice, while the former have the diamond lattice. These two lattices differ in the makeup of the basis that sits at each lattice site.

In figure 2.11, we illustrate the two lattices in the left panel of the figure. The basic cell is a face-centered cube (FCC), with atoms at the eight corners and one atom centered in each face. These atoms are shown as the various shades of red. Each lattice site has a basis of two atoms, one red and one blue in the figure. The tetrahedral coordination is indicated by the green bonds shown for the lower, left blue atom—these point toward the nearest red neighbors. Only four of the second basis atoms are shown as the others lay outside this FCC cell. This is not the unit

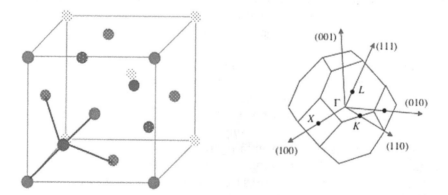

Figure 2.11. The zinc-blende lattice (left) and its reciprocal lattice (right).

cell, but is the shape most commonly used to describe this lattice. In diamond (and Si and Ge as well), the two atoms of the basis are the same, both C (or Si or Ge). In the compound materials, there will be one atom from each of the compounds in the basis; e.g., one Ga and one As atom in GaAs. We will see below that the four nearest neighbors means that we will have four exponentials in the Bloch sums (in (2.46) the Bloch sum was composed of vectors to the nearest neighbors, just as here).

In the right panel of figure 2.11, the Brillouin zone for the FCC lattice is shown. This is a truncated octahedron. Of course, it could also be viewed as a cube in which the eight corners have been lopped off. The important crystal directions have been indicated in the figure. The important points are the Γ point at the center of the zone, the X points at the centers of the square faces, and the L points at the center of the hexagonal faces. It is important to realize that these shapes stack nicely upon one another by just shifting the second cell along any two of the other axes. Hence, if one moves from Γ along the (110) direction, once you pass the point K, you will be in the top square face of the next zone. Thus, you will arrive at the X point along the (001) direction of that Brillouin zone. This will become important when we plot the energy bands later in this section.

The Hamiltonian matrix will be an 8×8 matrix, which can be decomposed into four 4×4 blocks. When we consider only interactions between the nearest neighbors, the form of these blocks becomes quite simple. The two diagonal blocks are both diagonal in nature, with the atomic s and p energies along this diagonal. One of these two blocks represents the A atom (red in figure 2.11), and the other represents the B atom (blue in figure 2.11). The other two blocks—the upper right block and the lower left block—are full rank matrices. However, the lower left block is the Hermitian conjugate (complex conjugate and transpose) of the upper right block, as this is required for the total Hamiltonian to be Hermitian. These two blocks represent the interactions of an orbital on the A atom with an orbital on the B atom; these blocks will contain the Bloch sums. That is, an A atom will have interactions with its four B neighbors. The Bloch sums contain the four translation operators from one atom to its four neighbors. If we take an A atom as the origin of the reference coordinates, then the four nearest neighbors are positioned according to the vectors

$$
\begin{aligned}
r_1 &= \frac{a}{4}(\boldsymbol{a}_x + \boldsymbol{a}_y + \boldsymbol{a}_z) \\
r_2 &= \frac{a}{4}(\boldsymbol{a}_x - \boldsymbol{a}_y - \boldsymbol{a}_z) \\
r_3 &= \frac{a}{4}(-\boldsymbol{a}_x + \boldsymbol{a}_y - \boldsymbol{a}_z) \\
r_4 &= \frac{a}{4}(-\boldsymbol{a}_x - \boldsymbol{a}_y + \boldsymbol{a}_z).
\end{aligned}
\tag{2.52}
$$

The first vector, for example, points from the lower left red atom to the blue shown in the left panel of figure 2.9, along with the corresponding bond. Now, the 8×8 matrix has the following general form

$$
\begin{bmatrix}
E_s^A & 0 & 0 & 0 & H_{ss}^{AB} & H_{sx}^{AB} & H_{sy}^{AB} & H_{sz}^{AB} \\
0 & E_p^A & 0 & 0 & H_{xs}^{AB} & H_{xx}^{AB} & H_{xy}^{AB} & H_{xz}^{AB} \\
0 & 0 & E_p^A & 0 & H_{ys}^{AB} & H_{yx}^{AB} & H_{yy}^{AB} & H_{yz}^{AB} \\
0 & 0 & 0 & E_p^A & H_{zs}^{AB} & H_{zx}^{AB} & H_{zy}^{AB} & H_{zz}^{AB} \\
& & & & E_s^B & 0 & 0 & 0 \\
& & & & 0 & E_p^B & 0 & 0 \\
& & & & 0 & 0 & E_p^B & 0 \\
& & & & 0 & 0 & 0 & E_p^B
\end{bmatrix}
\tag{2.53}
$$

with the lower left block being filled in by the appropriate transpose complex conjugate of the upper right block (which also reverses the AB to BA). We have also adopted a short hand notation of x, y, z for p_x, p_y, p_z [12]. We note that, if we reverse the two subscripts (when they are different), this has the effect of conjugating the resulting complex energy. Interchanging the A and B atoms inverts the coordinate system, so that these operations result in the number of Bloch sums reducing to only 4 different ones.

To begin, we consider the term for the interaction between the s states of the two atoms. The s states are spherically symmetry, so there is no angular variation of importance outside that in the arguments of the exponentials, so we don't have to worry about the signs in front of each term in the Bloch sum. Thus, this term becomes [12]

$$
H_{ss}^{AB} = \langle s^A | H | s^B \rangle (e^{ik \cdot r_1} + e^{ik \cdot r_2} + e^{ik \cdot r_3} + e^{ik \cdot r_4}).
\tag{2.54}
$$

In the Slater–Koster nomenclature, the matrix element is denoted as E_{ss}. This nomenclature is the same whether the calculation is first principles, where the Hamiltonian contains the many-body terms, or empirical, where the value is adjusted to fit experiment. By expanding each exponential into its cosine and sine terms, the sum can be rewritten, after some algebra, as

$$
B_0(\mathbf{k}) = 4 \left[\cos\left(\frac{k_x a}{2}\right) \cos\left(\frac{k_y a}{2}\right) \cos\left(\frac{k_z a}{2}\right) \right.
$$
$$
\left. - i \sin\left(\frac{k_x a}{2}\right) \sin\left(\frac{k_y a}{2}\right) \sin\left(\frac{k_z a}{2}\right) \right]
\tag{2.55}
$$

We note that the sum has a symmetry between k_x, k_y, and k_z, which arises from the fact that the coordinate axes are quite easily interchanged due to the spherical symmetry of these orbitals. We will see this term appear elsewhere as well. For example, let us consider the term for the equivalent p states

$$
H_{xx}^{AB} = \langle p_x^A | H | p_x^B \rangle (e^{ik \cdot r_1} + e^{ik \cdot r_2} + e^{ik \cdot r_3} + e^{ik \cdot r_4}) = E_{xx} B_0(\mathbf{k}).
\tag{2.56}
$$

Here, the same Bloch sum has arisen because all the p_x orbitals point in the same direction; the positive part of the wave function extends into the positive x (or y or z) direction. Since the x axis does not have any real difference from the y and z axes, (2.55) also holds for the two terms H_{yy}^{AB} and H_{zz}^{AB}. Thus, the entire diagonal of the upper, right block has the same Bloch sum. Now, let us turn to the situation for the interaction of the s orbital with one of the p orbitals, which becomes

$$H_{sx}^{AB} = \langle s^A | H | p_x^B \rangle (e^{i k \cdot r_1} + e^{i k \cdot r_2} - e^{i k \cdot r_3} - e^{i k \cdot r_4}) = E_{sx}^{AB} B_1(\mathbf{k}). \tag{2.57}$$

Now, we note that two signs have changed. This is because two of the p_x orbitals point away from the A atom, while the other two point toward the A atom. That is, each p orbital has a positive part on one side of the atom and a negative part on the other side of the atom. Hence, the central atom sees two positive parts of the p orbital and two negative parts. Thus, the sign changes. Hence, these two pairs of displacement operators have different signs. Following the convention so far, we have given the energy an appropriate nomenclature. The Bloch sum becomes

$$B_1(\mathbf{k}) = 4 \left[-\cos\left(\frac{k_x a}{2}\right) \sin\left(\frac{k_y a}{2}\right) \sin\left(\frac{k_z a}{2}\right) \right. $$
$$\left. + i \sin\left(\frac{k_x a}{2}\right) \cos\left(\frac{k_y a}{2}\right) \cos\left(\frac{k_z a}{2}\right) \right] \tag{2.58}$$

We can now consider the other two matrix elements for the interactions between an s state and a p state. These are created by using the obvious $k_x \to k_y \to k_z \to k_x$, which leads to

$$B_2(\mathbf{k}) = 4 \left[-\sin\left(\frac{k_x a}{2}\right) \cos\left(\frac{k_y a}{2}\right) \sin\left(\frac{k_z a}{2}\right) \right. $$
$$\left. + i \cos\left(\frac{k_x a}{2}\right) \sin\left(\frac{k_y a}{2}\right) \cos\left(\frac{k_z a}{2}\right) \right] \tag{2.59}$$

and

$$B_3(\mathbf{k}) = 4 \left[-\sin\left(\frac{k_x a}{2}\right) \sin\left(\frac{k_y a}{2}\right) \cos\left(\frac{k_z a}{2}\right) \right. $$
$$\left. + i \cos\left(\frac{k_x a}{2}\right) \cos\left(\frac{k_y a}{2}\right) \sin\left(\frac{k_z a}{2}\right) \right] \tag{2.60}$$

These Bloch sums will also carry through to the interactions between two p orbitals as well. For example, if we consider p_x and p_y, it is the p_z axis that is missing, and thus this leads to B_3, which has the unique k_z direction.

When we reverse the atoms, the tetrahedron is inverted, and the four vectors (2.51) are reversed. In the Bloch sum, this is equivalent to reversing the directions of the momentum, which is an inversion through the origin of the coordinates. This takes each B into its complex conjugate. Reversing the order of the wave functions in the matrix element would also introduce a complex conjugate to the energies, but these have been taken as real, so this doesn't change anything. However, we have to note that we do have to keep track of upon which atom the s orbital is located for the s-p matrix elements, as this mixes different orbitals from the two atoms. We can now write the off-diagonal block—the upper, right block of (2.53)—as

$$
\begin{bmatrix}
E_{ss}B_0 & E_{sx}^{AB}B_1 & E_{sx}^{AB}B_2 & E_{sx}^{AB}B_3 \\
E_{sx}^{BA}B_1 & E_{xx}B_0 & E_{xy}B_3 & E_{xy}B_2 \\
E_{sx}^{BA}B_2 & E_{xy}B_3 & E_{xx}B_0 & E_{xy}B_1 \\
E_{sx}^{BA}B_3 & E_{xy}B_2 & E_{xy}B_1^* & E_{xx}B_0
\end{bmatrix}.
\tag{2.61}
$$

As discussed above, the rest of the 8×8 matrix is filled in to make the final matrix Hermitian, with the lower, left block being the Hermitian conjugate of (2.61). The total 8×8 matrix now can be diagonalized to find the energy bands as a function of the wave vector $\mathbf{k}$.

Before doing the diagonalization, we can discuss some so-called sanity checks, as at certain points, such as the Γ point and the X point, the matrix will simplify. For example, at the Γ point, the 8×8 can be decomposed into four 2×2 matrices. Three of these are identical, as they are for the three p symmetry results, which retain their degeneracy. The fourth matrix is for the s symmetry results. This is an important result, as it means that at the Γ point, the only admixture occurs between like orbitals on each of the two atoms. Each of these smaller matrices is easily diagonalized. The admixture of the s orbitals leads to

$$
E_{1,2} = \frac{E_s^A + E_s^B}{2} \pm \sqrt{\left(\frac{E_s^A - E_s^B}{2}\right)^2 + 16E_{ss}^2}.
\tag{2.62}
$$

When the A and B atoms are the same, as in Si, then this reduces to $E_{1,2} = E_s \pm 4E_{ss}$. Similarly, the p admixture leads to the result

$$
E_{3,4} = \frac{E_p^A + E_p^B}{2} \pm \sqrt{\left(\frac{E_p^A - E_p^B}{2}\right)^2 + 16E_{xx}^2}.
\tag{2.63}
$$

Again, when the A and B atoms are the same, as in Si, then this reduces to $E_{3,4} = E_p \pm 4E_{xx}$. There is an important point here, The atomic s energies lie below the p energies. Yet, the bottom of the conduction band is usually spherically symmetry, or s like, while the top of the valence band is typically triply degenerate, which means p like. Hence, the bottom of the two bands (valence and conduction) are both s like, while the tops are p like. The top of the valence band is the lowest of the two p levels, so is likely derived from the anion in the compound, so that the top of the valence band is derived from anion p states. On the other hand, the bottom of the conduction band is the higher energy s level, so is generally derived from the cation s states.

When we fit the various coupling constants to experimental data, then this method has come to be known as the semi-empirical tight-binding method (SETBM) [12]. In figure 2.12, we plot the band structure for GaAs using just the 8 orbitals discussed here. The band gap fits nicely, but the positions of the L and X minima of the conduction bands are not correct. The L minimum should only be about 0.29 eV above the bottom of the conduction band, while the X minima lie about 0.5 eV above the bottom of the conduction band. It turns out that the empty s orbitals of the next level lie slightly above the orbitals used here. If the excited orbitals (s*) are added to give a 10 orbital model (called the $sp^3 s^*$ method, in which the Bloch sums remain the same as the s orbitals, but the coupling energies are different), then a much better fit to the experimental bands can be obtained [36]. This is shown in figure 2.13. Referring to figure 2.12, the bands in both of these figures are plotted along the line from L down the (111) direction to Γ, then out along the (100) direction to the X point, then jump to the second zone X point, from which they can return to Γ along the (110) direction, while passing through K. This is a relatively standard path, and one we use throughout this chapter and book. An alternative to using the excited states is to go to second neighbor interactions, which can also provide some corrections.

In figure 2.14, the energy bands for Si are plotted using the $sp^3 s^*$ approach. It may be seen that this is not a direct gap, as the minimum of the conduction band is along the line from Γ to X. The position is about 85% of the way to X, and this line is usually denoted Δ. Reference again to figure 2.11 shows that there are six such lines, so the minimum indicated in the figure is actually one of six such minima, one each along the six difference versions of the (100) line.

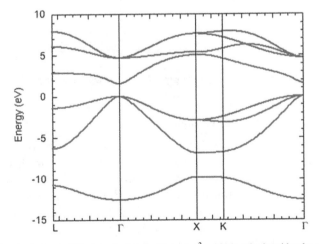

Figure 2.12. The band structure of GaAs calculated with just sp^3 orbitals calculated by the semi-empirical tight binding method.

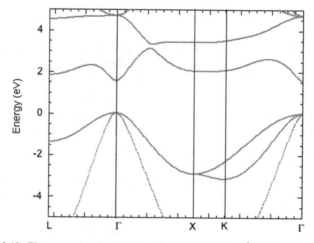

Figure 2.13. The energy bands of GaAs calculated with the sp^{3}s* 10 band approach.

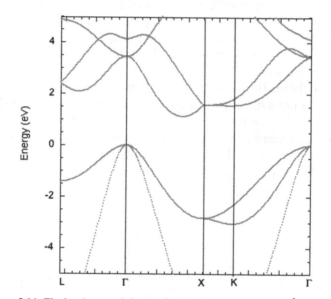

Figure 2.14. The bands around the gap for Si calculated using the sp^{3}s* method.

2.3.4 First principles and empirical approaches

As discussed earlier, one can use actual wave functions and the pseudo-potentials to evaluate the various energy parameters that have cropped up in this discussion. Or, one can use these energy parameters as fitting parameters to fit the bands to experimentally determined values, as was done in the previous section. One of the better approaches to first-principles tight-binding was formulated by Sankey [37]. A set of localized atomic orbitals, called *fireballs*, were created and coupled to an efficient exchange and correlation functional. This approach was extended to three center integrals to improve the behavior. Further, molecular dynamics—allowing

Table 2.1. Tight-binding parameters for select materials.

	E_s^A	E_s^B	E_p^A	E_p^B	E_{ss}	E_{xx}	E_{xy}	E_{sx}^{AB}	E_{sx}^{BA}
Si	−4.2	−4.2	1.72	1.72	−2.08	0.43	1.14	1.43	−1.43
GaAs	−2.7	−8.3	3.86	0.85	−1.62	0.45	1.26	1.42	−1.14
InAs	−3.53	−8.74	3.93	0.71	−1.51	0.42	1.1	1.73	−1.73
InP	−1.52	−8.5	4.28	0.28	−1.35	0.36	0.96	1.33	−1.35

the lattice to relax while computing the inter-atom forces—was used as well to get relaxed, or optimized, structures. A thorough review has recently appeared [38].

On the other hand, most people prefer the simpler, and more computationally efficient, route of the semi-empirical approach [12], where the parameters are varied so that the bands fit to experimentally determined gaps. But, there is always a question as to which experiments should be used for these measurements, and whose results should be accepted. The general road followed by computational physicists is to fit to optical measurements of inter-band transitions. But this adjusts gaps, as opposed to adjusting actual points in the band. My preference, as a transport person, is to fit the various extrema of the conduction band as well as the principle energy gap at Γ. Earlier I mentioned the positions of the X and L points in GaAs relative to the bottom of the conduction band at Γ. These have been determined experimentally, and are more important for semiconductor devices, as most of these devices utilize n-type material where the transport is by electrons in the conduction band. As major regions of the conduction band will be sampled in such devices used today, it is more important that the conduction band be correct, than that the optical gap at L be correct (although this gap can still be fit reasonably closely).

Let us consider some of the problems that arise in the empirical approach. We start with Si, where the top and bottom of the valence band are taken as 0 and −12.5 eV, respectively, with the first for convenience and the last coming from experiments [39]. The important s symmetry state that should be the bottom of the conduction band at Γ is not the doubly degenerate state in figure 2.14, but the single state above, which lies 4.1 eV above the top of the valence band [38]. Using (2.62), these values give $E_{ss} = 2.08$ eV and $E_s = -4.2$ eV. The difference $E_s - E_p$ is given as 7.2 eV by Chadi and Cohen [40], so that $E_p = 3$ eV. We now use (2.63) to find that $E_{xx} = 0.75$ eV. At this point, there are only two parameters left with which to fit all the other critical points in the band structure. This becomes a difficult task, and it is easy to see why extending the basis set to the excited states or to incorporate second-neighbor interactions becomes important. These will lead to further parameters with which to fit the other critical points in the Brillouin zone. Even with the added parameters for the compound semiconductors, it was obvious in figure 2.12 that a good fit was not obtained. Adding just the excited states gave enough new parameters that the good fit of figure 2.13 could be obtained. In tables 2.1 and 2.2, parameters for a few semiconductors are given. There are many sets of parameters in the literature, and another, different set is due to Teng *et al* [41].

Table 2.2. Some excited state parameters.

	E_{s*}^A	E_{s*}^B	E_{s*p}^{AB}	E_{s*p}^{BA}
Si	6.2	6.2	1.3	1.3
GaAs	6.74	8.6	0.75	1.25

Some have tried hard to find transferable parameters, that is parameters which work in both GaAs and InAs, as an example. One such set, which is also useful for strained material, is given by Tan *et al* [42]. Note that the diagonal elements E_{ss*} is usually ignored due to the orthogonality of the wave functions.

2.4 Momentum space methods

Clearly, the easiest way to approach the momentum space approach is to adopt the free electron wave functions, which we discussed in section 2.1. If the potentials due to the atoms are ignored, then this plane wave approach is exact, and one recovers the free electron bands described in figure 2.1. But, it is the perturbation of these bands by the atomic potentials that is important. The simplest case was described with figure 2.2. If the atomic potentials can be made relatively smooth, then the plane wave approach can be made to be relatively accurate, but the use of the word 'relative' here can be abused. The problem is how to handle the rapidly varying potential around the atoms, and also to incorporate the core electrons which modify this potential, but don't contribute (much) to the bonding properties of the semiconductors. Generally, the potential near the cores varies much more rapidly than just a simple Coulomb potential, and this means that a very large number of plane waves will be required. The presence of the core electrons, and this rapidly varying potential, can complicate the calculation, and make an approach which is solely composed of plane waves a very difficult task. One possible compromise was proposed by Slater [43], who suggested that the plane waves should be *augmented* by treating the core wave functions as those found by solving the isolated atom problem in a spherically symmetric potential. Then, the local potential around the atom was described by this same spherically symmetric potential up to a particular radius, beyond which the potential was constant. Hence, this *Augmented Plane Wave (APW) method* used what is called a *muffin-tin* potential (suggested by the relationship to a real muffin tin).

Still later, Herring [44] suggested that the plane waves should have an admixture of the core wave functions, so that by varying the strength of each core admixture, one could make the net wave function orthogonal to all of the core wave functions. By using these orthogonalized plane waves (or OPW method), the actual potential used in the band structure calculation could be smoothed (which leads to the pseudo-potential of section 2.2) and a smaller number of plane waves would be required. The OPW approach, like the APW method above, is one of a number of such cellular approaches in which real potentials are used within a certain radius of the atom, and a smoother (or no) potential is used outside that radius. The principle is that outside the so-called core radius we can use a plane wave representation, but we have to make these plane waves orthogonal to the core states. To begin with this, a set of core wave functions

$$\langle j, r_a | \psi_j(r - r_a) \rangle \tag{2.64}$$

is adopted, where the subscript j signifies a particular core orbital, and r_a represents the atomic position. This is then Fourier transformed as

$$\langle j, r_a | k \rangle = \int d^3r \psi_j(r - r_a) e^{ik \cdot r}. \tag{2.65}$$

Now, a version of a OPW plane wave can be constructed as

$$\psi_k = |k\rangle - \sum_{j,a} c_{j,a} |j, a\rangle \langle j, ak\rangle. \tag{2.66}$$

The constants $c_{j,a}$ are now adjusted to make the plane wave state orthogonal to each of the core wave functions. At the same time, the potential is smoothed by the core wave function, and we obtain the pseudo-potential in this manner. This is now the basis for the self-consistent, first-principles simulation, that depends upon generating the set of OPW functions and determining the proper set of interaction energies (Hartree, Hartree–Fock, LDA, and so on) and finding the correct pseudo-potential. This approach was apparently started by Phillips [45], and now there are many approaches (see [19]), but we will describe in this section the *empirical pseudo-potential method*. In this approach, which is fully in the same spirit as the empirical tight-binding method [12], one adjusts the pseudo-potentials to fit experimental measurements of many critical points in the Brillouin zone [46–49].

Now, it is important to recognize that the above approach is *three* dimensional. It involves the Fourier transform of a three dimensional structure. If we are going to use the momentum space approach for a one-dimensional or two-dimensional crystal, then we have to generate a full three-dimensional structure. For the one dimensional material, we create a two-dimensional lattice in the plane perpendicular to the one dimension of the material. For the two-dimensional layered material, we use multi-layers to generate the three dimensional structure. In each of these cases, however, the desired one-dimensional or two-dimensional structures must be separated sufficient far in space, that the replicas do not interact! That is, for e.g. layers of graphene, we need to space the layers further apart than they would occur in graphite so that the individual layers do not interact [50]. Similarly, for the TMDCs, multi-layers must be used, and by varying the separation one can study the transition from bulk to monolayer, whether using the empirical pseudo-potential method [35] or first-principles DFT [51]. In figure 2.15, we illustrate the pseudo-one-dimensional approach, where a xylyl-dithiol is stretched between two gold layers [52, 53] (in experiments, one of the gold layers would likely be a tip of the scanning tunneling microscope used to study the molecule [54]). The figure shows a single unit cell (or *super* cell), which is replicated in all three directions to generate the Fourier transforms, although only the unit cell is included in the computations. We have studied this with a momentum space DFT approach using the Vienna *ab-initio* Simulation Package (VASP) [55] as well as with two local orbital approaches,

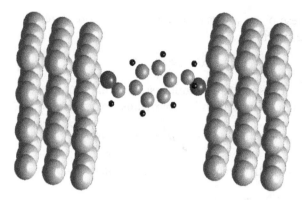

Figure 2.15. To simulate a quasi-one-dimensional structure one needs to create the pseudo-three-dimensional structure. Here, a xylyl-dithiol is positioned between two gold electrodes. This is one unit cell of the structure. Here, the green atoms are C, the blue atoms are S, the black atoms are H, and the gold atoms are Au.

Fireball [56] and Siesta [57]. VASP gives a slightly larger separation between the highest occupied molecular orbital and the lowest unoccupied molecular orbital, by about 1 eV (4 eV versus 3 eV). This difference is likely due to differences in the exchange and correlation functions used for the various calculations. One important point is that one must make sure that the gold work function is computed accurately by the self-consistent potential, otherwise the energy levels cannot be fully trusted [58]. To do this, the molecule is removed and the gold-air-gold structure is simulated. VASP achieves this, but it is harder to achieve with the local-orbital tight-binding approaches, although it has been done with Siesta.

2.4.1 The local pseudo-potential approach

Since we are working in momentum space, it is actually the Fourier transform of the pseudo-potential which will be of interest, since the use of a plane-wave basis essentially moves everything into the Fourier transform space (or momentum representation, quantum mechanically). Around a particular atom at site r_a, the pseudo-potential (2.25) may be written as

$$V_P(r - r_a) = \sum_G \tilde{V}_a(G)e^{IG\cdot(r-r_a)}. \qquad (2.67)$$

Here, G is the set of reciprocal lattice vectors, and is, in principle, an infinite set of such vectors. We will see later that a limited set can be introduced to reduce the computational complexity. But, of course, accuracy will improve as more reciprocal lattice vectors are used. The inverse of (2.67) now becomes

$$\tilde{V}_a(G) = \frac{1}{\Omega} \int d^3 r e^{-iG\cdot(r-r_a)} V_P(r - r_a). \qquad (2.68)$$

The quantity Ω is the volume of the unit cell in the crystal, as the reciprocal lattice vectors are defined by the unit cells of the crystal. It becomes convenient to work with the two atoms per unit cell of the zinc-blende or diamond crystals. In order to set a point of reference, we will reference the lattice vector to the mid-point between the two atoms, which is taken to be the origin. Then, we can write $r - r_{a,1} = t$, $r - r_{a,2} = -t$. Then, (2.67) becomes

$$V_P(r - r_a) = \sum_{\alpha=1,2} V_P(r - r_{a,\alpha})$$
$$= \sum_G [\tilde{V}_1(G)e^{iG\cdot t} + \tilde{V}_2(G)e^{-iG\cdot t}]e^{iG\cdot r}. \tag{2.69}$$

In the zinc-blende lattice and the diamond lattice, these two vectors refer to the positions of the two atoms of the basis set. Hence, $t = (a/8)(111)$, or one-eighth of the body diagonal distance of the face-centered cell (this is not the unit cell), while a is the length of the edge of this face-centered cell. At this point, it is convenient to introduce the symmetric and antisymmetric potentials as

$$V_S = \frac{1}{2}[\tilde{V}_1(G) + \tilde{V}_2(G)]$$
$$V_A = \frac{1}{2}[\tilde{V}_1(G) - \tilde{V}_2(G)], \tag{2.70}$$

so that

$$\tilde{V}_1(G) = V_S(G) + V_A(G)$$
$$\tilde{V}_2(G) = V_S(G) - V_A(G). \tag{2.71}$$

If we now insert the definitions (2.71) into (2.69), we can write the *cell* pseudo-potential as

$$V_c(G) = \tilde{V}_1(G)e^{iG\cdot t} + \tilde{V}_2(G)e^{-iG\cdot t}$$
$$= 2V_S(G)\cos(G \cdot t) + 2iV_A(G)\sin(G \cdot t). \tag{2.72}$$

Working with the Schrödinger equation for the pseudo-wave-function, we can develop the Hamiltonian matrix form. To begin, following (2.66) we define a plane wave function, at a particular momentum **k**, lying within the first Brillouin zone, as

$$|k\rangle = \sum_G C_G |k + G\rangle. \tag{2.73}$$

Now, introducing (2.72) and (2.73) into the Schrödinger equation (2.2) for this wave function, we find

$$\sum_G C_G \left[\frac{\hbar^2}{2m_0}(k + G)^2 + V_c(r) - E \right] |k + G\rangle = 0. \tag{2.74}$$

We now pre-multiply by the adjoint state $\langle k + G'|$, and do the inferred integration, so that the matrix elements between reciprocal lattice vectors G and G' are

$$\sum_G C_G \left\{ \left[\frac{\hbar^2}{2m_0}(k + G)^2 - E \right]\delta_{G,G'} + \langle k + G'|V_c(r)|k + G\rangle \right\} = 0. \tag{2.75}$$

Each equation within the curly brackets produces one row of the eigen-value matrix. Using (2.69) and (2.72), the off-diagonal elements can be evaluated as

$$\langle k + G' | V_c(r) | k + G \rangle = \frac{1}{\Omega} \int d^3r \sum_{G''} e^{-i(k+G')\cdot r} V_c(G'') e^{i(G''+k+G')\cdot r}$$
$$= \sum_{G''} V_c(G'') \delta(G'' + G - G') = V_c(G - G'). \tag{2.76}$$

Thus, we pick out one Fourier coefficient, depending upon the difference between G and G'. The diagonal contribution to this is $V_c(0)$, which provides only an energy shift of the overall spectrum. Usually, this is used to align the energy scale so that the top of the valence band lies at $E = 0$, although the reference energy can be placed anywhere. While (2.75) calls for an infinite number of reciprocal lattice vectors, typically a finite number is used, and this is large enough to allow some degree of convergence in the calculation. A common set is the 137 reciprocal lattice vectors composed of the set (and equivalent variations) of vectors (000), (111), (200), (220), (311), (222), (400), (331), (420), and (422) (all in units of $2\pi/a$). Among these sets, the magnitude squared $(G - G')^2$ takes values of 0, 3, 4, 8, 11, 12, 16, 19, 20, or 24 (in appropriate units). Normally, the off-diagonal elements are computed only for $(G - G')^2 \leqslant 11$ (or 12), as the Fourier amplitude for higher elements is quite small [48, 49], and going beyond these three does not appreciably affect the band structure [59]. An important point in considering the matrix elements is that the sines and cosines in (2.72) vanish for a number of the values given here. In the case of Si, for example, both atomic pseudo-potentials are equal, so all the sine terms vanish, and the potential has symmetric potential terms. In the zinc-blende structure, the sine term vanishes for $(G - G')^2 = 8$, while the cosine terms vanish for $(G - G')^2 = 4$ (this latter is also true in the diamond structure).

In figure 2.16, we show the band structure for Si, which has been computed using the local pseudo-potential described in this section. As remarked above, we have only three parameters to play with when we limit the Fourier coefficients as $(G - G')^2 \leqslant 11$. The values used for this are shown in table 2.3 for silicon and some III–V materials.

In figure 2.17, we plot the equivalent computation for GaAs. For this latter case, we have six parameters due to the presence of the asymmetric terms. As discussed at the beginning of the chapter, the fit has focused primarily on getting the conduction X and L valleys positioned correctly relative to the bottom of the conduction band. Still, one can notice that the minimum near X is actually in (toward Γ) from the actual X point, a result that is somewhat controversial. A startling feature of this band structure, relative to that for Si, is the large polar gap that opens in the valence band. This is seen in nearly all of the III–V materials. The parameters used here are also shown in table 2.3.

2.4.2 Adding nonlocal terms

It generally is found that the local pseudopotential has some problems in fitting to the available optical data. Some have suggested that one way to adjust e.g. the width

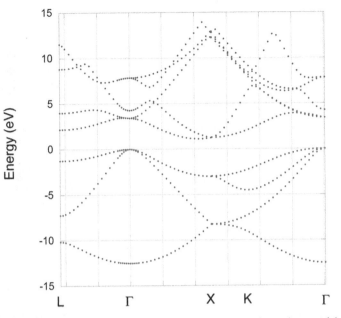

Figure 2.16. The band structure of Si computed with a local pseudo-potential.

Table 2.3. Local EPM parameters for select semiconductors.

	$V_S(3)$	$V_S(8)$	$V_S(11)$	$V_A(3)$	$V_A(4)$	$V_A(11)$
Si	−3.05	0.748	1.025			
GaAs	−3.13	0.136	0.816	1.1	0.685	0.172
InAs	−2.74	0.136	0.816	1.085	0.38	0.236
InSb	−2.47	0.	0.524	0.816	0.38	0.236

of the valence band is to add a term to the diagonal elements (the kinetic energy) by replacing the free electron mass with an adjusted value [60]. But, Pandey and Phillips [61] point out that there is no physical justification for this, and the sign is often wrong and thus does not provide the proper correction. These latter authors have, instead, suggested the use of a nonlocal pseudo-potential. They point out that in Ge, this provides an additional parameter which can represent repulsive effects from the 3d core states. (While the core states are removed from the pseudo-wave-functions and the pseudo-potential, it is known that they often will hybridize with the valence s-states, and will produce an effect (see, e.g., the discussion in [62]). Thus, adding the nonlocal corrections is a methodology to treat the effect of the d-states on primarily the valence electrons. But, this argument is weak, as one must question why it is used in Si, where there are no core d-states to worry about. The fact is, that it provides some additional angular behavior and provides an additional set of parameters with which to improve the fit between the computed band structure and the experimental data that is used to achieve the fit.

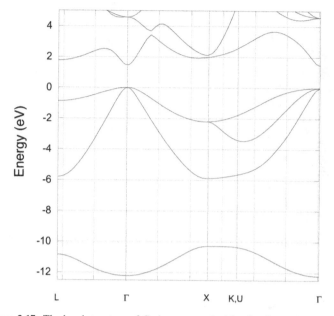

Figure 2.17. The band structure of GaAs computed with a local pseudo-potential.

To introduce a non-local (or angular variation) effect to the pseudo-potential, the local pseudo-potential itself is expanded by adding a non-local term which is expanded into spherical harmonics. Even with spherical symmetry, it is known that the solution to the Schrödinger equation involves angular variables and the angular solutions are normally expressed in terms of so-called spherical harmonics. The nonlocal part of the pseudo-potential for atom 1 may be written in terms of a projection operator P_1 onto the subspace Y_{lm} (m runs from $-l$ to l) of spherical harmonics with the same value of l as (here we follow the treatment of [61])

$$V_P^{NL}(\mathbf{r}) = \sum_l V_l(r)P_l, \tag{2.77}$$

where

$$V_l(r) = \begin{cases} A_i, & r < r_{0l} \\ 0, & r > r_{0l} \end{cases}. \tag{2.78}$$

Here, r_{0l} is an effective radius for the spherical harmonic within the spirit of the muffin-tin potential that was discussed above. Since each of the two atoms will contribute a nonlocal correction term to $V(r)$, as in (2.72), we can deal with each of the two terms separately, and then combine the results using (2.72). Here, however, we will use the factor of 1/2, discussed above to remove the additional factor of 2 in (2.72) and continue to interpret Ω as the volume of the unit cell, which is its definition in (2.68). The matrix element in (2.76) can now be computed from (2.77) as

$$\langle k + G' | V_c(r) | k + G \rangle = \frac{1}{\Omega} \int d^3r \sum_l e^{-i(k+G') \cdot r} V_l(r) P_l e^{i(k+G) \cdot r}. \tag{2.79}$$

To proceed, we use the common form of expanding exponentials in terms of Legendre polynomials as

$$e^{iK \cdot r} = \sum_l (i)^l (2l + 1) j_l(Kr) P_l(\cos \vartheta). \tag{2.80}$$

In this expression, the angle is that between the radius vector $\mathbf{r}$ and the momentum vector $\mathbf{K}$. As we have two exponentials, we will use primed and unprimed indices to correspond to the primed and unprimed $\mathbf{G}$ vectors that are in (2.79). The projection operator in (2.79) will select the appropriate Legendre polynomial and corresponding spherical Bessel function j_l, from the second exponential upon which it operates. Then, the matrix element can be rewritten as (we use the reduced notation $\mathbf{K} = \mathbf{k} + \mathbf{G}$ and $\mathbf{K}' = \mathbf{k} + \mathbf{G}'$)

$$\langle K' | V_P^{NL} | K \rangle = \frac{1}{\Omega} \int d^3r \sum_{l,\,l'} i^{l-l'} (2l + 1)(2l' + 1) \\ \times A_l P_l(\cos \vartheta) P_{l'}(\cos \vartheta') j_l(Kr) j_{l'}(K'r). \tag{2.81}$$

The angle ϑ' can be expanded in terms of the angle ϑ and the angle between the two momentum vectors, with the sine term integrating to zero under the three dimensional integral in the above equation. Hence, we have

$$\cos(\vartheta') = \cos(\vartheta)\cos(\vartheta_{KK'}). \tag{2.82}$$

As is generally the case when we want to simplify the computation, we are only interested in the values of $l = 0, 2$. The first value will give a low order correction, while the second case will account for the angular variation one might expect for a d-state. For $l = 0$, the integration over the angle is simple and gives a factor of 2. For the case of $l = 2$, we can expand the second Legendre polynomial and do the integration as [63]

$$\int_0^\pi P_2(\cos \vartheta) P_2(\cos \vartheta') d\vartheta = \int_0^\pi P_2(\cos \vartheta) P_2(\cos \vartheta) P_2(\cos \vartheta_{KK'}) d\vartheta \\ = \frac{2}{2l + 1} P_2(\cos \vartheta_{KK'}) \delta_{ll'}. \tag{2.83}$$

The solution of the remaining radial integration has been given by Pandey and Phillips [61] to be

$$\langle K' | V_P^{NL} | K \rangle = \frac{4\pi}{\Omega} \sum_l A_l P_l(\cos \vartheta_{KK'}) F(Kr, K'r), \tag{2.84}$$

with

$$F(x, x') = \begin{cases} \dfrac{r_{0l}^2}{x^2 - x'^2}[xj_{l+1}(x)j_l(x') - x'j_l(x)j_{l+1}(x')], & x \neq x' \\[12pt] \dfrac{r_{0l}^2}{2}\Big[j_l^2(x) - j_{l+1}(x)j_{l-1}(x)\Big], & x = x'. \end{cases} \qquad (2.85)$$

This result has also been found by Chelikowsky and Cohen [64], although they differ on the definition of the volume element discussed above. It is important to point out that the magnitudes in this latter function can be equal even in the off-diagonal elements, but the non-local correction is applied only to these off-diagonal elements as we discuss below.

There is nothing in this derivation to ascertain whether or not the diagonal element should be corrected with a non-local term. However, the diagonal element is a volume average (the zero Fourier coefficient), so that there should be no angular variation to couple a d-state correction. Moreover, it is clear in Pandey and Phillips [61] that the corrections are more important with the higher Fourier coefficients (they cite 8 and 11 for example). Indeed, these latter authors find that the equivalent local potential (this includes the correction for the nonlocal behavior), for Ge, has $V_S(3)$ increased (in magnitude) by only 4.4%, while $V_S(8)$ is increased by 220%, and $V_S(11)$ is increased by 30%. As a last point, an energy dependence of A_0 seems to have been introduced by Chelikowsky and Cohen [64]. When this is included, we have

$$A_0 = \alpha_0 + \beta_0 \frac{\hbar^2(\boldsymbol{K} \cdot \boldsymbol{K}' - k_F^2)}{2m_0}. \qquad (2.86)$$

Now, it has to be remembered that there is a value of each of the parameters for each of the atoms. However, for most semiconductors $\alpha_0 = 0$.

In figure 2.18, the non-local calculation for Si is compared with the local one of figure 2.16. The local calculation is shown in the red curves while the non-local one is shown in the blue curves. There are only small detailed differences between the two results. But, the fit is still empirical, so one might expect this to be the case, and the Fourier coefficients have had to be modified to achieve this fit when the non-local terms are included. For this fit, we used the values the values shown in table 2.4, along with $\alpha_0 = 3.5$.

In figure 2.19, the non-local calculation for GaAs is shown. Again, there are only small detailed differences of these curves from those of figure 2.14. Nevertheless, there has been a shift in the values of the various parameters in the calculation. For the fit shown here, we used the values shown in table 2.4. For this material, the principal non-local correction comes from the A_2 term, so that $A_0 = 0$. There are actually many sets of parameters in the literature; one such is Chelikowsky and Cohen [64].

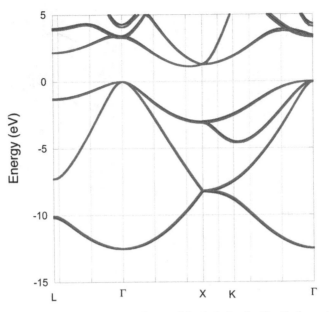

Figure 2.18. Comparison of the local (red) pseudopotential calculation for Si with the nonlocal (blue) one.

Table 2.4. Non-local EPM parameters for select semiconductors.

	$V_S(3)$	$V_S(8)$	$V_S(11)$	$V_A(3)$	$V_A(4)$	$V_A(11)$	$\beta_0{}^A$	$\beta_0{}^B$	$A_2{}^A$	$A_2{}^B$
Si	−3.05	0.5	0.6				0.2	0.2	0	0
GaAs	−2.98	0.21	0.816	1.3	0.68	0.136	0	0	1.7	5.0
InAs	−2.79	0.096	0.45	1.09	0.43	0.408	0.35	0.25	6.8	13.6
InSb	−2.47	−0.1	0.26	0.816	0.48	0.236	0.45	0.48	7.48	6.53

2.4.3 The spin–orbit interaction

It is well known that the quantum structure of atoms can cause the angular momentum of the electrons to mix with the spin angular momentum of these particles. Since the energy bands are composed of both the s- and p-orbitals of the individual atoms in the semiconductors, it has been found that the spin–orbit interaction affects these calculations as well. The spin–orbit interaction is a relativistic effect in which the angular motion of the electron interacts with the gradient of the confining potential to produce an effective magnetic field. This field then couples to the spin in a manner similar to the Zeeman effect. Early papers which used the OPW method clearly demonstrated that the spin–orbit interaction was important for the detailed properties of the bands [65, 66]. Not the least of these effects is the splitting of the three-fold bands at the top of the valence band, producing the so-called *split-off band*. This latter band lies from a few meV to a significant fraction of an eV below the top of the valence band in various

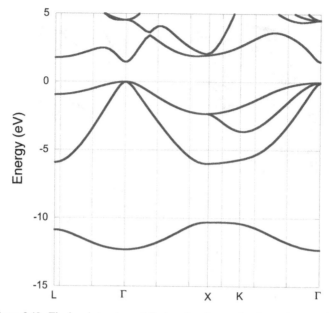

Figure 2.19. The band structure of GaAs using the non-local pseudo-potential.

semiconductors. Usually, the effect in the conduction band is smaller, but can still be significant.

The first inclusion of the spin–orbit interactions in pseudo-potential calculations for semiconductors is thought to be due to Bloom and Bergstresser [67], who extended the interaction Hamiltonian of Weisz [68] to compound materials. The spin–orbit interaction is known to be stronger in heavier atoms, so they considered the heavier materials InSb and CdTe. Subsequently, it has been applied to most of the major semiconductors. In general, the formulation of Weisz can be rewritten as [69]

$$\langle K', s'|H_{SO}|K, s\rangle = (K' \times K) \cdot \langle s'|\sigma|s\rangle \sum_l \lambda_l P_l'(\cos \vartheta_{KK'}) S(K' - K), \qquad (2.87)$$

where K and K' are the shifted wave vectors defined above (2.81), P_l' is the derivative of the Legendre polynomial, and S is the structure factor (our sine and cosine terms we used earlier in (2.72)). The term σ is a vector of the 2×2 Pauli spin matrices, so that the direction of the leading cross product picks out one Pauli term. Now, the wave function is more complicated. If we use the 137 plane basis discussed earlier, then there will be 137 basis states with spin up and another 137 basis states with spin down. So, each plane wave is now associated with a spin wave function and our matrix has doubled in rank. Since the bonding electrons we are interested in are only composed of the s- and p-states, we need keep only the $l = 1$ term in (2.87). We also have two atoms in our basis, and this will lead to even and odd values for the parameter λ_l. It is important to note that this version of the spin–orbit interaction is a *full-zone* correction. It is basically different than the spin–orbit interaction we deal with in the next section, where it is coupled to the $\boldsymbol{k}\cdot\boldsymbol{p}$ terms of the Hamiltonian, and treated as a perturbation correction to the band structure.

With the above changes, we can write (2.87) as [37, 70]

$$\langle \boldsymbol{K}', s'|H_{SO}|\boldsymbol{K}, s\rangle = -i(\boldsymbol{K}' \times \boldsymbol{K}) \cdot \langle s'|\boldsymbol{\sigma}|s\rangle$$
$$\times \{\lambda_S \cos[(\boldsymbol{K}' - \boldsymbol{K}) \cdot \boldsymbol{t}] + i\lambda_A \sin[(\boldsymbol{K}' - \boldsymbol{K}) \cdot \boldsymbol{t}]\}. \tag{2.88}$$

The two parameters are

$$\lambda_S = \frac{1}{2}(\lambda_1 + \lambda_2)$$
$$\lambda_A = \frac{1}{2}(\lambda_1 - \lambda_2), \tag{2.89}$$

where the 1 and 2 refer to atoms A and B in the unit cell, with

$$\lambda_1(K, K') = \mu B_n(K)B_n(K')$$
$$\lambda_2(K, K') = \alpha\mu B_m(K)B_m(K'), \tag{2.90}$$

where the subscripts n and m correspond to the row of the periodic table in which the atom resides, and μ and α are two fitting parameters. The functions B in (2.90) are determined by the core wave functions for the appropriate states as

$$B_n(K) = i\sqrt{12\pi}\, C \int_0^\infty j_n(Kr)R_n(r)dr. \tag{2.91}$$

In this equation, the j is a spherical Bessel function and the R is the radial part of the core wave function. Pötz and Vogl [70] have shown that the functions of (2.90) can be approximated by the relations

$$B_2 = \frac{1}{\left(1 + \kappa_2^2\right)^3}$$

$$B_3 = \frac{5 - \kappa_3^2}{5\left(1 + \kappa_3^2\right)^4} \tag{2.92}$$

$$B_4 = \frac{5 - 3\kappa_4^2}{5\left(1 + \kappa_4^2\right)^5},$$

where

$$\kappa_n = \frac{Ka_B}{\zeta_n}. \tag{2.93}$$

Here, a_B is the Bohr radius and ζ_n is the normalized (to the Bohr radius) radial extent of the core wave functions. Pötz and Vogl [70] have given values for the parameters for many of the tetrahedrally coordinated semiconductors. Our own simulations have used their values, but with an adjustment to the principal coupling parameter μ.

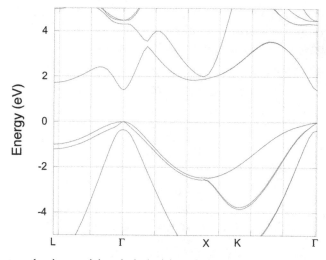

Figure 2.20. The energy bands around the principal minima of the conduction band for GaAs are shown for a non-local EPM, including the spin–orbit interaction.

In figure 2.20, the energy bands for GaAs, found by including the spin–orbit interaction are shown. Here, the coupling parameter $\mu = 0.0125$, somewhat stronger than that found by Pötz and Vogl [70]. In addition, the non-local parameter A_2^B has had to be increased to 8.5 in order to control the shifts of the bands. The opening of the split-off band is clearly seen, and the top of the valence band is now only doubly degenerate. However, one can see that these two bands also split quickly once one is away from the Γ point. The two bands that are degenerate at Γ are referred to as the heavy hole band (upper of the two) and the light hole band (lower of the two). There are other splittings in the bands that can be seen at various points in the zone. There is also a splitting of the conduction band away from the exact Γ point, but this is too small to be seen in the figure, as it is only a few meV. This latter splitting tends to be linear in k, so has to vanish at Γ.

2.5 The $k \cdot p$ method

While the spin–orbit interaction is easily incorporated within the pseudo-potential method, and most other computational approaches, it is often treated on its own as a perturbational approach, in which it is usually coupled with the $k \cdot p$ terms of the Hamiltonian. One reason for this approach is to find an analytical formula for the energy bands that also expresses the non-parabolic behavior discussed in section 2.1.2. We already have pointed out in that section, where we treated the nearly free electron model, that away from the actual minimum or maximum of the bands, the dispersion becomes non-parabolic. The non-parabolicity arises from the interaction between the wave functions of the two bands that would ordinarily (without the gap) be degenerate at the band edge (zone edge or zone center). This interaction remains even as the crystal potential opens the gap. However, the interaction decreases in amplitude as the gap between the two bands increases, and this causes the bands to not diverge quite as rapidly. At large values of momentum, away from the crossing, the bands return to the free electron bands, while an expansion in small momentum leads back to the parabolic relationship that has to be there at the band extremum. The spin–orbit interaction is a perturbative approach which couples various bands, and thus leads to a mixing of the basic s and p states, which alters the admixtures of the components as the wave vector, and hence the energy, vary.

It was established in section 2.1 above that the general solutions of the Schrödinger equation in a periodic potential are Bloch wave functions. It is these Bloch functions that will be utilized to set up the Hamiltonian for the $\mathbf{k} \cdot \mathbf{p}$ method. Here $\mathbf{k}$ is the wave vector describing the propagation of the wave and is related to the crystal momentum of the electron (in whichever band it is located). On the other hand, $\mathbf{p} = -i\hbar\nabla$ is the quantum mechanical momentum operator, which is related to motion in real space. If we put the vector form of the Bloch function (2.12) into the vector Schrödinger equation (2.2), we find that

$$\left[-\frac{\hbar^2}{2m_0}\nabla^2 - \frac{i\hbar^2}{m_0}\mathbf{k} \cdot \nabla + \frac{\hbar^2 k^2}{2m_0} + V(r) \right]u_k(r) = Eu_k(r), \qquad (2.94)$$

where the exponential term has been dropped as it is common to all terms in the equation and we have to remember that there is a Bloch function for each of the bands. This equation can be rewritten as

$$\left[-\frac{\hbar^2}{2m_0}\nabla^2 - \frac{i\hbar^2}{m_0}\mathbf{k} \cdot \nabla + V(r) \right]u_k(r) = \left(E - \frac{\hbar^2 k^2}{2m_0} \right)u_k(r). \qquad (2.95)$$

In this equation, we have used the operator form of the momentum, and the free electron term has been incorporated with the energy. The second term, in the square brackets in (2.95), is the so-called $\mathbf{k} \cdot \mathbf{p}$ term. As remarked above, this term is often treated by perturbation theory to build up the effective mass from a perturbation summation over all energy bands in the crystal. This is not necessary. If we use a limited basis set, a diagonalization procedure can be utilized to incorporate these terms exactly. The procedure we follow is due to Kane [71, 72]. In this approach, we use the four basis states that were used for the tight-binding approach in section 2.3.3. That is, we use four hybridized orbitals (which is different from the four states on each atom of the basis) for the single s and three p states. But, these will be doubled to add the spin wave function to each orbital, as explained further below. This is not the only approach possible, and a 16 band version can be found in Yu and Cardona [73].

We still must add the spin–orbit interaction terms to the Hamiltonian in (2.95). The spin–orbit interaction will split the triply degenerate (six-fold degenerate when spin is taken into account) valence bands at the maximum, which occurs at the zone center (the Γ point), as discussed in the last section. This splitting arises in nearly all tetrahedrally coordinated semiconductors. The interaction mainly splits off one (spin degenerate) state, leaving a doubly degenerate (four-fold with spin) set of levels that correspond to the light holes and the heavy holes. Thus, to account properly for this band arrangement, it is necessary to include the spin–orbit interaction in (2.95). This interaction leads to two additional terms that arise from coupling of the orbital crystal momentum (as well as the free electron momentum) and the spin angular momentum (motion of the electron spinning on its own axis) of the electrons. These two terms are

$$\frac{\hbar}{4m_0}\{[\nabla V(r) \times p] \cdot \boldsymbol{\sigma} + [\nabla V(r) \times k] \cdot \boldsymbol{\sigma}\}, \qquad (2.96)$$

where $\boldsymbol{\sigma}$ is the vector of Pauli spin matrices as used in the previous section. The second of these terms gives a term that is linear in k at the zone center and actually shifts the maxima of the valence band a negligibly small amount away from Γ in most compound semiconductors where there is no inversion symmetry of the crystal. It is usually a very small effect, and this term can be treated in perturbation theory. When coupled with effects from higher-lying bands, this term will be quite important in actually producing a mass different from the free-electron mass for the heavy-hole band, and these results are discussed below. This term, however, is ignored in the present calculations. The first term of (2.96) is called the k-independent spin–orbit interaction and is the major term that must be included in (2.95).

2.5.1 Valence and conduction band interactions

It is convenient to adopt a different set of wave functions than the normal set of orbitals. While one could use simply the s- and p-state atomic functions, the Hamiltonian that would result is somewhat more complex. By using a suitable combination of some of the states, it will appear in a simpler format. The basic sp^3 hybridization of the conduction and valence bands suggests the use of s, p_x (we will denote this as X), p_y (Y), and p_z (Z) wave functions. These are not proper Bloch functions, but they have the symmetry of the designated band states near the appropriate extrema and are the local cell part of the Bloch states. The four interactions of (2.95) and (2.96) leave the bands doubly (spin) degenerate, which eases the problem and size of the Hamiltonian matrix to be diagonalized. As mentioned above, we should take s and p hybrids for the two atoms in the basis separately, but the energies and wave functions discussed here are assumed to be appropriate averages. For the eight (with spin) basis states, the wave functions are now taken as the eight functions (the arrow denotes the direction of the electron spin in the particular state) [71, 72]

$$
\begin{array}{cc}
|iS\downarrow\rangle & |iS\uparrow\rangle \\[4pt]
\left|\dfrac{X - iY}{\sqrt{2}}\uparrow\right\rangle & \left|\dfrac{X + iY}{\sqrt{2}}\downarrow\right\rangle \\[8pt]
|Z\downarrow\rangle & |Z\uparrow\rangle \\[4pt]
\left|\dfrac{X + iY}{\sqrt{2}}\uparrow\right\rangle & \left|\dfrac{X - iY}{\sqrt{2}}\downarrow\right\rangle.
\end{array} \qquad (2.97)
$$

The four states in the left hand column forms one set of states, which are degenerate with the set of states in the right hand column. In evaluating the Hamiltonian matrix, which will be 8×8, the wave vector is taken in the z-direction. This selection is done for convenience (it is the direction normal to the circular pairing of the X and Y functions). One can use an arbitrary direction for the momentum, but the matrix is more complicated. On the other hand, the z-axis can be rotated to an arbitrary direction by a coordinate rotation, and this will change the Hamiltonian matrix appropriately.

Since the basis functions lead to doubly degenerate levels, the 8×8 matrix separates into a simpler block diagonal form, containing two 4×4 matrices, one for each spin direction. These are the diagonal blocks, and the 4×4 off-diagonal blocks are zero. Hence, each of the diagonal blocks will separate on its own, and has the form

$$
\begin{bmatrix}
E_s & 0 & kP & 0 \\
0 & E_p - \dfrac{\Delta}{3} & \sqrt{2}\dfrac{\Delta}{3} & 0 \\
kP & \sqrt{2}\dfrac{\Delta}{3} & E_p & 0 \\
0 & 0 & 0 & E_p + \dfrac{\Delta}{3}
\end{bmatrix}. \tag{2.98}
$$

The parameter Δ is a positive quantity, given by

$$
\Delta = \frac{3i\hbar}{4m_0^2} \left\langle X \left| \left(p_y \frac{\partial}{\partial x} - p_x \frac{\partial}{\partial y} \right) \right| Y \right\rangle, \tag{2.99}
$$

which gives the matrix elements of (2.96) with the p wave functions (the spin is taken in the p_z direction) normal to k. Empirically, this is the spin-orbit splitting energy that describes the energy difference between the split-off valence band and the top of the valence band, and is a measured quantity for most materials. The momentum matrix element P arises from the $\boldsymbol{k} \cdot \boldsymbol{p}$ term in (2.95) and is given by

$$
P = -\frac{i\hbar}{m_0}\langle S | p_z | Z \rangle \tag{2.100}
$$

E_s and E_p are an average of the atomic energy levels we have used earlier. The fourth line in the matrix (2.98) is an isolated level, and is the heavy-hole band. Since this isolated level is at the top of the valence band, it is necessary to set $E_p = -\Delta/3$, if we want to reference the zero of energy to the top of the valence band as done in the previous sections. This heavy hole band has an energy which is just the free electron curvature of the second term in (2.95). Unfortunately, this curvature has the wrong sign, a point to which we return later. Once this energy level choice is made, the characteristic equation for the determinant of the remaining 3×3 matrix is

$$
E'(E' - E_G)(E' + \Delta) - k^2 P^2 \left(E' + \frac{2\Delta}{3} \right) = 0, \tag{2.101}
$$

where the reduced energy E' is given by the reduced energy (the term in parentheses) on the right-hand side of (2.95) and we have set $E_G = E_s$ in recognition that the bottom of the conduction band is determined by the s-states, as discussed in section 2.3.

Small kP. If the size of the last term in (2.101) is quite small (e.g., the energy is near to the band extremum), the solutions will be basically just those arising from the products of the first term, with a slight adjustment for the kP term. For this case, the three bands are given by the three zeroes of the first term, which leads to

$$E_C = E_G + \frac{k^2 P^2}{3}\left(\frac{2}{E_G} + \frac{1}{E_G + \Delta}\right) + \frac{\hbar^2 k^2}{2m_0}$$

$$E_{lh} = -\frac{2k^2 P^2}{3E_G} + \frac{\hbar^2 k^2}{2m_0} \qquad (2.102)$$

$$E_{\Delta} = -\Delta - \frac{k^2 P^2}{3(E_G + \Delta)} + \frac{\hbar^2 k^2}{2m_0}.$$

The three solutions are for the conduction band, the light hole band, and the split off (spin orbit) band, respectively (the heavy-hole band has been discussed above, and we will return to it below). The free electron contribution has been restored to each energy from (2.95). Within this approximation, the bands are all parabolic for small values of k. We note that, in each case, the effective mass will be inversely proportional to the square of the momentum matrix element. For example, for the conduction band, we can infer the effective mass to be given by

$$\frac{1}{m_c} = \frac{1}{m_0} + \frac{2P^2}{3\hbar^2}\left(\frac{2}{E_G} + \frac{1}{E_G + \Delta}\right), \qquad (2.103)$$

and this mass will be considerably smaller than the free electron mass m_0. Similarly, masses can be inferred for the other two bands as well, but these values are all slightly different, although all depend on both the energy gap and the momentum matrix element. Near $k = 0$, the effects of the higher bands are relatively minor.

$\Delta = 0$. The Two-Band Model. If the spin–orbit interaction is set to zero, then the split off band becomes degenerate with the light-hole and the heavy hole bands at $k = 0$. The remaining interaction is just between the almost mirror image conduction and light hole bands. In this case, these two energies are given by

$$E_c = \frac{E_G}{2}\left[1 + \sqrt{1 + \frac{4k^2 P^2}{E_G^2}}\right] + \frac{\hbar^2 k^2}{2m_0}$$

$$E_{lh} = \frac{E_G}{2}\left[1 - \sqrt{1 + \frac{4k^2 P^2}{E_G^2}}\right] + \frac{\hbar^2 k^2}{2m_0}. \qquad (2.104)$$

The general form here is just the hyperbolic description found in (2.20) from the nearly-free electron model, except here the interaction energy is defined in terms of the momentum matrix element and the energy gap. We can use this to define an effective mass at the band edge (in the $k \to 0$ limit). If we expand the square root for the small argument limit, we find that

$$\frac{1}{m_c} = \frac{1}{m_0} + \frac{2P^2}{\hbar^2 E_G}$$

$$\frac{1}{m_{lh}} = -\frac{1}{m_0} + \frac{2P^2}{\hbar^2 E_G}. \tag{2.105}$$

The opposite signs on the free electron mass keep these two bands from being pure mirror images about the center of the energy gap.

$\Delta \gg E_G$, kP. In the case for which the spin orbit splitting is large, a somewhat different expansion of the characteristic equation can be obtained. For this case, the spin orbit energy is taken to be larger than any corresponding energy for which a solution is being sought, and the resulting quadratic equation can be solved as

$$E_c = \frac{E_G}{2}\left[1 + \sqrt{1 + \frac{8k^2P^2}{3E_G^2}}\right] + \frac{\hbar^2 k^2}{2m_0}$$

$$E_{lh} = \frac{E_G}{2}\left[1 - \sqrt{1 + \frac{8k^2P^2}{3E_G^2}}\right] + \frac{\hbar^2 k^2}{2m_0} \tag{2.106}$$

$$E_\Delta = -\Delta - \frac{k^2 P^2}{3(E_G + \Delta)} + \frac{\hbar^2 k^2}{2m_0},$$

and the spin–orbit split-off band has been brought forward from (2.102). We note here that the terms under the square-root differ from those in (2.104) just by the factor of 2/3. The conduction band and the light-hole band remain almost mirror images.

Hyperbolic Bands. It may be seen from (2.104) and (2.106) that the conduction band and the light hole band are essentially hyperbolic in shape. The relation between the momentum matrix element and the band edge effective mass changes slightly with the size of the spin orbit energy, but this change is relatively small, as the numerical coefficient of the k^2 term changes by only a factor of 2/3 as Δ goes from zero to quite large. The major effect in these equations is the hyperbolic band shape introduced by the kp interaction, and the variation introduced by the spin–orbit interaction is quite small other than the motion of the maximum of the split off band away from the heavy hole band. For this reason, it is usually decided to introduce the band edge masses directly into the hyperbolic relationship without worrying about whether or not the size of the spin orbit energy is significant. Then, the most common form seen for the hyperbolic bands is that of (2.104) with the effective masses given by (2.105). For most direct gap group III V compound semiconductors, the conduction band and light hole band effective masses are small and on the order of 0.01 to 0.1, so that the free electron term is essentially negligible. Since the momentum matrix element arises from the sp^3 hybrids, the results of (2.104) are that the masses scale with the band gap in an almost linear fashion, with modest variations from the momentum matrix element. Materials with narrow band gaps usually have very small values of the effective masses.

It is often useful to rearrange the mirror image bands, which are hyperbolic in nature, to express the momentum wave vector k directly in terms of the energy. This is easily done, using (2.104) and (2.105), with the results that

$$\frac{\hbar^2 k^2}{2m_c} = E\left(\frac{E}{E_G} - 1\right)$$
$$\frac{\hbar^2 k^2}{2m_{lh}} = -E\left(1 - \frac{E}{E_G}\right). \tag{2.107}$$

Here, we must remind ourselves that the energy of the conduction band goes from a value of E_G upwards, as the energy is measured from the valence band maxima. If we shift the zero of energy to the bottom of the conduction band, then the sign is changed in the two parentheses.

2.5.2 Examining the valence bands

The equations above have not been corrected for possible interactions with the higher lying bands, as these are beyond the approximations used in this section. One can go beyond this simple four-band approximation to a higher order approach. Indeed, if the heavy-hole band is to be corrected, then more bands are required with the interactions between them handled by perturbation theory rather than direct diagonalization. In general, these lead to the linear k terms mentioned above, which are important at very small values of k in the valence band. There are also quadratic terms in k that lead to variations of the masses beyond those of the hyperbolic bands, and quadratic cross terms that lead to warping of the band away from a spherically symmetric shape (at $k = 0$). These extra terms are traditionally handled by perturbation theory [74], typically by a 16 band formulation (eight bands which are spin split) [73]. The dominant effect of these terms, however, is on the light- and heavy hole bands, and these band by be written as [74]

$$E_h = -\left[Ak^2 \pm \sqrt{B^2 k^4 + C^2\left(k_x^2 k_y^2 + k_y^2 k_z^2 + k_z^2 k_x^2\right)}\right], \tag{2.108}$$

where the upper sign is for the light holes and the lower sign is for the heavy holes. The constants A, B, and C are known for many semiconductors. For Si, they take the values 4, 1.1, and 4.1, respectively. In addition, the quartic terms in the parentheses under the square root lead to significant angular variation of the bands—they are not spherically symmetric, a point raised in chapter 1.

At this point, it is useful to go through an example to show how the pseudo-potential bands are fit with the $k \cdot p$ perturbation approach and then the masses are extracted by the higher order perturbation of (2.108). For this, we consider the ternary alloy $AlAs_{0.16}Sb_{0.84}$. This alloy finds considerable usage as a cladding layer in infrared lasers grown on InAs [75], barrier material in high-electron mobility transistors using InAs [76], and the solar cell world for InAs absorbing layers [77]. In

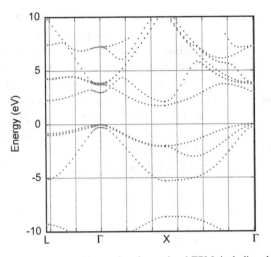

Figure 2.21. The energy bands of $AlAs_{0.16}Sb_{0.84}$ using the nonlocal EPM, including the spin–orbit interaction.

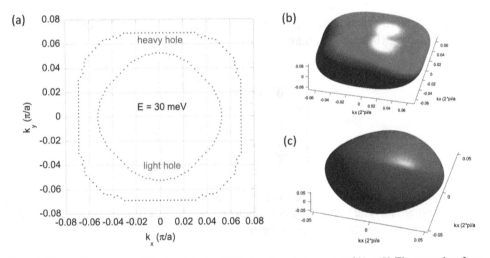

Figure 2.22. (a) The contours of the bands in the (001) plane for a hole energy of 30 meV. The warped surfaces can be inferred from these shapes. (b) Three dimensional energy surface of the heavy holes at an energy of 30 meV. (c) Three dimensional energy surface of the light holes at an energy of 30 meV. Panels (b) and (c) are reproduced courtesy of D Vasileska, with her permission.

figure 2.21, we show the energy band structure for this material as computed using the nonlocal EPM method described above, including the full zone spin–orbit interaction of section 2.4.3. The energy gap is indirect with the X minimum, slightly away from the X point as in Si, of 2.09 eV. The Γ and L minima are somewhat higher. The spin–orbit splitting at Γ is 0.58 eV.

Our interest is in the valence bands at this point. In figure 2.22, we plot the constant energy surfaces for the light- and heavy-hole energies. In panel (a), the constant energy surfaces are plotted in the $k_z = 0$ plane. In panels (b) and (c), the three dimensional constant energy surfaces are plotted for the heavy and light holes, respectively. These plots illustrate the nature of the non-spherical warping of the hole bands. The jitter in panel (a) arises from the discretion of the energy in the calculation, and has been smoothed in the other two panels. It is clear that determining an effective mass for either of these two bands depends upon just what crystal direction is chosen.

In arriving at the form of (2.108), Dresselhaus *et al* used the spin split three bands of the upper valence band [74]. Here, they used the p_x, p_y, and p_z orbitals directly. Introducing the $\mathbf{k} \cdot \mathbf{p}$ perturbation affects the three orbitals without reference to the spin–orbit interaction, and the 6×6 Hamiltonian splits into 3×3 matrices along the diagonals, with zeros for the off-diagonal 3×3 matrices. These matrices are given by

$$\begin{bmatrix} Lk_x^2 + M(k_y^2 + k_z^2) - E' & Nk_xk_y & Nk_xk_z \\ Nk_xk_y & Lk_y^2 + M(k_x^2 + k_z^2) - E' & Nk_yk_z \\ Nk_xk_z & Nk_yk_z & Lk_z^2 + M(k_y^2 + k_x^2) - E' \end{bmatrix} = 0 \quad (2.109)$$

The spin–orbit interaction adds a full 6×6 interaction matrix which couple the spins on each of the orbitals. This matrix is given by

$$H_{SO} = \frac{\Delta}{3} \begin{bmatrix} 0 & -i & 0 & 0 & 0 & 1 \\ i & 0 & 0 & 0 & 0 & -i \\ 0 & 0 & 0 & -1 & i & 0 \\ 0 & 0 & -1 & 0 & i & 0 \\ 0 & 0 & -i & -i & 0 & 0 \\ 1 & i & 0 & 0 & 0 & 0 \end{bmatrix} \begin{bmatrix} x\uparrow \\ y\uparrow \\ z\uparrow \\ x\downarrow \\ y\downarrow \\ z\downarrow \end{bmatrix} \quad (2.110)$$

The authors make an arbitrary choice of a basis at this point, in order to define the three parameters. They choose this basis to have the group notation Γ_{25}^+ for which the degenerate states at $k \sim 0$ transform as $\epsilon_1^+ \sim yz$, $\epsilon_2^+ \sim zx$, $\epsilon_3^+ \sim xy$.

$$L = \frac{\hbar^2}{m_0^2} \sum_{lj} \frac{\left|\langle \epsilon_1^+ |p_x| lj \rangle\right|^2}{E_G - E_l}$$

$$M = \frac{\hbar^2}{m_0^2} \sum_{lj} \frac{\left|\langle \epsilon_1^+ |p_y| lj \rangle\right|^2}{E_G - E_l} \quad (2.111)$$

$$N = \frac{\hbar^2}{m_0^2} \sum_{lj} \frac{\left|\langle \epsilon_1^+ |p_z| lj \rangle\right|^2}{E_G - E_l},$$

and Δ is given by (2.99) with the atomic orbitals replacing those in this latter equation. Here, l represents the band and j is a state in that band (the band here is the valence band).

The energy E' is defined just below (2.101). These three matrix elements relate to the three parameters in (2.108) as

$$A = \frac{1}{3}(L + 2M) + \frac{\hbar^2}{2m_0}$$

$$B = \frac{1}{3}(L - M) \tag{2.112}$$

$$C^2 = \frac{1}{3}[N^2 - (L - M)^2].$$

The A, B, C can be determined from the band edge effective masses in the various directions, and using (2.108) we can write these as [73]

$$k\|(100): \qquad \begin{array}{l} \dfrac{1}{m_{hh}} = \dfrac{2}{\hbar^2}(-A + B) \\[2ex] \dfrac{1}{m_{lh}} = \dfrac{2}{\hbar^2}(-A - B) \end{array} \tag{2.113}$$

and

$$k\|(111): \qquad \begin{array}{l} \dfrac{1}{m_{hh}} = \dfrac{2}{\hbar^2}\left[-A + B\left(1 + \dfrac{|C|^2}{3B^2}\right)\right] \\[3ex] \dfrac{1}{m_{lh}} = \dfrac{2}{\hbar^2}\left[-A - B\left(1 + \dfrac{|C|^2}{3B^2}\right)\right]. \end{array} \tag{2.114}$$

In figure 2.23, we fit the warped energy surfaces described by (2.108) to the heavy- and light-hole masses for these two directions. This allows us to estimate the masses, then to fit to the equations (2.113) and (2.114) to evaluate the various parameters. From this figure, it seems that we can obtain a reasonable fit to the pseudo-potential bands near the top of the valence band in a manner that reproduces the warped nature of these bands. This fit gives dimensionless values for $A \sim 8.6$, $B \sim 2.2$, $C \sim 1.6$, in units of eV-m^2, with the energy given as in (2.108). From the $k \cdot p$ perturbation, we estimate that the effective masses in the <111>, <100>, and <110> directions are 0.7, 0.65, 0.825 for the heavy holes, respectively, and 0.36, 0.373, and 0.33 for the light holes, respectively. From the A, B, C parameters, we can now find the average (over the angles of the wave vector), given by [73]

$$\frac{1}{m_{hh}} = \frac{2}{\hbar^2}\left[-A + B\left(1 + \frac{2|C|^2}{15B^2}\right)\right]$$

$$\frac{1}{m_{lh}} = \frac{2}{\hbar^2}\left[-A - B\left(1 + \frac{2|C|^2}{15B^2}\right)\right]. \tag{2.115}$$

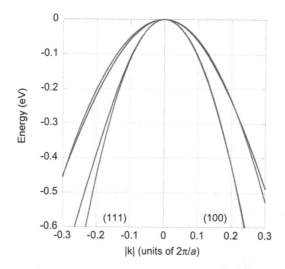

Figure 2.23. Fit of the Dresselhaus *et al* [74] perturbation approach to the EPM band structure near the top of the valence band. Here, the blue curves are the EPM results and the red curves are the Dresselhaus model. The top pair of curves is for heavy holes and the bottom pair of curves is for light holes. The positive k axis is along (100) while the negative k axis is along (111).

This gives values of 0.67 for the heavy holes and 0.37 for the light holes (all in units of the free electron mass; we will delve more heavily into the effective mass in section 2.6). It is these average masses that we will utilize in the density of states functions in chapter 4. In table 2.5, we present values for the Dresselhaus parameters for a variety of semiconductors (these have been normalized by dividing by $m_0/\hbar^2$).

2.5.3 Wave functions

The wave functions that form the basis set for the $\boldsymbol{k} \cdot \boldsymbol{p}$ method were given in (2.97). Once the Hamiltonian is diagonalized and the energies are found, the appropriate sums of the basis vectors give the new set of orthonormal wave functions that are constructed from the set of basic orbitals. The doubly degenerate wave functions that result from the diagonalization of the Hamiltonian, by the methods described above, may be written in the form [71, 72]

$$\psi_{i1} = a_i|iS\downarrow\rangle + b_i\left|\frac{X - iY\uparrow}{\sqrt{2}}\right\rangle + c_i|Z\downarrow\rangle$$

$$\psi_{i2} = a_i|iS\uparrow\rangle + b_i\left|-\frac{X + iY}{\sqrt{2}}\downarrow\right\rangle + c_i|Z\uparrow\rangle,$$

(2.116)

where the subscript i takes on the values c, lh, or Δ, while the values 1, 2 correspond to the two spin states. These give six of the wave functions. As previously, the wave vector is oriented in the z-direction. The heavy-hole wave functions are

Table 2.5. Dresselhaus parameters and hole masses [78].

	A	B	C	$<m_{hh}>/m_0$	$<m_{lh}>/m_0$
Si	−4.27	−0.63	4.93	0.537	0.153
Ge	−13.3	−8.57	12.78	0.33	0.043
AlAs	−4.03	−2.09	4.63	0.87	0.067
AlSb	−4.12	−2.09	4.71	0.81	0.066
GaP	−4.2	−1.97	4.6	0.67	0.17
GaAs	−6.98	−4.5	6.2	0.51	0.076
GaSb	−11.7	−8.19	11.07	0.28	0.05
InP	−6.28	−4.17	6.24	0.56	0.12
InAs	−19.7	−16.8	13.66	0.43	0.026
InSb	−35	−31.4	20.92	0.45	0.016

$$\psi_{hh1} = \left| \frac{X + iY}{\sqrt{2}} \uparrow \right\rangle$$

$$\psi_{hh2} = \left| \frac{X - iY}{\sqrt{2}} \downarrow \right\rangle. \tag{2.117}$$

Finally, the coefficients are written in terms of the energies of the various bands as

$$a_i = \frac{kP}{N}\left(E_i' + \frac{2\Delta}{3}\right)$$

$$b_i = \frac{\sqrt{2}}{3N}\left(E_i' - E_G\right) \tag{2.118}$$

$$c_1 = \frac{1}{N}\left(E_i' - E_G\right)\left(E_i' + \frac{2\Delta}{3}\right)$$

$$N^2 = a_i^2 + b_i^2 + c_i^2.$$

The last expression provides the normalization of the wave functions. For small kP, the conduction band remains s like, and the light hole and split off valence bands are composed of admixtures of the p-symmetry wave functions. In the hyperbolic band models, however, the conduction band also contains an admixture of p-symmetry wave functions. This admixture is energy dependent, and it is this energy dependence that introduces a similar (energy dependent) effect into the overlap integrals that will be calculated for the scattering processes in chapter 5. As with the details of the hyperbolic bands, the size of the spin orbit splitting makes only slight changes in the numerical factors, but these can be ignored as being of second order in importance.

2.6 The effective mass approximation

From the discussion in section 2.1, we know that the properties of an electron in one of the energy bands is not the same as a free electron. We know that the electron is a wave inside the crystal, a fact that was established by Davisson [79] and Thompson [80]. In fact, the energy band structure is a form of frequency filter, that only allows propagating waves in certain ranges of frequency or energy. While the idea of an effective mass arose in section 2.1, as well as above, from the energies for the bands near the extrema, we effectively would like to have a more general wave to find the *effective* mass at an arbitrary point in the energy band. This is what the effective mass approximation gives us. The intuitive way to achieve this is to recognize that what we call an electron in the semiconductor is not a real electron, but what we call a quasi-particle. This is because we are going to assign it certain properties, which we know real electrons have, but are different within the semiconductor. The important ones are the mass and, in the case of holes the charge. The most intuitive way to do this says that we will make the quasi-particle have the same momentum and energy as the real particle would have. Thus, its energy is given by the wave frequency as

$$E = \hbar\omega, \tag{2.119}$$

where the kinetic energy is the point in the band structure, measured from the convenient zero, for example at the minimum or maximum of the band. Then, we need the kinetic momentum to be the same as the wave momentum, where the kinetic momentum is assigned to the quasi-particle and the wave momentum comes from the band structure. This means that we enforce the relation

$$m^*v = \hbar k. \tag{2.120}$$

As for any normal wave, we know that the information, or the position of the quasi-particle in the present situation, moves with the group velocity of the wave, or

$$v = \frac{\partial\omega}{\partial k} = \frac{1}{\hbar}\frac{\partial E}{\partial k}, \tag{2.121}$$

which leads to the effective mass being

$$\frac{1}{m^*} = \frac{v}{\hbar k} = \frac{1}{\hbar^2 k}\frac{\partial E}{\partial k}. \tag{2.122}$$

Now, this is different from what one sees in a great many textbooks. In the following, we will describe how this result is actually the correct result, and then show how the result found in the textbooks is just a limiting expression valid only near the band extremum.

When the electron moves in the potential that arises from the atoms (and from the varieties of electron-electron potentials), it follows the energy band and thus is not stationary in time. On the other hand, if we connect it to the Bloch wave function, then it sits with a wave momentum and a corresponding energy at the momentum $\hbar k$, and can be thought of as being stationary at that momentum. In this connection, the detailed velocity and momentum is different from the velocity and momentum associated with the Bloch wave function. We often associate the latter with the idea of a quasi-particle with properties that are different than those of the free electron, even though the latter might well be the basis upon which the energy bands are constructed. At the end of the day, though, we need to give our quasi-particle an effective mass to complete the description and its response to external fields and potentials. The energy bands and the resulting motion arises from the Hamiltonian that is expressed in section 2.3, and which may be written as

$$H = \frac{p^2}{2m_0} + V(\mathbf{r}), \tag{2.123}$$

where the potential can be the pseudo-potential, or a modification of it to incorporate some form of the electron-electron energies. The quantity p is the conventional momentum operator from quantum mechanics, which is expressed as $p = -i\hbar\nabla$, in vector form. We want to examine how this potential interacts with the Bloch functions and gives rise to a quasi-momentum, which we can also call the crystal momentum [81]. The approach we follow is based upon ideas in [80], but more recently discussed by Zawadzki [82].

We note that the momentum operator which appears in the Hamiltonian (2.123) does not actually commute with this Hamiltonian. Rather, we find that

$$\frac{d\mathbf{p}}{dt} = -\frac{i}{\hbar}[\mathbf{p}, H] = -\nabla V(\mathbf{r}). \tag{2.124}$$

Similarly, the velocity of the electron in the crystal also does not commute with the Hamiltonian, and

$$\mathbf{v} = \frac{d\mathbf{r}}{dt} = -\frac{i}{\hbar}[\mathbf{r}, H] = \frac{\mathbf{p}}{m_0}. \tag{2.125}$$

The problem we have is that, since the energy is a constant of the motion, as we move through the spatially varying potential, the momentum (and hence the velocity) must vary spatially as well. Thus, these are oscillatory properties that vary as the electron moves through the various atomic potentials of the crystal. The oscillatory response, which is periodic in the crystal periodicity has been called a Bloch oscillation. But, these are not the properties that we wish to have for the quasi-particle to be used in e.g. transport calculations. So, to see how to approach this, let us step back a bit and examine the Bloch functions (2.12) a little further. Here, we will use them in the vector form.

The Bloch functions must be eigen-functions of the Hamiltonian, as we have used them to produce the energy bands in the last few sections. Hence, we must have

$$H\psi_{nk}(\mathbf{r}) = E_n(\mathbf{k})\psi_{nk}(\mathbf{r}) = E_n(\mathbf{k})e^{i\mathbf{k}\cdot\mathbf{r}}u_{nk}(\mathbf{r}), \tag{2.126}$$

where the part $u_{nk}(\mathbf{r})$ has the periodicity of the unit cell in the crystal. Now, the wave number k has some interesting properties as

$$\hbar k = \hbar\frac{2\pi}{\lambda} = \frac{h}{\lambda} = P, \tag{2.127}$$

where λ is the de Broglie wavelength, and P is the quasi-momentum, to be distinguished from the momentum operator in (2.123). The quasi-momentum arises from the wave part of the Bloch function, and will have some different properties from that of the momentum operator, or real momentum. Since the quantity $\hbar k$ is a relatively stationary quantity, the quasi-momentum must be a constant of the motion, and does not oscillate through the band with time. This follows as the Bloch function itself is stationary for a given k state. For example, following (2.127), the Bloch function also satisfies the eigenvalue equation

$$P\psi_{nk}(r) = \hbar k \psi_{nk}(r). \tag{2.128}$$

Quantum mechanically, a wave function can satisfy two eigenvalue equations only if the operators commute. Hence, the quasi-momentum must commute with the Hamiltonian in (2.123), and thus be a constant of the motion. There must be a relationship between this quasi-momentum and the momentum operator. So, let us write

$$P = p + i\hbar\gamma(r), \tag{2.129}$$

in which the last term must be determined. To do this, we apply this operator equation to the Bloch function itself, which leads to

$$\begin{aligned} P\psi_{nk}(r) &= (i\hbar\nabla + i\hbar\gamma)e^{ik\cdot r}u_{nk}(r) \\ &= \{\hbar k u_{nk}(r) + i\hbar[\gamma - \nabla(ln u_{nk}(r))]\}. \end{aligned} \tag{2.130}$$

In order for the eigenvalue equation (2.129) to be satisfied, the term in the square brackets must vanish, and this gives us

$$\gamma = \frac{1}{u_{nk}}\nabla u_{nk}. \tag{2.131}$$

If we now add an external potential to the crystal, the total Hamiltonian may be rewritten by including this extra term as

$$H_T = \frac{p^2}{2m_0} + V(r) + U_{ext}(r). \tag{2.132}$$

With this new Hamiltonian, we find that the quasi-momentum varies with the external potential as

$$\frac{dP}{dt} = -\frac{i}{\hbar}[P, H_T] = -\frac{i}{\hbar}[P, U_{ext}] = -\nabla U_{ext}, \tag{2.133}$$

where we have used (2.129) and (2.130) and the last term of (2.129) commutes with the external potential. Thus, the quasi-momentum is changed only by the non-periodic external force. We relate the quasi-momentum to our quasi-particle electron through the Bloch state, so that our quasi-particle is accelerated only by the external fields and its average motion in the absence of these fields is a constant of the motion.

Let us now return to the velocity. As is well known in quantum mechanics, we may express the time variation of an operator, which is not explicitly a function of time via the Heisenberg notation as

$$v(t) = e^{iHt/\hbar}ve^{-iHt/\hbar}. \tag{2.134}$$

We now take the expectation of this operator using the Bloch wave functions. This leads to

$$\langle v(t) \rangle = \langle \psi_{nk}|e^{iHt/\hbar}ve^{-iHt/\hbar}|\psi_{nk}\rangle$$
$$= \langle \psi_{nk}|e^{iE_nt/\hbar}ve^{-iE_nt/\hbar}|\psi_{nk}\rangle = \langle \psi_{nk}|v|\psi_{nk}\rangle, \tag{2.135}$$

which is a time independent result. Hence, the average velocity is constant (in the absence of external fields) due to the eigenvalue equation (2.126) and the properties of the Bloch functions. If we put the full Bloch function into (2.126) and find the resulting eigenvalue equation for just the cell periodic part, we find

$$Hu_{nk}(r) = \left[\frac{(p + \hbar k)^2}{2m_0} + V(r)\right]u_{nk}(r) = E_n(k)u_{nk}(r). \tag{2.136}$$

With this expansion of the Hamiltonian, we can write

$$\frac{\partial E_n(k)}{\partial k} = \frac{\partial H}{\partial k} = \frac{\hbar}{m_0}(p + \hbar k). \tag{2.137}$$

The expectation value of this last term, evaluated with the cell periodic parts of the Bloch function, gives

$$\frac{\partial E_n(k)}{\partial k} = \frac{\hbar^2}{m_0}\langle u_{nk}|-i\nabla + k|u_{nk}\rangle$$
$$= \frac{\hbar^2}{m_0}\langle u_{nk}|-i\nabla + k|e^{-ik\cdot r}\psi_{nk}\rangle \tag{2.138}$$
$$= -\frac{i\hbar^2}{m_0}\langle u_{nk}|e^{-ik\cdot r}\nabla|\psi_{nk}\rangle = \langle \psi_{nk}|p|\psi_{nk}\rangle.$$

This latter expression can be combined with (2.125) to give the average velocity in terms of the energy bands as

$$\langle v(t) \rangle = \frac{1}{\hbar}\frac{\partial E_n(k)}{\partial k}. \tag{2.139}$$

Thus, the average quasi-particle velocity is given by the point derivative of the energy as a function of quasi-momentum.

As the average velocity of the quasi-particle is given by the derivative of the energy with respect to the quasi-momentum, this means that the average velocity of the quasi-particle must be directly related to the quasi-momentum, and this allows us to define this relationship via an effective mass:

$$P = \hbar k = m^* \langle v(t) \rangle. \tag{2.140}$$

This *must* be regarded as the fundamental definition of the effective mass for the quasi-particle electron. It is different for every band, just as the Bloch function and the quasi-momentum are different for every band (and every momentum k). But, as mentioned, this is dramatically different from what is commonly found in text books. Here, we make the definition based upon the existence of the quasi-momentum and the average velocity, which describe the motion of our quasi-particle arising from the Bloch function.

It is important to note that, if we combine (2.140) with (2.139), we can write an expression for the mass in terms of the energy bands, as

$$\frac{1}{m^*} = \frac{v}{\hbar k} = \frac{1}{\hbar^2 k} \frac{\partial E_n(k)}{\partial k}. \tag{2.141}$$

for a spherically symmetric band.

To connect with the equation found in many textbooks, we note that there is a problem at the bottom of the band, where both the derivative and k vanish (k is defined from the point of the band minimum). In this situation, we find that one must use L'Hospital's rule and take the derivatives of the numerator and denominator with respect to k, and this leads to

$$\frac{1}{m^*} = \lim_{k \to 0} \left(\frac{1}{\hbar^2} \frac{1}{\partial k / \partial k} \frac{\partial^2 E_n(k)}{\partial k^2} \right) \sim \frac{1}{\hbar^2} \frac{\partial^2 E_n(k)}{\partial k^2}. \tag{2.142}$$

But, it must be remembered that this is a limiting form, valid only at the bottom of the conduction band (or, equivalently, at the top of the valence band). However, this latter form is the one found in most textbooks without any discussion of its range of validity. It is worth noting that most of the common derivations of this last expression have some errors in them that obscure these limitations, a fact that necessitates the present approach. It is worth pointing out that the first derivative mass (2.141) has long been known as the *cyclotron* mass [83], as cyclotron resonance measures a cross-sectional area of the energy surface, and this is related to the above mass.

Now, let's examine what happens here for the energy dependent mass in the hyperbolic bands. We insert the conduction band mass from (2.105) into the hyperbolic band (2.104) to obtain (we now reference the energy of the conduction band from the minimum of this band)

$$E_c = \frac{E_G}{2} \left[\sqrt{1 + \frac{2\hbar^2 k^2}{m_c E_G}} - 1 \right]. \tag{2.143}$$

We could also have used (2.106), but the numerical factors in front of P^2 would have changed in (2.105). Applying (2.141) to find the energy dependent mass gives us

$$\frac{1}{m^*} = \frac{1}{\hbar^2 k} \frac{\partial E_n(\boldsymbol{k})}{\partial k} = \frac{1}{\hbar^2 k} \left(\frac{\hbar^2 k}{m_c} \right) \frac{1}{\sqrt{1 + \dfrac{2\hbar^2 k^2}{m_c E_G}}} \qquad (2.144)$$

and

$$m^* = m_c \left(1 + \frac{2E}{E_G} \right). \qquad (2.145)$$

That is, the mass gets heavier as the energy increases away from the minimum of the conduction band. This is exactly the behavior we found in (2.23) and results from the fact that moving away from the band extrema reduces the interaction between the bands and the curve begins to trend toward the free electron behavior. Here, this is far away, but the result is that the mass gets larger as we increase the momentum. This is often referred to as a relativistic effect, and the general kp correction is often called the relativistic correction, due to the connection between this approach and the Dirac equation, in which the static mass term corresponds to the band gap in semiconductors [84, 85].

2.7 Dielectric scaling theory

As we have discussed above, the tetrahedrally coordinated semiconductors are basically those with the diamond and zinc-blende structures. The layer compounds are somewhat different. But, if we look at the real space and/or the momentum space band calculations, it is clear that the similarities between the materials are greater than their differences. The band structure arises primarily from their outer shell s and p orbitals, that combine to form the four tetrahedral bond orbitals. The differences in the band structures arise from quite subtle differences in the various parameters. One may well ask whether there is a general theory which describes these subtle differences sufficiently well to provide a scaling concept between materials. As this is a philosophical question, the answer probably relies upon the acceptance of the observer. Nevertheless, there have been attempts to do just this and develop a scaling theory.

Phillips [86] and Van Vechten [87], in particular, have developed what is called a dielectric theory of electronegativity that is based upon the similarities among the tetrahedrally coordinated semiconductors. This theory describes not only the relative iconicity and dielectric constants but also gives an excellent description for the ionization potential and interband transition energies. Notably, it is just these quantities that are needed to describe the optical properties and effective masses of the alloys, to be discussed in section 2.8. There have been other approaches, notably due to Harrison [88], to develop general theories for the variations between different materials in this group. The key factor is that basic measurements of certain optical transitions can be used to determine the interaction matrix elements in the tight-binding approach, as well as in the empirical pseudo-potential approach, and this will elaborate on the scaling rules to produce a generally complete interpretive theory for a given material. The interpolative nature of the dielectric theory allows an excellent approach without having to resort to extensive computer calculations. In this section, we will discuss the nature and predictions of this dielectric theory.

The dielectric method assumes a basic or universal model of the semiconductor by adopting a two-band description of the conduction and valence bands. The lower band is the bonding valence band, while the upper band is the anti-bonding conduction band. Between these two bands, there exists an average energy gap W_G. This is *not* the band gap between the minimum of the conduction band and the maximum of the valence band. This is a gap between the average energy of the conduction band and the average energy of the valence band. This average gap is defined as

$$W_G = \sqrt{E_h^2 + C^2}, \qquad (2.146)$$

where E_h is the so-called homopolar energy and C is the heteropolar energy. It is assumed that this average gap can be determined from the real, high frequency dielectric constant $\varepsilon_{r\infty}$, that will be discussed again in chapter 3. This dielectric constant is free from any polarization do to the lattice, and is determined by the interaction energy among the valence electrons; e.g., their valence plasma frequency ω_P. This high frequency dielectric constant is then related to the average energy gap and the valence plasma frequency through the basic form of the Penn dielectric constant [89]

$$\varepsilon_{r\infty} = 1 + \frac{(\hbar\omega_P)^2 DA}{W_G^2}, \qquad (2.147)$$

where the DA are corrections from this theory, and the parameters are given by

$$A = 1 - B + \frac{B^3}{3}$$
$$B = \frac{W_G}{4E_F} \qquad (2.148)$$
$$\omega_P^2 = \frac{Ne^2}{m_0\varepsilon_0},$$

and D is a number $\geqslant 1$, that is a correction factor to account for the slight interaction of core d electrons with the valence electrons (if they exist). This latter effect can produce a 10%–20% correction to the average number of 4 valence electrons per atom. The Fermi energy here is for the total number of valence electrons, measured from the bottom of the valence band and is typically close to the actual energy width of the valence band. In table 2.6, values for these parameters are given for a number of semiconductors.

The quantity W_G is not measurable directly from any of the optical properties, but is a true average between the bonding and anti-bonding orbitals. It is composed of contributions from the homopolar (group IV nature) and heteropolary

Table 2.6. Some relevant parameters for the dielectric theory.

	$\varepsilon_{r\infty}$ [78]	E_F (eV) [78]	$\hbar\omega_P$ (eV)	D
C	5.7	21.03	31.2	1.0
Si	11.9	12.34	16.6	1.0
Ge	16.2	12.6	15.6	1.26
GaAs	10.9	12.5	15.6	1.24
GaP	9.11	12.99	16.5	1.15
GaSb	14.44	12.0	13.9	1.31
InAs	12.25	12.69	14.2	1.35
InP	9.61	11.42	14.8	1.27
InSb	15.68	11.71	12.7	1.42
AlAs	8.16	11.95	15.5	1.1
HgTe	15.2	10.1	12.8	1.3
CdTe	7.1	10.0	12.7	1.3

(non-group IV nature) energies. These two contribution represent the symmetric and anti-symmetric parts of the atomic potentials from the two atoms in the unit cell, a separation as was done in the empirical pseudo-potentials in (2.70). The heteropolar potential C is in principle related to the valence difference of the two atoms in the unit cell and can be determined, in principle, from this fact, as

$$C = s\left(\frac{Z_A}{r_A} - \frac{Z_B}{r_B}\right)\exp\left[-\frac{k_s}{2}(r_A + r_B)\right],\qquad(2.149)$$

where k_s is the linearized Fermi–Thomas screening wave vector, and r_A and r_B are the covalent radii of the two atoms. The factor s represents a correction factor of approximately 1.5 (in addition to $e^2/4\pi\varepsilon_0$ in MKS units) to account for the fact that the Fermi–Thomas theory generally overestimates the screening. Obviously (2.149) is no more than the difference in the Coulomb energy of the two atoms, screened by the valence electrons, but this formulation for C suggests that C in some sense is a measure of the iconicity of the bond (difference from straight covalent bonding).

In a scaling theory, such as the dielectric method, one would expect the strength to lie in the ability to fit variations among the various compounds. Within this method, there are seven basic postulates or rules for implementing the scaling behavior. These are:

1. Any direct energy gap E_i, in the absence of d-level perturbations, is given in the general form of (2.145) as

$$E_i = \sqrt{E_{i,h} + C^2},\qquad(2.150)$$

where $E_{i,h}$ is related to the appropriate gap for the corresponding homopolar material (which is usually referenced to Si). (Throughout this section, the subscript h will be used to denote the homopolar value of the gap.) The origin of this assumption is the recognition that at any critical point in the Brillouin zone, the opening of a gap should be expressable in a form similar to that in section 2.2, with the size of that gap varying only due to any difference in the two atomic potentials. Hence, the idea is that the basic gap that opens is largely due to the tetrahedral nature of the homopolar material, like Si. Differences from this are due to the introduction of C in the heteropolar materials.

2. The values for $E_{i,h}$ and all other homopolar variables scale with the inverse power u_i of the interatomic distance through the relations

$$E_{i,h} = E_{i,h}^{Si}\left(\frac{d}{d_{Si}}\right)^{u_i}. \tag{2.151}$$

3. The ionization potential I_h, which is the energy of the top of the valence band relative to the vacuum level, also obeys (2.150). Thus, the ionization potential is treated like any other optical gap according to rule 1. The homopolar value of this gap scales as (2.151). This postulate is a little hard to justify, since the top of the valence band is usually composed of atomic p orbitals, and these usually do not scale well. Nevertheless, this energy is found to obey some scaling parameters.

4. The energy at the top of the valence band at X (the zone boundary along any of the (100) axes) is independent of the heteropolar energy C. This symmetry point is usually denoted X_4 in Si and X_5 in the compounds. This suggestion is thus that the energy at this point scales with (2.151), but does not depend upon any ionic contribution to the bonds. The scaling here is of a different nature than that at the zone center, where the ionic contribution leads to rule 3.

5. The energy at the top of the valence band at L (at the zone edge along the (111) directions) is usually found to lie midway between the values at the zone center and at the X point. This level is usually denoted L_3 in all the tetrahedrally coordinated materials. Hence, this energy may be written as

$$E_{h,L_3} = \frac{I_h + E_{h,X}}{2}. \tag{2.152}$$

6. The splitting of the conduction band X levels into the X_3 and X_1 states, as they are usually denoted, in the compounds is generally given by a simple value related to the iconicity. This splitting is usually of the order of 0.14 C. This splitting arises from the polar nature of the material, so it is expected to scale with the heteropolar energy C.

7. Finally, the perturbative effect of the d levels on the s-like states of greatest interest, those at Γ and L in the conduction band, which are the minima of the conduction band at these points, is expressed by decreasing the optical gap transitions from those expected from (2.150). The d states thus perturb the energy gaps at Γ, denoted the E_0 gap, and at L, denoted the E_1 gap. The d

states apparently do not perturb the gap at X, denoted the E_2 gap. The effect of the d states on the Γ and L gaps can be expressed as a modification of (2.150) to

$$E_i = [E_{i,h} - (D - 1)\Delta E_i]\sqrt{1 + \left(\frac{C}{E_{i,h}}\right)^2}. \qquad (2.153)$$

Obviously, here, the subscript only runs over the two gaps of interest. The parameter ΔE_i scales with d as in (2.151) and is a function only of this interatomic spacing. The interaction with the d levels leads to an increase in the number of electrons that contribute to the bonds and this tends to broaden the valence band. This broadening thus leads to a narrowing of the gaps. Some values of these parameters, computed for Si and Ge, are given in table 2.7, and for the polar compounds in table 2.8 below.

2.7.1 Silicon and germanium

The valence and conduction band structure for Si were shown in figures 2.13 and 2.18 for calculations via the SETBM and non-local empirical pseudo-potentials. The lowest energy gap is between the conduction band minima, located about 85% of the way along (100) to the X point, and the top of the valence band at Γ. The optical transition between these two points is indirect and therefore must involve a phonon (the wave number of the optical wave is much smaller than the values of k found in solids, so that the optical transition must be vertical on a band structure plot). Because of the symmetry of the (100) axes, there are six equivalent ellipsoids of constant energy in the conduction band, and these have their principal axes oriented along the (100) directions. We will return to these ellipsoids in chapter 4, where they are essential in determining the density of states. The lowest direct transition at Γ, although it is a transition from the valence band to the second conduction band, which is the proper dominantly s-state from the SETBM theory. It is this state that converts to the conduction band minima in the compound semiconductors.

Germanium differs from Si only in the fact that the minimum of the conduction band lies at the 8 L points in the Brillouin zone. Hence, there are four equivalent

Table 2.7. Properties of the homopolar gaps.

Parameter	Si	Ge	u_i
I_h	−5.17	−4.9	−1.31
X_4	−8.63	−8.14	−1.4
$E_{0,h}$	4.1	3.64	−2.75
$E_{1,h}$	3.6	3.27	−2.22
$E_{2,h}$	4.5	4.06	−2.38
ΔE_0	0	10.6	−2.0
ΔE_1	0	3.88	−2.0

Table 2.8. Principal parameters in the group III–V compounds.

	AlAs	GaP	GaAs	GaSb	InP	InAs	InSb
$E_{h,0}$	3.67	4.06	3.67	2.69	3.31	3.07	2.52
ΔE_0	10.9	12.67	10.9	7.81	8.75	7.92	0.27
E_0	2.96	2.75	1.42	0.7	1.35	0.35	0.2
I_h	−4.91	−5.15	−4.91	−4.42	−4.67	−4.5	−4.1
I	−6.02	−6.12	−5.7	−4.9	−5.74	−5.27	−4.7
X_5	−8.51	−8.62	−8.51	−8.0	−8.42	−8.34	−7.94
$E_{h,2}$	4.09	4.6	4.09	3.39	3.74	3.5	2.96
E_2	5.2	5.55	5.01	3.99	5.01	4.44	3.74
L_3	−7.25	−7.37	−7.11	−6.45	−7.08	−6.81	−6.32
$E_{h,1}$	3.29	3.57	3.29	2.76	3.03	2.85	2.43
ΔE_1	4.07	4.69	4.07	2.75	3.33	2.74	2.07
E_1	4.36	3.89	3.13	2.41	3.17	2.61	2.16
$\Delta E_{\Gamma X}$	−1.36	−0.16	0.49	0.19	0.98	1.02	0.32
$\Delta E_{\Gamma L}$	−0.85	−0.11	0.3	0.16	0.5	0.72	0.36
m_c/m_0	0.145		0.063	0.045	0.072	0.022	0.013
$P^2/2m_0$	18.5		21.1	14.9	17.4	15.6	15.2
C	3.62	3.3	2.9	2.1	2.74	2.74	2.1
D	1.07	1.11	1.24	1.24	1.45	1.45	1.43

ellipsoids; each ellipsoid has e.g. one half in the zone at the (111) point and one half in the zone at the ($\bar{1}\bar{1}\bar{1}$) point (the bar over the 1 indicates a negative value). Some of the values for the various parameters for the dielectric theory for Si and Ge are given in table 2.6.

2.7.2 Group III–V compounds

As expected from the entire concept of scaling, the band structures for the III–V compounds differ in detail from those of Si and Ge very little. There are two main differences, one of which is that the minima of the conduction bands will vary between the L, X, and Γ with the various parameters. The second major difference is an opening in the valence band, well below the maximum of this band, that is known as the polar gap. This polar gap can be seen in figure 2.17, where it splits off the mainly s-state that forms the bottom of the valence band. With little change, mainly in the direct band gap, figure 2.17 could have been for InAs, InSb, GaSb, or InP instead of GaAs. All are direct gap semiconductors. Moreover, each of these has the various conduction band minima arranged as Γ, L, X in terms of their position in energy from the valence band maximum. A number of parameters for the III–V materials are listed in table 2.8 [86].

One line in table 2.7, for $P^2/2m_0$, requires some discussion. We met this already in section 2.5.1, where it connects the $\boldsymbol{k} \cdot \boldsymbol{p}$ theory to the band edge effective mass (particularly in the conduction band). There, it was divided by the energy gap to obtain the band-edge mass (see (2.142) for the form of the energy). This suggests that the band edge effective mass scales with this factor and the energy gap through the first line of (2.105). Certainly, as you look at the conduction band masses in the above table, it is clear that the latter scales with the energy gap, which suggests that $P^2/2m_0$ should be relatively constant across the table. It doesn't quite fit this concept, but there are no order of magnitude variations for the parameter. The variation arises from the fact that the momentum matrix element P involves the cell periodic part of the Bloch function and this has variations among various materials. Nevertheless, it is close enough to being constant that it can be used as a scaling parameter of sorts.

2.7.3 Some group II–VI compounds

The basic band structure of the zinc-blende II–VI compounds, such as CdTe, differs little from the structure of GaAs. Only HgTe has a distinct difference. In the latter material, the lowest conduction band (the lowest anti-bonding level) actually lies below the highest valence band (the highest bonding level), so that the character of these two energy levels is reversed. For this reason, the material is said to have a negative energy gap. When alloys of HgTe and CdTe are prepared, the resultant band gap can be brought very close to zero (it passes through zero in the simplest approximation, but second-order effects such as an interaction between the bands is thought to prevent a value exactly equal to zero in practice). Thus materials with very small band gaps can be prepared. For this reason, these materials have found extensive usage in far infrared detectors.

However, these materials have another problem. In many cases of the II–VI, the ionicity is very high. Generally, we think of the ionic fraction of the bond being related to the ratio $C/\sqrt{E_h^2 + C^2}$. As the ionicity increases, the metastable zinc-blende lattice becomes unstable, and can collapse into the wurtzite structure [85]. This is one form which is similar to a hexagonal lattice in the basal plane, but is elongated along the z-axis [90]. (This is not a close-packed hexagonal lattice!)

2.8 Semiconductor alloys

2.8.1 The virtual crystal approximation

We should now be comfortable with the idea that the zinc-blende lattice is composed of two interpenetrating face centered cubic structures, one for each atom in the basis. Thus, in GaAs for example, one face centered cubic structure is made up of the Ga atoms while the second is composed of the As atoms. These two FCC cells refer to the A atoms and the B atoms. We can extend this concept to the case of pseudo-binary alloys, such as GaInAs or GaAlAs, which are supposed to be formulated by a smooth mixing of the two constituents on the A sub-lattice (e.g., GaAs and AlAs in GaAlAs or GaAs and InAs in GaInAs). In such $A_x B_{1-x} C$ alloys, all of the sites of one face centered cubic sublattice are occupied by type C atoms (confusingly, these C atoms are on the B sub-lattice), but the sites of the A sub-lattice are shared by the atoms of type (again confusingly) A and type B in a random fashion subject to the conditions

$$N_A + N_B = N_C = N$$

$$x = \frac{N_A}{N} \equiv c_A$$

$$1 - x = \frac{N_B}{N} \equiv c_B.$$

(2.154)

In this arrangement, a type C atom may have all type A neighbors or all type B neighbors, but on the average has a fraction x of type A neighbors and a fraction $1-x$ of type B neighbors. In effect, the structure is a face centered cubic structure of mixed AC and BC molecules, complete with interpenetrating molecular bonding. This structure composes what is called a pseudo-binary alloy with the properties determined by the relative concentrations of A and B atoms. In true pseudo-binary alloys, it should be possible to scale the properties by a smooth extrapolation between the two end-point compounds, but this isn't always the case.

It is also important to note that alloying on the B sub-lattice is also possible. Hence, materials like GaAsP, a mixture of GaAs and GaP, can exist, with equivalent changes in (2.154). So, the ability to alloy on either of the two sub-lattices means that a very many new materials with a wide range of properties can be created. The ability to create these alloys can be important in the search for a specific set of properties desired. For example, both $In_{0.53}Ga_{0.47}As$ and $In_{0.52}Al_{0.48}As$ are lattice matched to InP, but have dramatically different band gaps. This is useful for heterostructures, as we will see in a later section of this chapter.

In recent years, quaternary alloys have also appeared as $A_xB_{1-x}C_yD_{1-y}$ (the most common example is the quaternary InGaAsP, used in infrared light emitters). Here, C and D atoms share the sites on the B sub-lattice, while the A and B atoms share the A sub-lattice as described above. This new compound is still considered a pseudo-binary compound composed of a random mixture of two ternaries $A_xB_{1-x}C$ and $A_xB_{1-x}D$, which are only somewhat more complicated than the simple ones discussed in the preceding paragraph. Still, it is assumed that a true random mixture occurs so that the properties can easily be interpolated from those of the constituent compounds. That is, the randomness of the alloy prevents any correlation to occur among the various atoms, other than that of the crystal structure. Then any general theory of pseudo-binary alloys can be applied equally well to the quaternaries as to the ternaries. If these compounds are truly smooth mixtures, the alloy theory will hold, but if there is any ordering or correlation in the distribution of the two constituents, deviations from the alloy theory should be expected. For example, $In_xGa_{1-x}As$ may be a smooth alloy composed of a random mixture of InAs and GaAs. However, if perfect ordering were to occur, particularly near $x = 0.5$, the crystal structure would not be a zinc-blende lattice, but would be a chalcopyrite—a super-lattice on the zinc blende structure with significant distortion of the unit cell along one of the principal axes. In the latter case, changes are expected to occur in the band structure due to Brillouin zone folding about the elongated axis (the lattice period is now twice as long, which places the edge of the Brillouin zone only one half as far from the origin in the orthogonal reciprocal space direction). For many years, it has been assumed that the ternary and quaternary compounds formed of the group IIIV compounds were true random alloys. In recent times, however, it has become quite clear that this is not the case in many situations. We return to this below, as it gives some insight into deviations expected from the random alloy theory.

Consider a pseudo-binary alloy in which the AC and BC molecules are randomly placed on the crystal lattice. Attention will be focused on ternaries, but the approach is readily extended to quaternaries. The contribution to the crystal potential for the A and B atoms may be written as

$$V_{AB}(\mathbf{r}) = \sum_A V_A(\mathbf{r} - \mathbf{r}_A) + \sum_B V_B(\mathbf{r} - \mathbf{r}_B), \tag{2.155}$$

where $\mathbf{r}_i$ defines the lattice site of the particular sub-lattice on which the A and B atoms are randomly sited. This part of the total crystal potential may now be decomposed into symmetric and antisymmetric parts, although this separation is different from that introduced in the pseudo-potential earlier in this chapter. The former is the 'virtual-crystal' potential, and the latter is a random potential, whose average is presumed to be sufficiently small that it can be neglected, but often provides so-called alloy scattering. This decomposition is just

$$
\begin{aligned}
V_S(\mathbf{r}) &= c_A \sum_A V_A(\mathbf{r} - \mathbf{r}_A) + c_B \sum_B V_B(\mathbf{r} - \mathbf{r}_B) \\
V_A(\mathbf{r}) &= \sum_A V_A(\mathbf{r} - \mathbf{r}_A) - \sum_B V_B(\mathbf{r} - \mathbf{r}_B),
\end{aligned}
\tag{2.156}
$$

where c_A and c_B are defined in (2.153). For the pseudo-potential, these potentials are added as the A atom contribution to (2.70). The virtual-crystal potential, which is the symmetric potential, is a smooth interpolation between the potential for the A-C crystal and that for the B-C crystal. The random part can contribute either to scattering of the carriers or to bowing of the energy levels in the mixed crystal. Bowing of a band gap means a deviation from the linear extrapolation. Normally, this bowing is toward a gap that is narrower than predicted by the virtual-crystal approximation. That is, the bowing can lower the gap which indicates an increased stability of the random alloy. If there is a regularity to V_A, or to V_B, so that they possess a significant amplitude in one of the Fourier components, it will make a significant impact on the Bloch functions and on the band structure. Thus one definition of a random alloy is that it is one in which the anti-symmetric potential is sufficiently random that none of the Fourier components are excited to any great degree. This means that the anti-symmetric potential must be aperiodic in nature.

The experimental measurements of the band gap variation for a typical alloy can be expressed quite generally as

$$E_G = x E_{G,AC} + (1 - x) E_{G,BC} - x(1 - x) E_{\text{bow}}. \tag{2.157}$$

Table 2.9. Some bowing parameters of ternary alloys [78, 92].

Alloy	Γ	L	X
AlGaAs	0.438	0.5	0.16
AlGaSb	0.4	0.21	0.0
AlInP	0.47	0.0	0.0
AlInAs	0.7	0.0	0.0
GaInP	0.65	0.05	0.2
GaInAs	0.405	0.5	0.08
GaInSb	0.413	0.4	0.33
GaAsP	0.186	0.0	0.211
GaAsSb	1.2	1.1	1.1
InAsP	0.32	0.22	0.22
AlAsSb	0.8	0.28	0.28

The general form of (2.157) is found in nearly all alloys; for example, there is a linear term interpolating between the two endpoint compounds that represents the virtual crystal approximation (the first two terms of this equation), and a negative bowing energy [coefficient of the $x(1 - x)$ term] that represents the contribution from the uncorrelated anti-symmetric potential. An equivalent form of (2.156) exists for alloying on the B sub-lattice. Moreover, such a form exists for all of the valleys of the conduction band, referenced to the top of the valence band. In table 2.9, some of these bowing parameters are given for the various sets of valleys.

In the quaternary compounds, it is necessary to extrapolate the band gap and lattice constants from those of the ternaries. There are many possible ternary materials; their number is roughly the number of binaries raised to the 3/2 power. Usually, however, they are grown lattice matched to some binary substrate. In alloys, a rule known as Vegard's law stipulates that the lattice constant will vary linearly between the values of the two endpoints [91]. For example, the edge of the face-centered cube in GaAs, InAs, and InP are found to be approximately 5.653 Å, 6.058 Å, and 5.869 Å, respectively [78]. Thus the lattice constant of $Ga_xIn_{1x}As$ is

$$a_{\text{InGaAs}} = 6.058 - 0.405x, \qquad (2.158)$$

and this is lattice matched to InP for $x = 0.466$. Thus this composition of the alloy may be grown on InP without introducing any significant strain in the layer.

2.8.2 Alloy ordering

As discussed above, it is quite possible that these alloy compounds are not perfectly random alloys, but in fact possess some ordering in their structure. The basis of ordering in otherwise random alloys lies in the fact that the ordered lattice, whether it has short range order or long range order, may be in a lower energy state than the perfectly random alloy. In a random alloy $A_xB_{1-x}C$, the average of the cohesive energy will change by

$$E_{\text{coh}} = E_{\text{coh}}^{BC} + x\left(E_{\text{coh}}^{AC} - E_{\text{coh}}^{BC}\right), \tag{2.159}$$

within the virtual crystal approximation. While the A-C compound is losing energy, the B-C compound is gaining energy, and this energy comes from the expansion or contraction of the lattice of the two end compounds (and the variation this produces in the energy structure). For example, in $In_xGa_{1-x}As$, the cohesive energy is the average of those of GaAs and InAs, but the gain of energy in the expansion of the GaAs lattice is exactly offset by that absorbed in the compression of the InAs lattice, at least within the linear approximation of the virtual-crystal approximation.

If any short range order exists, however, this argument no longer holds. Rather, the ordered GaAs regions undergo a loss of energy as their bonds are stretched in the alloy, while the ordered InAs regions gain energy as the bonds are compressed (here gain of energy is to be interpreted in the sense that the crystal is compressed and the equilibrium state now has a lower energy state). Since one may assume that the cohesive energy varies as $1/d^2$ in the simplest theory (where d is an interatomic distance). This behavior is found for most other interaction energies in the crystal. Hence, a net change of cohesive energy in the semiconductor compounds is a very simple calculation. However, the lattice constant, and hence d, varies linearly from one compound to the other due to Vegard's law. Yet, the cohesive energy varies with the inverse square of the change in the lattice constant. Thus, it is not guaranteed that the change in cohesive energy will follow the simple linear law given by (2.159).

The valence band actually contains just the $8N$ (where N is the number of unit cells in the structure) electrons in equilibrium. As one alloys two compounds, the absolute position of this band can move, yielding a change in the average energy of the bonding electrons. This is ignored when we take the top of the valence band as the zero of energy. In fact, the absolute energies of one compound relative to another becomes important in the alloy's stability. A decrease in the average energy of the valence band, or an increase of the cohesive energy, both indicate that ordering in the alloy is energetically favored. It is apparent that there are some alloys in which ordering is energetically favored. The data on GaAlAs is mixed, but even if it occurs, it would be only at low temperatures. In this case, only realistic total energy calculations can shed much light on the stability of the random alloy. The experimental situation has not been effectively investigated except for a few special cases.

In the case of InGaAs, InGaSb, and InAsP, all indications suggest that the alloy will favor phase separation and ordering at room temperature. Indeed, this tendency to order in the InGaAs and InAsP compounds may produce the well-known miscibility gap in the InGaAsP quaternary alloy that is found in the range $0.7 < y < 0.9$. The actual nature of any ordering that occurs can be quite subtle, however. As an example, in pioneering experiments with x-ray absorption fine structure (XAFS) measurements on InGaAs, Mikkelsen and Boyce [93] found that apparently the GaAs and InAs nearest neighbor bond lengths remain nearly constant at the binary values (the covalent radii) for all alloy compositions. The average cation anion distance follows Vegard's law and increases by 0.174 Å. The cation sub-lattice strongly resembles a virtual crystal (this is the sub-lattice in which alloying occurs), but the anion sub-lattice is very distorted due to the foregoing tendency. The distortion leads to two As-As (second-neighbor) distances which differ by as much as 0.24 Å, and the distribution of the observed second neighbor distances has a Gaussian profile about the two distinct values. The distortion of the anion sub-lattice is clearly beyond the virtual crystal approximation, and such a structure can be accommodated in a model crystal that resembles closely a chalcopyrite distortion, and can in fact explain the observable bowing of the band structure. If these observations are carried over to other semiconductor alloys, it is likely that, in alloys in which the alloying is occurring between two atoms of greatly differing size, the nearest-neighbor distance will probably prefer to adopt the binary value.

Zunger *et al* [94–97] have taken these theoretical ideas much further in an investigation of the alloying of group III–V semiconductors. In most of the arguments above, we have only looked at the average compression/expansion of the overall crystal lattice of the two binary constituents and have not included the tendency for the average nearest neighbor distance to remain at the binary value. For this to occur, there must be a relaxation of the common constituent sub-lattice within the unit cell as well as a possible charge transfer between the various common atoms on the non-alloyed sub-lattice. In fact, these authors find that the latter factors are the dominant ones in alloy ordering. They have investigated the tendency to order by adopting a total energy calculation using the non-local pseudopotential method. They calculate the total energy for a given composition of alloy and then vary the atomic positions to ascertain the lowest energy state. This has proven to be a very powerful approach.

Let us consider the above arguments of Zunger *et al* for the manner in which ordering may occur. For an $A_x B_{1-x} C$ alloy, the four cations of type A and B per face centered cubic cell can assume five different ordered nearest neighbor arrangements around the C atom: $A_4 C_4$, $A_3 B C_4$, $A_2 B_2 C_4$, $A B_3 C_4$, and $B_4 C_4$. These are denoted as $n = 0, 1, 2, 3, 4$ arrangements. Obviously, n indicates the number of B atoms in the cluster. If the solid is perfectly ordered with these arrangements (which correspond directly to $x = 0, 0.25, 0.5, 0.75$, and 1.0, respectively), the lattice structures are zinc-blende only for $n = 0$ or 4. For the other compositions, the ordered crystal structure is known as either luzonite or famatinite for $n = 1$ or 3, and either CuAuI or chalcopyrite for $n = 2$. The choice of the particular crystal structure is dominated by whichever is the lowest energy configuration for the crystal. In any case, it is now thought that a disordered or random alloy must be a statistical mixture of these various crystal structures. This suggests that highly-ordered alloys can generate new types of super-lattices with very short periods. Indeed, experiments have observed the highly-ordered $x = 0.5$ structure in both GaAsSb [98] and GaAlAs [99]. In the former case, both the CuAuI structure and the chalcopyrite structure are observed, while only the CuAuI structure seems to be found in the latter case. In addition, the famatinite structure has been observed in the InGaAs alloy for $x = 0.25$ and 0.75 [100]. It is clear that alloys of the binary semiconductors can be anything but random in nature and may be quite different from the simple virtual crystals they are made out to be. Mbaye *et al* [97] have calculated the phase diagrams for several alloys utilizing the total energy method discussed above, and find that random alloys, ordered alloys, and miscibility gaps can all occur and that the strain can actually stabilize the ordered stoichiometric compounds.

2.9 Hetero-structures

One of the principal reasons for developing alloy semiconductors is the desire to tailor the various band gaps to desired levels appropriate to create hetero-structures of dissimilar materials that have desired properties. For example, one such device is the resonant tunneling diode in the GaAs/AlGaAs system, in which two GaAlAs barriers surround a single thin GaAs layer to create a quantum well [101, 102]. This structure is then surrounded by a pair of GaAs layers. The quantum well will have an bound energy level that acts as the resonance in the tunneling through the device. Another type of device is the high-electron mobility transistor [103], in which a GaAs layer plays the role of the Si in a MOSFET and a layer of GaAlAs plays the role of the oxide. In this case, however, the device is a normally-on device and a metal gate is used to deplete the electron density sitting in the GaAs at the interface between the two materials.

When the two dissimilar materials come together, the various bands usually do not line up with one another, not the least because the band gaps are not the same. The difference between the two band gaps is broken into a conduction band offset and a valence band offset. In type-I hetero-structures, the smaller band gap lies within the larger band gap. In type-II hetero-structures, the two band gaps are staggered. It is also possible for one band gap to lie completely below the valence band maximum of the other material. So, determining how these bands line up is a serious problem that must be solved before the hetero-structure becomes useful.

The earliest theories of these band lineups generally suggested that the electron affinities be used to align the bands. The electron affinity of a material is a surface property that gives the energy difference between the bottom of the conduction band and the vacuum level (hence the difference between the ionization energy and the band gap). However, there is no real rationale why this should work. The conduction bands have subtle differences between semiconductors due to the variations of the parameters that go into calculating the bands, and there is almost no reason to explain the difference between direct and indirect band materials. Where does one chose to evaluate the electron affinity? Should this be done at the zone center or at the minimum of the conduction band, which may not be at the zone center. In addition, the electron affinity is a surface property, and the surface may have differences from the bulk, as we will see in section 2.10 below. Consider for example the fact that the electron affinity, and the work function measuring the energy difference between the Fermi level and the vacuum level, and the ionization energy, all represent a potential barrier at the surface of the semiconductor. This potential barrier prevents the electrons from leaving the bulk material. Such a potential barrier must be supported by a charge dipole, supposedly created by some electrons moving outward from the bulk, leaving positive charge behind. This is the surface dipole. When two dissimilar materials are brought together, the two individual surface dipoles are replaced by a interface dipole, which is somehow a combination of the two surface dipoles. This interface dipole depends crucially on the nature of the interface at the hetero-structure junction. Even different growth conditions lead to differences in the dipole. There is no obvious way to determine this dipole.

This may seem to be concentrating on an unimportant point, but it turns out that one cannot effectively design different hetero-structures, especially when they may contain hundreds of layers of different materials, without knowing how the bands line up at each and every interface. Consequently, a great deal of effort has been expended on this problem, but by and large the answer is no closer to being found. The difficulty in achieving the answer resides in some of the above discussion, but is amplified for example in the AlGaAs/GaAs case. Here, the band offsets, and the nature of the interfaces, are different when GaAs is grown on top of AlGaAs as opposed to AlGaAs being grown on top of GaAs. In practice, the actual band offset must be determined experimentally, and this may be difficult as well.

2.9.1 Some theories on the band offset

The problem of interest is shown in figure 2.24, where two dissimilar semiconductors are joined at a type I hetero-junction. The problem is to ascertain how the difference in the two band gaps is distributed between the conduction and valence bands. It is complicated when the free carrier concentrations are different in the two materials, since the Fermi level must be at a single energy in both semiconductors when no current flows through the device. When it is different on the two sides, band bending (as shown) must occur to align the Fermi levels. The range over which this bending occurs depends upon the doping in the two materials. But, the band offsets are independent of this band bending, and rather are thought to depend upon the

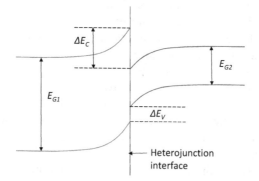

Figure 2.24. Pictorial view of the conduction and valence bands and their lineup and bending at the hetero-junction interface. The curvature is due to charge transfer across the interface, the forced lineup of the Fermi energy on the two sides, and Gauss' law making the fields continuous across the interface.

interface dipole. The dipole leading to the band bending is distinct from this other dipole. So, many different theories have appeared to explain how these band offsets are obtained. We cannot cover them all, but will deal with some of the more known ones.

One of the first was the so-called common anion rule [104]. In semiconductor hetero-junctions between materials like GaAs and AlAs, the anion (in this case As) is common between the two semiconductors. Since we found in section 2.2.3 that the top of the valence band is predominantly formed from the anion p-states, the common anion rule suggests that the top of the valence band should be continuous across the interface; e.g., $\Delta E_V \sim 0$. However, from (2.63) we know that there is an interaction energy between the p-states on the two atoms in the unit cell, here the E_{xx} energy. This interaction is likely different between the two semiconductors due to the effect of the different cations. Yet, it was suggested that one could use (2.63) with the known values of E_{xx} for the two semiconductors to determine the valence band offset. The problem lies in the fact that the bonding of the As atoms at the interface will be different if the GaAs side is terminated with Ga atoms or with As atoms. The bonding differences, and the amount of charge that lies in the bond charge located in the tetrahedral bond between neighboring atoms affects this. This latter charge affects a shift in the core levels of the atoms and affects the interfacial dipole, which hasn't been determined in this approach. For example, the As inner $3p$ level is known to lie some 40.2 eV below the vacuum level in AlAs, and some 0.6 eV lower in GaAs. This difference has to arise from the interface dipole representing the differences in the overall pseudo-potentials in the two materials. As mentioned before, there is no real method of calculating this interface dipole, nor is there any method to estimate how well it is screened by the valence electrons. As a result, we can say that the common anion rule was an interesting and pleasing suggestion, but in the end it is not a theory that can be used for designing reliable heterostructures.

Frensley and Kroemer [105] recognized that the dipole calculation would be the most important part of estimating the valence-band offset between the two materials. They suggested that one could estimate the potential in the open areas of the zinc-blende and diamond structures, relative to the vacuum level, from pseudo-potential calculations for the individual materials. Then, one could use the difference of this potential from the two materials to estimate the dipole strength. Although this potential gives a good average of the potential in the material, there really is no reason to expect that it occupies an exalted level at an interface. Yet, the values obtained for the band offsets are as good as any other approach and better than most.

A somewhat more attractive approach is the charge neutral level suggested by Tersoff [106]. Although controversial when it was introduced, this approach has gained a number of followers, since the predictions seem to be in good agreement with some experiments. Here, it is assumed that there exists a *charge neutral* energy level in the gap of the semiconductor. This level certainly exists, and is usually the position of the Fermi level in intrinsic semiconductors. We will in fact use charge neutrality to determine this energy in chapter 4 were we discuss semiconductor statistics. Tersoff used pseudo-potential calculations to evaluate the imaginary band structure where the k is imaginary[1] [107] to determine the position of this level. He then argued that this level should be aligned on either side of the hetero-junction, and doing so would minimize the interfacial dipole charge. By doing this, the valence band offset would arise from the relative position of the two valence bands and the charge neutral energies. In contrast, one could ask just why the dipole charge should be minimized. The dipole arises from the nature of the bonding and doesn't seem to be a quantity that can be maximized or minimized. Then, if the charge neutral point is as important to the band offsets, one would expect that the Fermi level could be pinned at this point by some mechanism, especially if we have to invoke the constancy of the Fermi level across the hetero-junction. But, this isn't the case. As remarked above, the charge neutral approach is used to find the Fermi energy in the intrinsic semiconductor. This is called the intrinsic Fermi level. And, it is known that this level certainly doesn't pin the Fermi energy. Adding a single donor or acceptor to the intrinsic material will move the Fermi energy by about one-half of the energy gap (at low temperature). Thus, it is clear that any charge neutral energy has no special properties.

The problem of band lineup becomes even more complicated when one of the two materials is a ternary alloy (or, worse, a quaternary alloy). It is not at all clear that the ternaries are really random alloys, and the nature of the alloy becomes more unclear at a hetero-structure interface. It is known that the actual interface is rougher, and the As $3p$ levels are shifted differently, when GaAs is grown on top of AlGaAs than when AlGaAs is grown on top of GaAs. Whether this is a problem arising from the alloy, or is a problem from the known affinity of Al for any oxygen in the system, is just not known. For the latter growth sequence, we do know that approximately 63% of the band gap difference goes to the conduction band offset with the remainder going to the valence band offset.

2.9.2 The MOS heterostructure

We normally don't think of the metal-oxide-semiconductor (MOS) structure in terms of hetero-structures. But, in fact, SiO_2 is a wide-band-gap semiconductor [108]. So, it is absolute correct to think of the SiO_2-Si structure as a hetero-junction, if for no other reason than that the MOSFET is the most studied and produced device in the history of the world. The field effect transistor (FET) is actually the oldest known form of transistor, as it was originally patented by Lilienfeld in 1926 [109]. But, it was not until 1959 that a working device was demonstrated [110], as control of the surface was a serious problem which held up development until after the bipolar transistor. Within a few years, the MOSFET was the preferred device for the integrated circuit due both to its planar technology and its generally lower power dissipation. The rest, so they say, is history. However, it is not generally appreciated

---

[1] The normal band structure treated in this chapter assumes a propagating wave where k is real. One can also assume k is imaginary and find strongly localized states which do not support propagating waves. These states are evanescent, but can be important for indirect optical transitions that are assisted by phonons. The evanescent states are used e.g. for tunneling in quantum mechanics.

that it is also a good device in which to study mesoscopic physics. The quantum Hall effect was discovered in a Si MOSFET [111].

The MOSFET is usually provided with a gate that can control the Si surface by a voltage. Normally, a n-channel device utilizes a p-doped layer or undoped layer of Si adjacent to the oxide. Then, a positive voltage applied to the gate induces electrons in a thin layer next to the oxide, as shown in figure 2.25. The inversion channel electrons reside between the potential barrier introduced by the oxide-semiconductor interface and the confining potential represented by the conduction band bending in the silicon. A cut of this confinement is shown in the figure. In the classical model, the density in the conduction band is related to the separation of the Fermi energy from the conduction band edge as (this will be shown in detail in chapter 4)

$$n = N_c \exp\left(-\frac{E_C - E_F}{k_B T}\right), \tag{2.160}$$

where N_c is the effective density of states, a number on the order of 10^{19} cm^{-3}. As shown in the figure, the conduction band edge is a function of position, so this makes the density a function of position, and a maximum at the oxide-semiconductor interface. From (2.160), it is clear that we can write the decay of the density into the semiconductor in the form

$$n(x) = n(0)e^{-x/L}, \tag{2.161}$$

where L is a characteristic decay length. In fact, the exponential decay is given by defining the surface potential $\varphi_s(x)$ as the variation of the conduction band away from its bulk equilibrium value. The surface electric field is given as

$$E_s \sim \frac{en(0)L}{\varepsilon_r \varepsilon_0}. \tag{2.162}$$

If we take an inversion density of 5×10^{11} cm^{-2} at room temperature, then we find that the effective thickness of the inversion layer is about 3.3 nm.

Quantum mechanically, we need to ask what the corresponding de Broglie wavelength is for an electron at the Fermi energy in the inversion layer. Suppose

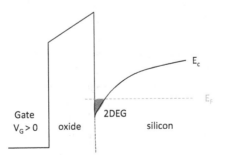

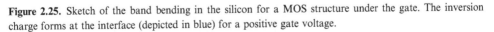

Figure 2.25. Sketch of the band bending in the silicon for a MOS structure under the gate. The inversion charge forms at the interface (depicted in blue) for a positive gate voltage.

we assume that the average energy of the carrier is just the thermal energy. Then, the de Broglie wavelength is given as

$$\lambda_d = \frac{h}{p} \sim \frac{h}{\sqrt{2m_c k_B T}}. \tag{2.163}$$

As we have discussed above, the minimum of the conduction band lies along the line from Γ to X in the Brillouin zone, and is located about 85% of the way to X. Because of the symmetry of the Brillouin zone, there are six equivalent minima. Each of the six ellipsoids have a longitudinal axis and two transverse axes, and corresponding values for the mass. In Si, it is generally felt that the effective mass values are $m_L = 0.91m_0$, $m_T = 0.19m_0$. Using the transverse mass for transport along the channel, we find that this wavelength is ~18 nm. Now, there is just no way to stuff an 18 nm electron into a 3.3 nm hole. Hence, the potential must be solved in a self-consistent manner. This results in the charge being peaked a nm or so away from the actual interface.

Silicon MOSFETs are usually fabricated with the surface normal along a (0,0,1) direction. Then, the quantization has a beneficial result of splitting the six ellipsoids into two sets. One pair of ellipsoids has the longitudinal mass normal to the interface, and this gives a lower energy set of quantum levels. The other four ellipsoids show the transverse mass in the direction normal to the interface and, as this is the smaller mass, will have higher lying quantum levels for motion normal to the surface. The advantage is that the two-fold set of valleys now show the smaller transverse mass in the transport direction, which gives a higher mobility. The introduction of strain about a decade ago in the industry was done for the same reason—to separate these valleys and gain a higher mobility. The doping has an effect on the potential, as band bending depletes the substrate so that there is a contribution to the surface field from this charge. Thus, in modern MOSFETs, we must replace the surface field of (2.162) with [112]

$$E_s = \frac{e}{\varepsilon}\left(N_A L + \frac{n_s}{2}\right). \tag{2.164}$$

Here, N_A is the acceptor concentration, L is our length from above, n_s is the inversion sheet concentration, and the total dielectric constant is used. The factor of ½ arises from the fact that the electric field appears on both sides of the inversion charge, while it only appears on the oxide side of the bulk depletion charge. So the results in figure 2.25 are doping dependent. With the momentum normal to the interface quantized, the transport is constrained to lie in the plane of the interface. Hence, these electrons form what is known as a quasi-two dimensional electron gas (2DEG) [112]. The mobility of these electrons is limited by scattering from a variety of sources, but primarily by scattering from ionized impurities, such as the acceptors in the bulk Si (these scattering processes will be discussed in chapter 5).

2.9.3 The GaAs/AlGaAs heterostructure

The most popular material system for hetero-junction devices is the GaAs/AlGaAs heterostructure system. It is important to point out that AlGaAs is a totally random alloy in principle, so that there is no clustering or precipitation of various compounds within the crystal. One reason for the ternary is that the band gap increases as the percentage of Al increases, thus one can create a series of quantum barriers and wells with multiple layers of the two compounds. However, AlAs is an indirect material, and the alloy becomes indirect at around $x = 0.55$. So, one usually stays with the more Ga-rich alloys, especially for optical applications, so that the band gap is direct at the Γ point. Another important point is that the lattice constant of both AlAs and GaAs are nearly the same. Hence, the alloy can be grown on GaAs with almost zero strain in the crystal, and this results in almost atomically sharp interfaces [113].

The band gap of AlGaAs is larger than that of GaAs, and this difference must be taken up by band bending at the interface. Part of the band discontinuity is taken up in the conduction band and part in the valence band, as shown in figure 2.24. Currently, it is felt that the conduction band discontinuity is about 63% of the total energy band discontinuity [114]. Prior to bringing the two materials together (conceptually), the Fermi level will be set in each material by the corresponding doping. Usually, the GaAs is undoped so that the Fermi level is near mid-gap and thought to be set by a deep trap level in this material. Once the interface is formed, however, there must be a single Fermi level that is constant throughout the material, as no current is flowing. This leads to the band bending as shown in the figure, and the band discontinuities provide certain offsets that are shown in the figure. In the early days, it was common to uniformly dope the GaAlAs, but this is no longer done. Rather, a single layer of dopant atoms is placed in the GaAlAs a distance d_{sb} from the interface [115]. Regardless of the doping method, electrons near the interface will move to the lower energy states in the quantum well on the GaAs side of the interface. For the δ-doping case (a single layer of dopants), all of the electrons will move to the GaAs. In the uniform doping situation, only a small fraction will move. This technique of getting the electrons into the GaAs is known as modulation doping [116]. In this approach, the actual ionized dopants are set some distance from the electrons, so that the Coulomb scattering potential is weakened. Moreover, the electrons themselves work to screen this scattering potential. These effects lead to very high mobilities for the electrons in the GaAs. In fact, mobilities above 10^7 cm^2 V^{-1} s^{-1} can be obtained for the electrons in the GaAs at low temperatures [117] and this is thought to be limited by scattering from the dopants themselves [118]. The usual dopant for the GaAlAs is silicon, which acts as a donor. However, it can also form a complex which leads to a trap level, known as the DX center [119]. This trap can be avoided if the composition of the GaAlAs is kept below about 0.25. So this sets additional limits on the heterostructure.

The GaAs/GaAlAs heterostructure is typically grown on a GaAs semi-insulating substrate. First, a superlattice formed of thin layers of GaAlAs and GaAs is grown. This has the double effect of smoothing the surface and trapping dislocations within the superlattice. Then, a thick undoped GaAs layer is grown, followed by the GaAlAs layer. For the latter layer, the growth is interrupted to deposit the dopant layer a desired distance from the interface, and then the additional ternary is grown to the desired thickness. For the highest mobility layers, d_{sb} can be more than 20 nm thick. The thicker this set-back layer is, the higher the mobility will be and the lower the inversion layer density will be. Finally, a heavily doped GaAs 'cap' layer is grown on the surface, which serves two purposes. First, it prevents unwanted oxidation of the GaAlAs surface. Second, it provides a layer to which it is easier to make ohmic contacts. The preferred growth method for high quality material is with molecular beam epitaxy, a growth technique achieved in ultra-high vacuum. The atomic constituents are provided by heated sources, known as Knudsen cells, in which the atomic species are individually vaporized and shutters are used to turn the sources on and off. Careful control of the deposition rates and the substrate temperature can lead to atomic layer epitaxy and the ultimate control of the growth. However, this process leads to low growth rates, well below a micron per hour, so it is not conducive to thick structures. This growth process, and slow grow rate, allows for the design of semiconductor multi-layers with very precise control of the overall structure for quite specific applications. Quantum well structures are created by sandwiching a GaAs layer between two GaAlAs layers, and the bound states in the well can be precisely tuned by careful control of the thickness of the GaAs layer and the effective barriers formed by the band offsets arising from a precisely controlled alloy composition. This process has led to a wide variety of emitters and detectors of radiation over a wide spectral range.

2.9.4 Other materials

If we consider just the binary III–V materials, there are already a large number of possible heterostructures that can be grown. In figure 2.26, we plot the energy gaps as a function of the lattice constant for the group IV and III–V compounds. Also shown are some rough connector lines for a few ternary alloys. One popular substrate is InP, not the least because it is lattice matched to $In_{0.53}Ga_{0.47}As$ which has an energy gap well suited to match the minimum dispersion in quartz fibers, a match very important for long distance fiber communications. The quaternaries InGaAsP or InGaAsSb provide the ability to both lattice match InP as a substrate and to vary the band gap over a very wide range. The ability to thus chose a material to provide desired characteristics is quite important in these materials. InAs has become important for THz HEMTs as well as with GaSb for spin applications.

The group III nitrides provide another set of materials with a wide range of attributes. For example, wurtzite GaN has a band gap of 3.28 eV and an a-plane lattice constant of 0.316 nm. In AlN, these values are 6.03 eV and 0.311 nm, while in InN, these values are 0.7 eV and 0.354 nm. Thus, alloys of these materials can span the entire visible range, and GaN-based systems have found a home in blue and blue-green lasers. They are also being pursued for high power HEMTs. While one normally tries to lattice match the various layers in a heterostructure, a controlled mismatch can be used quite effectively. The group III nitrides tend to be

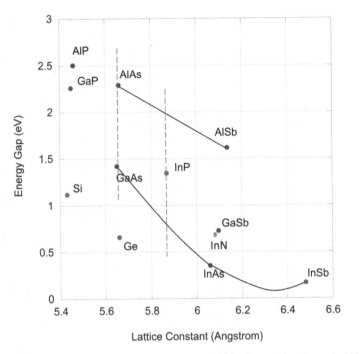

Figure 2.26. Band gap as a function of lattice constant for a variety of materials. The vertical blue lines show the range of materials that can be lattice matched to GaAs or InP.

ferroelectric, which means that they have a built-in polarization in the lattice. The discontinuity in this polarization at an interface can be used effectively to induce the inversion layer charge to form without the need for dopants [120].

2.10 Surfaces

The surface is the ultimate hetero-junction, in which the semiconductor is matched to the true electron vacuum. Generally, the most perfect surface is considered to be one that has been cleaved in high vacuum so that no 'pollutants' are absorbed onto the cleaved surface. The cleavage plane in diamond semiconductor is generally the [111] plane, while in the zinc-blende semiconductors it is usually the [110] surface, as these surfaces are charge neutral in the latter case (both A and B atoms lie in the surface plane in equal numbers). There are many tools available for studying the surface, such as electron emission spectroscopy with x-rays (XPS), ultraviolet photons (UPS), and especially with modern angle resolved images (ARPES). One can also use x-ray and electron scattering. The surface can also be imaged and studied with the scanning tunneling microscope and the atomic force microscope. So, we have been able to learn a great deal about various surfaces of semiconductors over the past several decades.

The projection of the bulk unit cell onto the surface presents a somewhat different view of the crystal structure. Moreover, it is often the case that the surface unit cell adopts a very different structure than what is expected by projection of the bulk unit cell. In general, if the atoms at or near the surface move around a bit, but the surface unit cell remains that of the bulk cell projection, it is said that the surface has *relaxed*. On the other hand, if the surface unit cell changes, then the surface is said to be *reconstructed*. Most semiconductors will have such changes.

The surface unit cell depends upon the orientation of the crystal itself. If the surface is a (100) surface, then the surface unit cell is just the face of the cube—four atoms forming a square with a fifth atom located in the center of the square (the face-centered atom from the cube). For a (110) surface plane, the surface cell is a slice through the cube which forms a rectangle of $a \times \sqrt{2}\,a$ in size, where a is the edge of the cubic cell. In addition, the second atom of the basis sits in this rectangle, centered on the short dimension but spaced $\sqrt{3}\,a/4$ from the closer pair of corner atoms. In Si and most zinc-blende materials, this (110) surface structure undergoes relaxation in which the four corner atoms move slightly out of the plane, while the second atom moves into the surface by about one-half of an angstrom. There is also some distortion in the next layer under the surface.

With the Si (100) surface, one can examine the above structure of a square with five atoms and realize that it is composed of smaller squares with four atoms, each spaced $\sqrt{2}\,a/2$ from each other. Each of these Si atoms has two bonds extending into the bulk, and two bonds sticking out from the surface. What happens is that these bonds tend to attract to neighboring atoms to form surface dimers. This leads to a reconstruction of the surface where the surface unit cells is 2 bulk cells in one direction and 1 bulk cell in the perpendicular direction, known as a 2 × 1 reconstruction. There are a great many different relaxations and reconstructions possible at the surfaces of various materials, in fact far too many to discuss fully in this limited space. But, those found in Si are a good introduction to at least the terminology.

2.11 Some nanostructures

In the beginning of section 2.4, we have already talked about one sort of nano-structure—the molecule placed between two Au electrodes. The molecule is of course quite small, and illustrates the problem of such small devices. Of course, now there is considerable interest in nanowires and nano-ribbons, and these small structures create new problems of the energy band structure. To rephrase what was said at the beginning of section 2.4, if we are going to use the momentum space approach for a one-dimensional or two-dimensional crystal, then we have to generate a full three-dimensional structure. For the one dimensional material, we create a two-dimensional lattice in the plane perpendicular to the one dimension of the material. The structure shown in figure 2.15 is, in fact, just one unit cell in the transverse lattice, and is assumed to be one unit cell in the direction of the molecule as well. Hence, the new unit (super-)cell for this momentum space approach has a great deal more atoms than that for the bulk material. For the two-dimensional layered material, we use multi-layers to generate the three dimensional structure. In each of these cases, however, the desired one-dimensional or two-dimensional structures must be separated sufficient far in space, that the replicas do not interact!

That is, for e.g. layers of graphene, we need to space the layers further apart than they would occur in graphite so that the individual layers do not interact [50]. Similarly, for the TMDCs, multi-layers must be used, and by varying the separation one can study the transition from bulk to monolayer, whether using the empirical pseudo-potential method [35] or first-principles DFT [121].

In the real space approach, one does not have to create the super-cell for the lattice, but one does have to use a great deal more atoms. The reason we don't need the super-cell is that the periodicity is generated from the Bloch sums (with the vectors to neighboring atoms), not from the Fourier transform of the crystal. Hence, the major periodicity arises from the direction of the nanowire or the direction of the nano-ribbon, where the latter is more common for the two-dimensional materials. In this section, we want to illustrate the real-space approach to determining the band structure of these materials. This band structure will be much more complicated, simply because there is a much larger number of atoms in the unit cell.

In the momentum space approach, the super-cell produces a so-called zone folding effect. The lateral dimension of the super-cell has a Fourier momentum that is much smaller than the normal Brillouin zone size, and this added periodicity leads to folding of the band structure back over this smaller momentum vector. Hence, many more bands exist in the new, smaller Brillouin zone.

In the real-space approach, we have periodicity in only one direction for e.g. a nanowire. Hence, the size of the nanowire governs how may atoms there are in each transverse layer, and the number of layers that arise in the super-cell of the nanowire along the growth direction multiplies the total number of atoms. So, even in this case, we are dealing with a much larger Hamiltonian, and the solution is more complicated.

2.11.1 $In_{0.53}Ga_{0.47}As$ nanowires

Nanowires have been studied for several decades now. They can be fabricated by top down approaches, as in other semiconductor devices, or they can be grown as self-forming structures by various types of epitaxy [122]. Nanowires have become of interest in modern semiconductors for their promise of enhanced capabilities. This is true for MOSFETs [123] as well as in photovoltaics [124]. As we mentioned just above, a good way to determine the band structure of a nanowire is to use the real-space semi-empirical tight binding approach of section 2.3.3. This approach has been extended from that given in section 2.3.3 to include the contribution of d-states that contribute to the valence band [125], providing a $sp^3d^5s^*$ basis set. In order to use this approach, the atomic structure of the nanowire needs to be constructed as it relies upon the atomistic orbitals of the various atoms. This is shown in figure 2.27 for a 4 nm diameter nanowire of $In_{0.53}Ga_{0.47}As$ that is clad by a 2 nm layer of InP. The nanowire is oriented in the (111) direction, and this leads to the hexagonal faceting of the ternary material. The band diagram is calculated for momentum vectors along the wire axis and in the 1D Brillouin zone. The Hamiltonian for this is constructed from the full atomistic basis of the layers shown in the figure. The basis set is constructed assuming a random alloy which leads to an effective InGa atom in

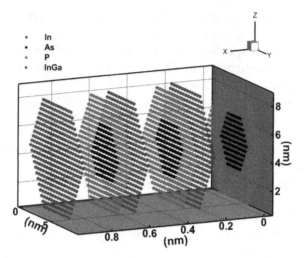

Figure 2.27. Unit cell structure of a 4 nm hexagonal $In_{0.53}Ga_{0.47}As$ nanowire with a 2 nm cladding layer of InP. Reprinted, with permission, from Hathwar R *et al* [125], copyright 2019 by IOP Publishing.

each layer of the nanowire. The full zone spin–orbit interaction is included in the interactions represented in the Hamiltonian.

Now, one should ask: 'What happens to the dangling bonds around the outside of the nanowire?' The standard approach is to assume that each unbound electron is attached to a H atom. This satisfies all the covalent bonds required and makes the edges more or less unreactive. Usually, the H atoms are not included in the Hamiltonian under the assumption that they are inactive in the energy structure and keep the dangling bonds from interfering with the calculation.

The first 400 conduction bands of the one-dimensional band structure are shown in figure 2.28. As $In_{0.53}Ga_{0.47}As$ is a direct gap material in the bulk, the minimum of the conduction band remains at the Γ point in the Brillouin zone. The first two conduction bands, each showing a spin splitting away from $k = 0$, are well separated in energy. It can be seen that the second band is more complicated and is actually composed of several closely spaced bands. Successive bands become even more complicated and eventually merge into what looks like a continuum of states. Indeed, in these higher bands, the density of states begins to appear like the bulk density of states (chapter 4).

2.11.2 Graphene nano-ribbons

The energy spectrum of graphene nano-ribbons depends upon the nature of their edges [29]. Usually, for computational considerations, these edges are either zig-zag or armchair. If we refer to figure 2.8, if the surface is terminated with the left-side of the structure, one sees the zig-zag edge. On the other hand, if the surface is along the top surface, we see the arm-chair edge (the two lower atoms are the seat, and the arms are the adjacent upper atoms). Calculations show that both types of edge produce a zero gap at the edges, hence both are metallic at zero order. However, the

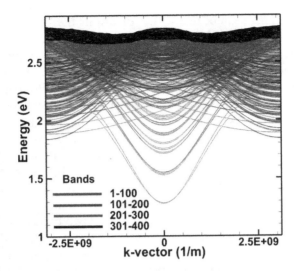

Figure 2.28. The one-dimensional band structure of the hexagonal nanowire illustrated in figure 2.27. The momentum direction of the one dimensional Brillouin zone is along the nanowire axis. Only the first 400 bands are shown. Reprinted, with permission, from Hathwar R *et al* [125], copyright 2019 by IOP Publishing.

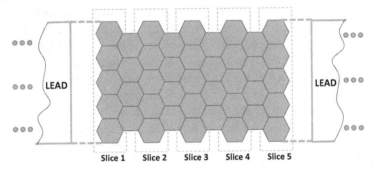

Figure 2.29. An illustrative drawing of a graphene nano-ribbon. The green dashed lines give the unit cell along the direction of the nano-ribbon. The structure uses arm-chair edges and each unit cell contains two rows of atoms. Adapted from Liu *et al* [128].

zig-zag edge produces a band of zero-energy modes (at the Dirac point) that is absent in the arm-chair case. More detailed calculations show that the metallic edge state in the arm-chair case is opened by interactions so that a gap in the spectrum is observed [126]. Hence, with arm-chair edges, the energy spectrum now is dominated by the bulk energy structure which has a gap due to the quantization arising from the finite width of the nano-ribbon.

Here, we will focus on a graphene ribbon with an armchair edge. In figure 2.29, we illustrate a typical (narrow, for illustrative reasons) graphene quantum wire scheme with two leads at the left and right ends [127]. In order to produce a common unit cell along the wire, we use two columns of carbon atoms for each slice. This will make the slice Hamiltonian have a dimension that is twice the number of atoms in each column, but this is necessary as we note that adjacent columns have different numbers of atoms. There can be many ways of writing the slice Hamiltonian and the coupling Hamiltonian. What is common in different approaches is that they all reflect the nature of the graphene atomic lattice and hexagonal structure. For simplicity and solely for illustrative purposes, we use three atoms in a column to show one of procedures of constructing the slice Hamiltonian, as depicted in figure 2.30. As shown, we consider the hopping energy between nearest neighboring carbon atoms to be γ_0, which is the value of the overlap integral for wave functions on two adjacent atoms as used in (2.46). From figure 2.30, we see that there are six atoms in the transverse unit cell, so we can write the slice Hamiltonian in the form of a 6 × 6 matrix, which is sparse in nature. The main diagonal is zero, and the off-diagonal elements correspond to the four blue bonds. To understand, it is convenient to break this larger Hamiltonian down into four 3 × 3 matrices. The diagonal terms correspond to the two bonds on the left side and the two bonds on the right. Say we number the three atoms in the left column as 1, 2, 3 (from the top), then the coupling is between 1 and 2 and 2 and 3, so there are non-zero elements (all given by γ_0) as 12, 21, 23, 32. The upper right matrix couples the left column to the right column and there are connects for 1–4 and 3–6. The entire 6 × 6 matrix is Hermitian, so the other connections are easy to fill. Between each slice is another 6 × 6 matrix representing the inter-slice coupling. We note from figure 2.30 that atom 2 in the central slice couples to atom 5 to the left, and atom 5 in the central slice couples to atom 2 on the right. So, the inter-slice coupling matrix has only a single non-zero entry [127].

There are some interesting aspects of the nano-ribbons in graphene. We note that the slice edge has two atoms, one of which is the A atom and the other the B atom from the basic bulk unit cell of the graphene layer. Each of these atoms has a Fourier component from the K and K' points of the Brillouin zone. The two different points in the Brillouin zone produce phase shifts in the atom Bloch functions, so that they also produce different waves along the atoms in each column of the slice discussed above. These waves now produce composites at both $K - K'$ and $K + K'$. The difference between the two points leads to a modulation across the nano-ribbon, whereas the sum vector is along the ribbon and can be ignored. But, the transverse

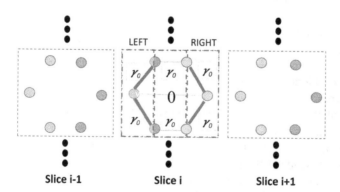

Figure 2.30. Pictorial representation of the coupling of the two slices per unit cell of the nano-ribbon. Adapted from Liu *et al* [128].

wave produces a density modulation that is called a 'Bloch beat.' From (2.40), we can write

$$\Delta K = K - K' = \frac{4\pi}{3\sqrt{3}\,a}. \tag{2.165}$$

The period of the standing wave is given by

$$\Delta L = \frac{2\pi}{\Delta K} = \frac{3\sqrt{3}\,a}{2} \sim 0.369 \text{ nm}. \tag{2.166}$$

As the spacing vertically between the atoms (say 1 and 2) is one-half the unit cell vector, this distance can be found from (2.39) as

$$t = \frac{|a_1|}{2} = \frac{\sqrt{3}}{2}a \sim 0.123 \text{ nm}. \tag{2.167}$$

The ratio of these two numbers is precisely 3. So, if the number of atoms in the column of figure 2.30 is a multiple of 3, there will be actual zeroes of the density every 3 atoms. If the number is not exactly a multiple of 3, then the wave is not commensurate with the lattice and the wave minimas will not be zero [128]. These waves have actually been imaged by scanning tunneling microscopy, and can produce interesting standing waves in the nano-ribbon [129]. The standing wave structure is also called a Kékule pattern, which formed of the six member ring of atoms, called a Clar sextet [130] in the chemistry community. This is a well characterized effect in poly-aromatic hydrocarbons, but is noted in electronics as well [131]. They readily show up in pseudo-potential simulations of graphene nano-ribbons [50]. But, these waves introduce a significant property of these nano-ribbons, and that is the strong dependence of the properties on the number of atoms in the ribbon width. When the number is $3p$, that is a multiple of 3, there is a single strong Kékule pattern at the edges. Then the number is $3p + 1$, one gets a pair of Kékule patterns, while when it is $3p + 2$, multiple Kékule patterns form. What is important is that the nature of the gap that opens at the edges depends upon this number of atoms, and this will be important later when conductivity and scattering are discussed.

References

[1] Nagaoka H 1904 *Philos. Mag.* ser. 6 **7** 445
[2] Rutherford E 1911 *Philos. Mag.* ser. 6 **21** 669
[3] Bohr N 1913 *Philos. Mag.* ser. 6 **26** 1
[4] Sommerfeld A 1916 *Ann. Phys.* **51** 1
[5] Slater J C 1960 *Quantum Theory of Atomic Structure* vol 1 (New York: McGraw-Hill)
[6] Hartree D R 1928 *Proc. Cambridge Phil. Soc.* **24** 89
[7] Fock V 1930 *Z. Physik.* **61** 126
    Fock V 1930 *Z. Physik.* **62** 795
[8] Hohenberg H and Kohn W 1964 *Phys. Rev.* **136** B864
[9] Kohn W and Sham L J 1965 *Phys. Rev.* **140** A1113

[10] Hellmann H 1935 *J. Chem. Phys.* **3** 61
[11] Lennard-Jones J E 1929 *Trans. Faraday Soc.* **25** 0668
[12] Slater J C and Koster G F 1954 *Phys. Rev.* **94** 1498
[13] Phillips J C and Kleinman L 1959 *Phys. Rev.* **116** 287
[14] Dirac P A M 1967 *The Principles of Quantum Mechanics* 4th edn (Oxford: Oxford University Press)
[15] Merzbacher E 1970 *Quantum Mechanics* 2nd edn (New York: Wiley)
[16] Brillouin L 1953 *Wave Propagation in Periodic Structures* (Toronto: Dover)
[17] Ziman J M 1963 *Electrons and Phonons* (Oxford: Oxford University Press)
[18] Martin R M 2004 *Electronic Structure* (Cambridge: Cambridge University Press)
[19] Kohanoff J 2006 *Electronic Structure Calculations for Solids and Molecules* (Cambridge: Cambridge University Press)
[20] Hedin L 1965 *Phys. Rev.* **139** A796
[21] Städele M, Majewski J A, Vogl P and Göring A 1997 *Phys. Rev. Lett.* **79** 2089
[22] Kittel C 1986 *Introduction to Solid State Physics* 6th edn (New York: Wiley)
[23] Mathew P T and Fang F 2018 *Engineering* **4** 760
[24] DeMarco K R, Bekker S and Vorobyov I 2019 *J. Physiol. Lond.* **597** 679
[25] Shamin S, Weber B, Thompson D W, Simmons M Y and Ghosh A 2016 *Nano Lett.* **16** 5779
[26] Saffarzadeh A, Demir F and Kirczenow G 2018 *Phys. Rev. B* **98** 115436
[27] Vukajlovic-Plestina J *et al* 2019 *Nat. Commun.* **10** 869
[28] Hathwar R, Saraniti M and Goodnick S M 2016 *J. Appl. Phys.* **120** 044307
[29] Castro Neto A H, Guinea F, Peres N M R, Novoselov K S and Geim A K 2009 *Rev. Mod. Phys.* **81** 109
[30] Li S-L, Tsukagoshi K, Orgiu E and Samori P 2016 *Chem. Soc. Rev.* **45** 118
[31] Wallace P R 1947 *Phys. Rev.* **71** 622
[32] Novoselov K S, Geim A K, Morozov S V, Jiang D, Katsnelson M I, Grigorieva I V, Dubonos S V and Firsov A A 2005 *Nature* **438** 197
[33] Khoshnevisan B and Tabatabaean Z S 2008 *Appl. Phys.* A **92** 371
[34] Kobayashi K and Yamauchi J 1995 *Phys. Rev. B* **51** 17085
[35] Cappelluti E, Roldán R, Silva-Guillén J A, Ordejón P and Guinea F 2013 *Phys. Rev. B* **88** 075409
[36] Vogl P, Halmerson H P and Dow J D 1983 *J. Phys. Chem. Sol.* **44** 365
[37] Sankey O F and Niklewski D J 1989 *Phys. Rev. B* **40** 3979
[38] Lewis J P *et al* 2011 *Phys. Status Sol.* B **248** 1989
[39] Madelung O (ed) 1996 *Semiconductors–Basic Data* (Berlin: Springer)
[40] Chadi D J and Cohen M L 1975 *Phys. Stat. Sol.* B **68** 405
[41] Teng D, Shen J, Newman K E and Gu B-L 1991 *J. Phys. Chem. Sol.* **52** 1109
[42] Tan Y P, Povolotskyi M, Kubis T, Boykin T B and Klimeck G 2015 *Phys. Rev. B* **92** 085301
[43] Slater J C 1937 *Phys. Rev.* **51** 846
[44] Herring C 1940 *Phys. Rev.* **57** 1169
[45] Phillips J C 1958 *Phys. Rev.* **112** 685
[46] Brust D, Phillips J C and Bassani F 1962 *Phys. Rev. Lett.* **9** 94
[47] Brust D 1964 *Phys. Rev.* **134** A1337
[48] Cohen M L and Phillips J C 1965 *Phys. Rev.* **139** A912

[49] Cohen M L and Bergstresser T K 1966 *Phys. Rev.* **141** 789

[50] Fischetti M V *et al* 2013 *J. Phys. Cond. Matter.* **25** 473202

[51] Kaasbjert K, Thygesen K S and Jacobsen K W 2012 *Phys. Rev.* B **85** 115317

[52] Speyer G, Akis R and Ferry D K 2003 *J. Phys. Conf. Ser.* **38** 25

[53] Speyer G, Akis R and Ferry D K 2005 *IEEE Trans. Nanotechnol.* **4** 403

[54] Xu B and Tao N J 2003 *Science* **294** 571

[55] https://vasp.at/

[56] http://Fireball-DFT.org

[57] https://departments.icmab.es/leem/siesta/

[58] Speyer G, Akis R and Ferry D K 2006 *J. Vac. Sci. Technol.* B **24** 1987

[59] Kane E O 1966 *Phys. Rev.* **146** 556

[60] Chelikowsky J, Chadi D J and Cohen M L 1973 *Phys. Rev.* B **8** 2786

[61] Pandey K C and Phillips J C 1974 *Phys. Rev.* B **9** 1552

[62] Persson C and Zunger A 2003 *Phys. Rev.* B **68** 073205

[63] Erdélyi A (ed) 1981 *Higher Transcendental Functions* (Malabar, FL: Krieger Publishing)

[64] Chelikowsky J R and Cohen M L 1976 *Phys. Rev.* B **14** 556

[65] Falicov L M and Cohen M H 1963 *Phys. Rev.* **130** 92

[66] Liu L 1962 *Phys. Rev.* **126** 1317

[67] Bloom S and Bergstresser T K 1968 *Sol. State Commun.* **6** 465

[68] Weisz G 1966 *Phys. Rev.* **149** 504

[69] De A and Pryor C E 2010 *Phys. Rev.* B **81** 155210

[70] Pötz W and Vogl P 1981 *Phys. Rev.* B **24** 2025

[71] Kane E O 1957 *J. Phys. Chem. Sol.* **1** 249

[72] Kane E O 1966 *Semiconductors and Semimetals* ed R K Willardson and A C Beer (New York: Academic), vol 1 75

[73] Yu P Y and Cardona M 2001 *Fundamentals of Semiconductors* 3rd edn (Heidelberg: Springer) sec 2.6

[74] Dresselhaus G, Kip A F and Kittel C 1955 *Phys. Rev.* **98** 368

[75] Christol P, El Gazouli M, Bigenwald P and Joulié A 2002 *Physica* E **14** 375

[76] Lin K P, Lee T-C, Li M-Y, Lee C-P and Lin Y-M 2014 *Electron. Lett.* **50** 1018

[77] Whiteside V R *et al* 2019 *Semicond. Sci. Technol.* **34** 025005

[78] Madelung O 1996 *Semiconductors—Basic Data* (Berlin: Springer)

[79] Davisson C and Kunsman C H 1921 *Sci.* **54** 522

[80] Thomson G P and Reid A 1927 *Nature* **119** 890

[81] Smith R A 1961 *Wave Mechanics of Crystalline Solids* (London: Chapman and Hall)

[82] Zawadzki W 2013 *Acta Phys. Polon.* A **123** 132

[83] Kittel C 1963 *Quantum Theory of Solids* (New York: Wiley), 227

[84] Ferry D K 2018 *An Introduction to Quantum Transport in Semiconductors* (Singapore: Pan Stanford) ch 11

[85] Ferry D K and Nedjalkov M 2018 *The Wigner Function in Science and Technology* (Bristol: IOP Publishing) sec. 12.5

[86] Phillips J C 1973 *Bonds and Bands in Semiconductors* (New York: Academic)

[87] Van Vechten J A 1968 *Phys. Rev.* **182** 891

[88] Harrison W A 1980 *Electronic Structure and the Properties of Solids* (San Francisco, CA: W. H. Freeman)

[89] Penn D 1962 *Phys. Rev.* **128** 2093

[90] Slater J C 1965 *Quantum Theory of Molecules and Solids* (New York: McGraw-Hill) vol 2

[91] Vegard L 1921 *Z. Phys.* **5** 17

[92] Vurgaftman I, Meyer J R and Ram-Mohan L R 2001 *J. Appl. Phys.* **89** 5815

[93] Mikkelson J C and Boyce J B 1983 *Phys. Rev. Lett.* **49** 1412

[94] Zunger A and Jaffe E 1984 *Phys. Rev. Lett.* **51** 662

[95] Srivastava G P, Martins J L and Zunger A 1985 *Phys. Rev.* **B 31** 2561

[96] Martins J L and Zunger A 1986 *Phys. Rev. Lett.* **56** 1400

[97] Mbaye A, Ferreira L and Zunger A 1986 *Appl. Phys. Lett.* **49** 782

[98] Jen H R, Cherng M J and Stringfellow G B 1986 *Appl. Phys. Lett.* **48** 782

[99] Kuan T S, Kuech T F, Wang W I and Wilkie E L 1985 *Phys. Rev. Lett.* **54** 201

[100] Nakayama H and Fujita H 1986 *Inst. Phys. Conf. Ser.* **79** 289

[101] Tsu R and Esaki L 1973 *Appl. Phys. Lett.* **11** 562

[102] Ferry D K 2015 *Transport in Semiconductor Mesoscopic Devices* (Bristol: IOP Publishing)

[103] Mimura T, Joshin K, Hiyamizu S, Hikosaka K and Abe M 1981 *Jpn. J. Appl. Phys.* **20** L598

[104] McCaldin J O, McGill T C and Mead C A 1976 *Phys. Rev. Lett.* **36** 56

[105] Frensley W R and Kroemer H 1977 *Phys. Rev.* **B16** 2642

[106] Tersoff J 1986 *Phys. Rev. Lett.* **56** 2755

[107] Chang Y-C and Shulman J N 1982 *Phys. Rev.* **B 25** 3975

[108] Hughes R C 1975 *Phys. Rev. Lett.* **35** 449

[109] Lilienfeld J E 1926 Method and apparatus for controlling currents US Patent 1475175

[110] Atalla M M, Tannenbaum E and Scheibner E J 1959 *Bell Sys. Tech. J.* **38** 749

[111] von Klitzing K, Dorda G and Pepper M 1980 *Phys. Rev. Lett.* **45** 494

[112] Ando T, Fowler A and Stern F 1982 *Rev. Mod. Phys.* **54** 437

[113] Suzuki Y, Seki M and Okamoto H 1984 *16th Congress on Solid State Devices Materials, Kobe, 1984* (Tokyo: Business Center Academy Society of Japan), 607

[114] Heiblum M, Nathan M I and Eizenberg M 1985 *Appl. Phys. Lett.* **47** 503

[115] Schubert E F and Ploog K 1965 *Jpn. J. Appl. Phys.* **24** L608

[116] Dingle R, Störmer H L, Gossard A C and Wiegmann W 1978 *Appl. Phys. Lett.* **33** 665

[117] Pfeiffer L, West K W, Störmer H L and Baldwin K W 1989 *Appl. Phys. Lett.* **55** 1888

[118] Lin B J F, Tsui D C, Paalanen M A and Gossard A C 1984 *Appl. Phys. Lett.* **45** 695

[119] Mooney P M 1990 *J. Appl. Phys.* **67** R1

[120] Ambacher O *et al* 1999 *J. Appl. Phys.* **85** 3222

[121] Kaasbjerg K, Thygesen K S and Jacobsen K W 2012 *Phys. Rev.* **B 85** 115317

[122] Samuelson L 2003 *Mater. Today* **6** 22

[123] Fang W W *et al* 2007 *IEEE Electron Dev. Lett.* **28** 211

[124] Hathwar R, Zou Y, Jirauschek C and Goodnick S M 2019 *J. Phys. D* **52** 093001

[125] Lee S, Oyafuso F, Allmen P V and Klimeck G 2004 *Phys. Rev.* **B 69** 045316

[126] Son Y-W, Cohen M L and Louie S G 2006 *Nature* **444** A453

[127] Liu B, Akis R and Ferry D K 2014 *J. Comput. Electron.* **13** 950

[128] Brey L and Fertig H A 2006 *Phys. Rev.* **B 73** 235411

[129] Park C *et al* 2011 *Proc. Nat. Acad. Sci.* **108** 18622

[130] Clar E 1972 *The Aromatic Sextet* (New York: Wiley)

[131] Ezawa M 2006 *Phys. Rev.* **B 73** 045432

IOP Publishing

Semiconductors (Second Edition)
Bonds and bands
David K Ferry

Chapter 3

Lattice dynamics

The study of acoustic waves propagating in solids, particularly semiconductors, has been around for a great many years. The earliest such studies followed the normal theory of deformable solids. When we expand to include the actual motion of the atoms within the solid, then we must use the adiabatic theory studied in the chapter 1, if we have any hope of solving a tractable problem. With this approach, we can attempt to follow the motion of the atoms without worrying about the presence of the electrons and their coupling to the atoms. There are many reasons to study the motion of the atoms. Perhaps the most important is to learn how the semiconductor responds to mechanical forces applied to it, such as pressure. But, our interest is mainly how the motion of the atoms leads to scattering of the electrons by this motion.

In this chapter, our first task is to develop the idea of waves that can exist in a simple one-dimensional chain of atoms. This effort, and its extension to a lattice with a basis makes connection to the ideas of the Brillouin zone of the chapter 2. It is the quantization of these modes that will lead to the ideas of phonons and the structure of these phonons. Scattering of the electrons by the lattice is treated as the absorption or emission of a phonon by the electron, and we will deal with this in chapter 5. Following the quantization ideas, we turn to treat the simple deformable solid theory for acoustic waves, as these can be used to study properties of the crystal interatomic forces. In this approach, the description of the solid is one of a continuous media represented by the solid volume.

We then turn to a discussion of the methods one can use to calculate the dispersion relations for the phonons in a particular crystal structure. In essence, the phonon dispersion relations are the lattice dynamic equivalent of the energy bands we determined for the electrons in the lattice. The lattice is common to both treatments, and it is this lattice with its periodic properties that set the Brillouin zone, so that the same zone is common to both the phonon dispersion and the electron bands. Finally, we discuss the anharmonicity of the lattice, where we go beyond the simple harmonic oscillator approach used to treat the phonon quantization and dispersion.

3.1 Lattice waves and phonons

The motion of the various atoms in the crystalline solid is much like the motion of the electrons, with the important exception that the atoms are forced *on the average* to remain in their equilibrium atomic positions which define the lattice. The lattice is, of course, a three-dimensional system. However, when the wave is along one of the principal axes of the lattice, one passes a regular array of atoms in a one-dimensional chain as one moves through the crystal. Hence, the one-dimensional chain is quite important in real solids as well as being very intuitive for understanding the nature of the lattice waves and the resulting phonons. While a simple model, it is easily extended to the typical motion for an entire atomic plane perpendicular to the wave motion.

3.1.1 One-dimensional lattice

We consider a one-dimensional chain of atoms that constitutes such a chain. At rest, the atoms are separated a distance a, as shown in figure 3.1 (this is, of course, the same as figure 2.3). Each atom has a mass M, and all the atoms are assumed to be identical (otherwise, it would not be a lattice). The goal here is to solve for the waves, and their dispersion relations which can exist in this lattice chain. As in the previous chapter, it is clear that this dispersion exists in a reciprocal lattice which defines a Brillouin zone. We write an equation for a particular atom, say s, in the chain. In reality, all of the atoms will be moved slightly, with the amplitude of the motion varying along the chain. It is this variation of the amplitude that constitutes the wave in this lattice. We consider that the forces between the atom can be represented by an everyday spring connected between each pair of atoms. As the atom moves, relative to its neighbors, the spring on one side will be extended while that on the other side will be compressed. It is these 'springs' which lead to the forces that return the atom to its equilibrium position.

As with the electronic case in section 2.3, we need only consider the forces between nearest neighbor atoms. We take these forces in the quadratic limit, which is the linear limit where the force is only a linear function of the displacement of the atom. Then, one can immediately write down the differential equation for the motion of the sth atom as

$$M\frac{d^2u_s}{dt} = F_s = C(u_{s+1} - u_s) + C(u_{s-1} - u_s), \tag{3.1}$$

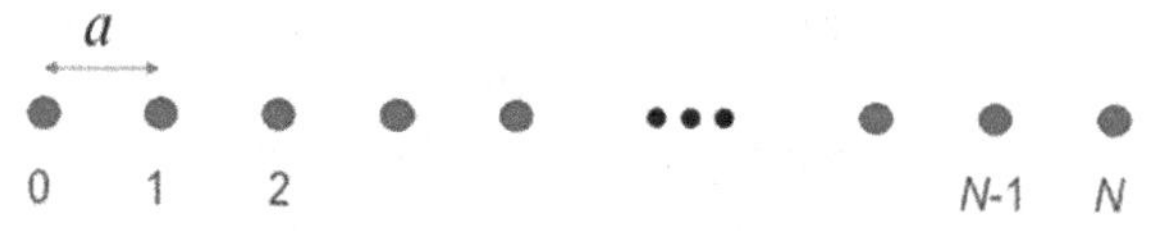

Figure 3.1. A one-dimensional chain of atoms, for which we will discuss the atomic motion. Here, the atoms will now be allowed to move around these equilibrium positions.

where u_s is the amplitude of the motion of the particular atom. The constant C is the force constant for the springs that connect one atom to its neighbor. Our current interest is in waves which propagate in this lattice. We will describe this wave as

$$u_s \sim e^{i(qx-\omega t)}, \tag{3.2}$$

where q is the wave number (we want to distinguish it from that for the electron). Further, we also assert that the idea of Bloch waves extends to the atomic motion, so that the shift between one atom and its neighbor can be described by an appropriate displacement operator

$$u_{s\pm 1} = e^{\pm iqa}u_s. \tag{3.3}$$

Using (3.2) and (3.3), we can rewrite (3.1) as

$$-M\omega^2 u_s = C(e^{iqa} + e^{-iqa} - 2)u_s. \tag{3.4}$$

Since the amplitude of motion drops out, we have left just the required dispersion relation between frequency and wave number for the wave. This is given as

$$\omega^2 = \frac{2C}{M}[1 - \cos(qa)] = \frac{4C}{M}\sin^2\left(\frac{qa}{2}\right). \tag{3.5}$$

It is clear from this result that all appropriate values of the frequency are found by taking q within the first Brillouin zone, just as for electrons, and we define the zone exactly as in chapter 2:

$$-\frac{\pi}{a} < q \leqslant \frac{\pi}{a}. \tag{3.6}$$

We further note that the right-hand side of (3.5) is positive definite, and while we may take the square root of both sides, the positive square root must be chosen. This follows because the energy in the lattice vibrations must be positive. The dispersion curves are shown in figure 3.2 for this one-dimensional result.

For small values of the wave vector q, the sinusoid may be expanded with its linear approximation. The frequency is now linearly related to the wave vector through

$$\omega = \sqrt{\frac{C}{M}}\, qa. \tag{3.7}$$

This is a familiar form for elastic waves, in that the frequency is a linear function of the wave number q. The velocity of the wave is given by the corresponding group velocity, and, as this is a low frequency wave of the lattice, it is called the *sound velocity*:

$$v_s = \frac{\partial \omega}{\partial q} = a\sqrt{\frac{C}{M}}. \tag{3.8}$$

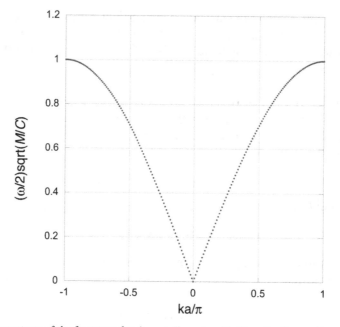

Figure 3.2. The spectrum of the frequency for the one-dimensional lattice vibrations, with a single atom per unit cell. The slope near $q \sim 0$ gives the acoustic sound velocity.

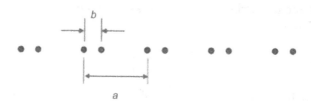

Figure 3.3. The diatomic lattice has two atoms per unit cell. Here we assume that they have different masses, and the spacings vary as well.

In fact, by measuring the sound velocity of an acoustic wave through the lattice, (3.8) may be used to determine information about the force constant C. This wave is called an acoustic wave (hence the sound velocity) and the wave length is quite long, measuring many hundreds of atomic spacings.

3.1.2 The diatomic lattice

Let us now consider the slightly more difficult problem of a diatomic linear chain in which there is a basis of two atoms, as shown in figure 2.5, and repeated in figure 3.3. The two atoms of the basis, one blue and one green, will be assumed to have different masses, M_1 and M_2, respectively. To accommodate this structure, we will designate the blue atoms with values of s that are even and the green atoms with values of s that are odd. As discussed previously, the lattice constant is given as a, while b is one of the nearest neighbor distances (the other is $a - b$). To ease the recognition within the equations, we will also designate the displacement of the blue atoms by u_s, as before. However, we will denote the displacement of the green atoms by w_s. While one may think that the distance between the atoms should be equal, this is not the case in many materials, and particularly in the semiconductors in which we are interested. In these materials, a glance at figure 2.9 will show that the structure along lines such as (111) clearly show the gaps indicated in figure 3.3. Naturally, we will assume that the springs between the atoms have different lengths, but are characterized by the same spring constant. Then, we can write the equations, in analogy to (3.1) as

$$M_1 \frac{d^2 u_s}{dt^2} = C(w_{s+1} + w_{s-1} - 2u_s)$$
$$M_2 \frac{d^2 w_{s-1}}{dt^2} = C(u_s + u_{s-2} - 2w_{s-1})$$
(3.9)

We assume that both of the two atomic displacements u and w propagate as waves, according to (3.2), so that we may rewrite the above equations as

$$-M_1 \omega^2 u_s = C(w_{s+1} + w_{s-1} - 2u_s)$$
$$-M_2 \omega^2 w_{s-1} = C(u_s + u_{s-2} - 2w_{s-1})$$
(3.10)

We now introduce the displacement operators, according to (3.3), and we can rearrange the equations as

$$(2C - M_1 \omega^2) u_s = C w_{s-1}(e^{iqa} + 1)$$
$$(2C - M_2 \omega^2) w_{s-1} = C u_s(e^{-iqa} + 1)$$
(3.11)

The dispersion relation is found by diagonalizing the resulting determinant of the above equations. Since there is no forcing function, the determinant must vanish, and

$$\begin{vmatrix} (2C - M_1 \omega^2) & -C(1 + e^{iqa}) \\ -C(e^{-iqa} + 1) & (2C - M_2 \omega^2) \end{vmatrix} = 0.$$
(3.12)

The solution to the determinant is then

$$\omega^4 - 2C\left(\frac{M_1 + M_2}{M_1 M_2}\right)\omega^2 + \frac{2C^2}{M_1 M_2}[1 - \cos(qa)] = 0.$$
(3.13)

It is clear that if we let the two masses be equal, we still will not recover the simpler dispersion relation of the last section. This is because of the differences in distances between the atoms. Our lattice must still have a basis with two atoms per unit cell. Hence, the diatomic nature of the lattice dictates that we get two roots of (3.13).

To see the nature of the solutions that arise from the dispersion relation (3.13) for this diatomic lattice, let look at some limiting cases. First, we examine the situation for $q = 0$, for which the last term in (3.13) vanishes. Then, the two solutions give

$$\omega^2 = 0$$

$$\omega^2 = 2C\left(\frac{M_1 + M_2}{M_1 M_2}\right). \tag{3.14}$$

The first of these solutions is just the acoustic mode discussed in the last section. The second solution, however, is a higher frequency mode, and is called the optical mode, given by

$$\omega_o = \sqrt{2C\left(\frac{M_1 + M_2}{M_1 M_2}\right)}. \tag{3.15}$$

This frequency has a reduced mass given by the geometric mean of the two atomic masses, and represents the coupling of the two atoms. This mode represents the wave displacement of the green atom of mass M_2 relative to that of the blue atom M_1. Hence, the two chains of different atoms are displaced relative to each other. Even for $q > 0$, the two sub-lattices vibrate relative to each other with both chains in motion. The appearance of the reduced mass in (3.15) is a sign of a normal mode that arises from the coupled oscillations of the two individual chains of atoms.

Now, let us take the short wavelength limit, where $q = \pi/a$. Then, (3.13) becomes

$$\omega^4 - 2C\left(\frac{M_1 + M_2}{M_1 M_2}\right)\omega^2 + \frac{4C^2}{M_1 M_2} = 0. \tag{3.16}$$

Again, there are two solutions, given as

$$\omega_1 = \sqrt{\frac{2C}{M_1}}$$
$$\omega_2 = \sqrt{\frac{2C}{M_2}}. \tag{3.17}$$

Each of these two frequencies involve only a single mass. That is, the first frequency is vibration of the blue atoms with the green atoms at rest. The second frequency is the reverse—the vibration of the green atoms with the blue atoms at rest. Which of the two frequencies is highest depends upon the size of the two masses. The higher frequency oscillation is that of the lightest mass, while the lower frequency oscillation is taken as that of the heaviest mass. The higher frequency is associated with the optical modes, while the lower frequency is associated with the acoustic modes. Obviously, if the masses are equal, the two modes are degenerate, and it is difficult to ascertain which chain is vibrating. In figure 3.4, we plot the two modes throughout the first Brillouin zone for the particular case in which $M_1 = 2M_2$. For this much difference in the two masses, the gap at the zone edge is fairly large. The upper mode in the figure (the blue curve) is the optical branch while the lower mode is the acoustic branch.

The results for these two lattices in this section and the previous one point out an important point. We find that one branch of the spectrum appears for each atom in

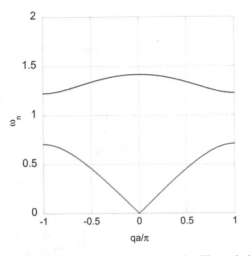

Figure 3.4. Plot of the two modes of the phonons in the diatomic lattice. The optical modes are shown in blue, while the acoustic modes are shown in red. This is the case for $M_1 = 2M_2$, and the frequency is scaled by $\omega_n = \sqrt{3M_2/4C}\,\omega$.

the basis, or for each atom in the unit cell of the crystal. We can extend this to three dimensions, where we will find three branches for each atom in the unit cell. Thus, in the zinc-blende or diamond lattices we expect to find six modes, three of which are acoustic modes and three of which are optical modes. But, for more atoms per unit cell, the number of acoustic modes is limited to three, as there can be only three ways in which the atoms can vibrate in phase. So, if we have more than two atoms per unit cell, all but three of the modes must be optical modes. So, if there are P atoms per unit cell, then there will be three acoustic modes and $3(P - 1)$ optical modes. Of the three acoustic modes, only one can be a longitudinal mode, in which the atom motion is parallel to the wave number q. The other two modes must be transverse modes, in which the atom motion is perpendicular to the wave number. This carries over to the optical modes as well, as one third of the optical modes can be longitudinal, with the remaining two-thirds being transverse modes.

3.1.3 Quantization of the one-dimensional lattice

One often talks about phonons as a concept and usage lies in discussion of condensed matter physics. A question that one might ask is how to connect the idea of a phonon to the lattice dynamics and motion of the atoms. Nearly everyone is familiar with quantization of the harmonic oscillator, as it should be familiar from introductory quantum mechanics. However, its connection to the atomic motion is perhaps not so clear. In this section, we want to show how this connection is made in terms of normal modes and Fourier transforms. We will do this in one dimension for clarity, but it is quite easily extended to three dimensions.

We begin with the Hamiltonian terms that describe only the lattice vibrations and the atomic motion, as described in (1.5). The inter-atomic potential, which provides the springs and spring constants of the preceding paragraphs will be expanded to second order, and this provides a quadratic approximation to the real potential. Once this simplified picture is achieved, then the equations are Fourier transformed and the idea of normal modes introduced. In this picture, the Hamiltonian can now be written as a sum over the various Fourier modes. The sum over lattice vibrations becomes a summation over a set of modes, which are then easily shown to be equivalent to harmonic oscillators. The latter are then described by their creation and annihilation operators, which correspond to the creation and/or annihilation of a single phonon. So, when we talk about a phonon, we are talking about a particular momentum q and energy $\hbar\omega_q$ that describe a single mode of atomic motion. We recall that the momentum is quantized by the periodicity of the lattice, and there are N discrete values for the momentum, where N is the number of unit cells in the lattice. In addition, there are $3P$ modes of atomic motion, where P is the number of atoms per unit cell. In Fourier transform space of energy vs. momentum, there will be $3PN$ different harmonic oscillators. So, when we talk about a phonon, we refer to a very particular excitation of one of this large number of harmonic oscillators. That is, the phonon is a quasi-particle of the atomic motion, and if an electron is to interact with it, we must describe the exact energy and momentum of that mode in which the quasi-particle exists.

As mentioned above, the part of the total Hamiltonian that is related to the atomic motion is given in (1.5), where it was achieved by use of the adiabatic approximation. In this approximation, the electrons move so rapidly that they follow the slow atomic motion adiabatically. Hence, we need only concern ourselves with the motion of the atoms. The lattice Hamiltonian is given as

$$H_L = \sum_j \frac{P_j^2}{2M_j} - \sum_{r,s\neq r} \frac{Q_r Q_s}{4\pi\epsilon_0 X_{rs}}$$
$$= \sum_j \frac{P_j^2}{2M_j} - \sum_{i\neq j,j} V|r_i - r_j| \tag{3.18}$$

In general, we consider that the atoms each have an equilibrium position, about which they have a periodic displacement. Hence, we can write the instantaneous position in terms of the equilibrium position as

$$r_j = R_{j0} + x_j. \tag{3.19}$$

The potential should average over time to a constant given by the equilibrium positions of the atoms. We can then expand this potential in a Taylor series about these equilibrium positions. The first-order terms must vanish, since the positions would not be equilibrium if these terms were non-zero. That is, the equilibrium positions must be in local minima of the potential, and the first derivatives about this local minimum will vanish. We then keep only the second-derivative terms as they are the most important ones. Thus, we expand (3.18) as

$$\sum_j \frac{P_j^2}{2M_j} - \frac{1}{2}\sum_{i\neq j,j} \frac{\partial^2 V}{\partial r_j \partial r_i} x_j x_i + \dots. \tag{3.20}$$

The result (3.20) is still in a mixed representation, and we need to replace the momenta by their spatial equivalents in a semi-classical sense, using

$$\sum_j \frac{P_j^2}{2M_j} \longrightarrow \sum_j \frac{M_j}{2}\left(\frac{\partial x_j}{\partial t}\right)^2, \tag{3.21}$$

so that we reach the classical form

$$H = \sum_j \frac{M_j}{2}\left(\frac{\partial x_j}{\partial t}\right)^2 + \frac{1}{2}\sum_{i\neq j,j} \frac{\partial^2 V}{\partial r_j \partial r_i}x_j x_i. \tag{3.22}$$

The Fourier transform of the atomic motion may be written as (now in one dimension)

$$x_j = \frac{1}{\sqrt{N}}\sum_q u_q e^{i(qr_j - \omega t)}. \tag{3.23}$$

In three dimensions, there would also be a polarization vector to distinguish between the various longitudinal and transverse modes. But, we ignore this here. We now use (3.23) in the potential term first, and this becomes

$$\begin{aligned}
\sum_{i\neq j,j} \frac{\partial^2 V}{\partial r_j \partial r_i}x_j x_i &= \frac{1}{N}\sum_{i\neq j,j}\sum_{q,q'} u_q u_{q'} \frac{\partial^2 V}{\partial r_j \partial r_i}e^{i(qr_i + q'r_j)} \\
&= \frac{1}{N}\sum_{i\neq j,j}\sum_{q,q'} u_q u_{q'} \frac{\partial^2 V}{\partial r_j \partial r_i}e^{iq(r_i - r_j) + i(q+q')r_j}
\end{aligned}. \tag{3.24}$$

We have added and subtracted a term in the exponent to explicitly reach the last term, and this can be summed over as

$$\sum_j e^{i(q+q')r_j} = \sum_{\text{cells}} e^{i(q+q')R_{j0}}\sum_{x_j} e^{i(q+q')x_j}, \tag{3.25}$$

where the last sum just runs over the positions of the atoms within the unit cell. The first sum runs over the basic lattice sites, and thus represents closure of the system and vanishes unless the coefficient of the exponential vanishes. When this occurs, (3.25) just gives us $N\delta(q+q')$. Then, summing over one of the momenta just gives the result that one momentum is the negative of the other. We add to this by defining the effective force constant is defined to be

$$C_q = \sum_j \frac{\partial^2 V}{\partial r_j \partial r_i}e^{iq(r_i - r_j)}. \tag{3.26}$$

Since we cannot tell one atom from another, the effective force constant is really independent of the index i, and we can write the Hamiltonian explicitly within the Fourier space as

$$H = \sum_q \left(\frac{M_q}{2}\frac{du_q}{dt}\frac{du_{-q}}{dt} + C_q u_q u_{-q}\right). \tag{3.27}$$

Here, the mass is the average mass in the unit cell, and again is really independent of the momentum wave number. But, we now see that the Hamiltonian is a sum over the Fourier modes, and each mode has an equation that is similar to a harmonic oscillator. This similarity is made more visible if we write the effective force constant as

$$C_q = \frac{1}{2}M\omega_q^2.$$

(3.28)

With this, we consider each Fourier mode as a harmonic oscillator. The phonon energy for that mode represents the spacing in the harmonic oscillator energy levels corresponding to that mode. Hence, creating a phonon in this particular mode, described by its wave number q, raises the population within that mode by one unit, increasing its energy. This increase of energy represents the increase of motion of that mode in real space. If we return to the momentum as P_q in the first term, as in (3.21), then quantization of this harmonic oscillator follows from requiring that

$$[u_q, P_q] = u_q P_q - P_q u_q = i\hbar.$$

(3.29)

Now both the atomic motion and its derivative are subject to the quantization condition.

As with the normal quantum approach to the harmonic oscillator, it is common to introduce creation and annihilation operators. In transport theory where we have scattering of the electrons by the phonons, these operators correspond to the emission of a phonon by the electron thus creating an excitation of that particular harmonic oscillator (determined by the wave number q). Correspondingly, the absorption of a phonon by the electron corresponds to the annihilation of a phonon in that particular harmonic oscillator. Thus, these processes correspond to the transfer of energy between the electron gas and the lattice itself. This is the subject of chapter 5. The creation and annihilation operators are defined in terms of the mode amplitude and momentum by

$$a_q^\dagger = \sqrt{\frac{M}{2\hbar\omega_q}}\left(\omega_q u_{-q} - i\frac{P_q}{M}\right)$$
$$a_q = \sqrt{\frac{M}{2\hbar\omega_q}}\left(\omega_q u_q + i\frac{P_q}{M}\right)$$

(3.30)

where the first of these is the creation operator and the last is the annihilation operator. It is relatively easy to show, using (3.29), that these operators satisfy the commutator relationship

$$\left[a_q, a_{q'}^\dagger\right] = \delta_{qq'}.$$

(3.31)

And, just as for the normal harmonic oscillator, we can write the Hamiltonian (3.27) in a simpler form using these operators as

$$H = \sum_q \hbar\omega_q \left(a_q^\dagger a_q + \frac{1}{2} \right). \tag{3.32}$$

The product of operators in the parentheses is known as the number operator, and, in equilibrium, the number of excitations in the harmonic oscillator is given by the Bose–Einstein distribution

$$E_{q,n} = \hbar\omega_q \left(n_q + \frac{1}{2} \right).$$
$$n_q = [e^{\hbar\omega_q/k_B T} - 1]^{-1} \tag{3.33}$$

Further properties of these operators and the harmonic oscillators themselves can be found in any good quantum mechanics textbook, such as [1].

3.2 Waves in deformable solids

One of the standard ways of determining the force constants, at least for the acoustic modes, is to use externally excited acoustic waves, which then propagate through the crystal, which is treated as a deformable solid body. Measuring the velocity of the acoustic waves at a given frequency of excitation allows us to determine the force constant from the velocity of propagation of the wave. The excited wave can be either a longitudinal or transverse mode, depending upon the transducer used. By considering the crystal as a deformable body, we are really treating it as a continuous medium rather than a collection of atoms. In this sense, it is treated as a homogeneous, although anisotropic, medium, and we deal with the long wavelength (small q) acoustic modes. The excitation is sufficiently small that Hooke's law remains valid, so that the strain is directly proportional to the stress within the solid. In recent years, this has again become relatively important due to situations where a hetero-structure has two materials with slightly different lattice constants. Knowing the elastic properties of either material, it is possible for determine how much strain is introduced in the crystal due to the lattice mismatch and how this strain changes the band structure.

The unstressed crystal may be defined in terms of three orthogonal axes, which we align with the usual rectangular coordinates, so that the unit vectors are defined by $\mathbf{a}_x, \mathbf{a}_y, \mathbf{a}_z$. After a small, but homogeneous deformation of the lattice is applied by the external stress, the coordinate system, and the unit vectors are deformed into a new set which are described by $\mathbf{a}_x', \mathbf{a}_y', \mathbf{a}_z'$. These new axes may quite generally be written in terms of the old set via

$$\mathbf{a}_x' = (1 + \epsilon_{xx})\mathbf{a}_x + \epsilon_{xy}\mathbf{a}_y + \epsilon_{xz}\mathbf{a}_z$$
$$\mathbf{a}_y' = \epsilon_{yx}\mathbf{a}_x + (1 + \epsilon_{yy})\mathbf{a}_y + \epsilon_{yz}\mathbf{a}_z. \tag{3.34}$$
$$\mathbf{a}_z' = \epsilon_{zx}\mathbf{a}_x + \epsilon_{zy}\mathbf{a}_y + (1 + \epsilon_{zz})\mathbf{a}_z$$

where the factors ϵ_{ij} define the deformation of the crystal due to the forces applied. While the original unit vectors were of unit length, the new ones will not be so. That is, the new vector in the x-direction now has length squared given by

$$\mathbf{a}'_x \cdot \mathbf{a}'_x = (1 + \epsilon_{xx})^2 + \epsilon_{xy}^2 + \epsilon_{xz}^2 \approx 1 + 2\epsilon_{xx} + \cdots \tag{3.35}$$

to lowest order in the small quantities. This leads to $\mathbf{a}'_x \sim 1 + \epsilon_{xx}$. Thus, the change in length is given to first order just by the deformation constant in that direction.

On an atomic basis, the distortion of the crystal leads to an atomic movement as well. If the atom was initially at r, after the distortion it will be moved to r'. Thus, we may define a displacement vector as

$$u = r' - r = x\left(\mathbf{a}'_x - \mathbf{a}_x\right) + y\left(\mathbf{a}'_y - \mathbf{a}_y\right) + z\left(\mathbf{a}'_z - \mathbf{a}_z\right)$$
$$= u_x\mathbf{a}_x + u_y\mathbf{a}_y + u_z\mathbf{a}_z \tag{3.36}$$

The deformations, and the definitions of the distortion waves, can define the general strain constants e_{ij}. These are defined in terms of the deformations through

$$e_{ij} = \mathbf{a}'_i \cdot \mathbf{a}'_j = \epsilon_{ij} + \epsilon_{ji} = \frac{\partial u_i}{\partial r_j} + \frac{\partial u_j}{\partial r_i}$$
$$e_{ii} = \mathbf{a}'_i \cdot \mathbf{a}'_i - 1 = \epsilon_{ii} = \frac{\partial u_i}{\partial r_i} \tag{3.37}$$

With the presence of the strain in the crystal, the lengths change and therefore the volume will also change. The new volume is given by

$$V' = \mathbf{a}'_x \cdot \left(\mathbf{a}'_x \times \mathbf{a}'_z\right) \approx 1 + \epsilon_{xx} + \epsilon_{yy} + \epsilon_{zz} + \cdots, \tag{3.38}$$

from which we can define the *dilation* of the crystal as

$$\Delta = \frac{V' - V}{V} = \epsilon_{xx} + \epsilon_{yy} + \epsilon_{zz}. \tag{3.39}$$

In general, one normally applies a stress to the crystal, which leads to strain appearing within the crystal. We have discussed these strain components, but not yet introduced the stress to the argument. As we can see from the definitions above, the strain is a second rank tensor. Similarly, the stress will also be a second rank tensor. But also, we see from the definitions in (3.37) that the off diagonal elements are symmetric in their subscripts, which means that our definition of the strain is non-rotational, so that there are no components which arise from e.g. the curl of a vector. The importance of this symmetry is that instead of the nine components of the deformation, the strain tensor has only six components, and this number will be reduced with further symmetry arguments later for our tetrahedrally-coordinated semiconductors. To see how this will occur, we note that the cubic symmetry means that we really cannot distinguish between the x-, y-, and z-axes. Hence, this will lead to symmetry effects. Hooke's law relates the stress tensor to the strain tensor, so we expect the relation to be a fourth rank tensor, so that $T_{ij} = C_{ijkl}e_{kl}$. This would lead us to a C matrix with 81 elements. But, with the reduction above, we expect only 36 elements, and we will write this with a new short-hand notation. We describe the six independent elements of the stress as

$$T_1 = T_{xx}, \quad T_2 = T_{yy}, \quad T_3 = T_{zz},$$
$$T_4 = T_{xy}, \quad T_5 = T_{yz}, \quad T_6 = T_{zx}. \tag{3.40}$$

This now defines a new, six element vector, and we can redefine the strain elements by this same notation. The 36 elements of the C matrix characterize Hooke's law, and are termed the elastic stiffness constants. As we will see, these are the spring constants we used for the atomic motion earlier. We may write this relation as

$$T_i = \sum_j C_{ij} e_j. \tag{3.41}$$

This last set of six equations is the most useful for general purposes. The forces that are applied from the acoustic transducers introduce the stress to the crystal, and this is connected to the strain through (3.41).

In a cubic crystal, as we mentioned above, it is impossible to determine which axis is the x, or y, or z. Thus, we can set $C_{11} = C_{22} = C_{33}$ by this symmetry. The crystal also possesses three-fold rotational symmetry about the set of (111) directions (the cube diagonals), and these rotations take $x \to y \to z \to x$. When one writes down the energy in the crystal, there will be terms such as $e_{ij}e_{kl}$. Since the energy is a scalar, these latter terms cannot have any preferred direction, and the above rotational symmetry then requires that these terms are unchanged by these rotations. Hence, this requires that $C_{14} = C_{15} = C_{16} = 0$, and similarly for equivalent terms, since these terms connect a compressional stress to a shear strain. Moreover, this also requires that $C_{44} = C_{55} = C_{66}$, since a static solid is considered which cannot rotate under the shear stress. Finally, rotation about any of the principle axes leaves the crystal unchanged, which requires the C matrix to be symmetrical. With this, and the equivalence of the principle axes, we find that $C_{12} = C_{13} = C_{23}$. This now leaves us with just three independent stiffness constants, so that the relation (3.41) is reduced to

$$
\begin{bmatrix} T_1 \\ T_2 \\ T_3 \\ T_4 \\ T_5 \\ T_6 \end{bmatrix}
=
\begin{bmatrix}
C_{11} & C_{12} & C_{12} & 0 & 0 & 0 \\
C_{12} & C_{11} & C_{12} & 0 & 0 & 0 \\
C_{12} & C_{12} & C_{11} & 0 & 0 & 0 \\
0 & 0 & 0 & C_{44} & 0 & 0 \\
0 & 0 & 0 & 0 & C_{44} & 0 \\
0 & 0 & 0 & 0 & 0 & C_{44}
\end{bmatrix}
\begin{bmatrix} e_1 \\ e_2 \\ e_3 \\ e_4 \\ e_5 \\ e_6 \end{bmatrix}. \tag{3.42}
$$

Thus all the shear strains are related to the shear stresses by a single constant, C_{44}, while the compressional stresses and strains are related by just a pair of constants. One of these, the diagonal component, relates strain that results from stress along the same axis, while the second, off-diagonal one relates the shear resulting from a stress, that deforms the cube. This deformation results in a stretching along the y- and z-axes for stress applied in the x-direction.

While we use the reduced notation in (3.42), it is important to remember that the stress and strain tensors are properly second-rank tensors. The force, which itself is a vector, arises as the divergence of the stress tensor. That is

$$F_i = \sum_j \frac{\partial T_{ij}}{\partial r_j}. \tag{3.43}$$

This can now be used to compute the equations of motion for the three components of the displacement within the crystal. In our homogeneous medium approximation, the mass density ρ is the mass per unit volume of the crystal. Then, we can write the general equation as

$$\rho \frac{\partial^2 u_i}{\partial t^2} = F_i = \sum_j \frac{\partial T_{ij}}{\partial r_j}. \tag{3.44}$$

The equations may be formulated quite easily using (3.40) to replace the stress terms on the right hand side of (3.43), and (3.41) to equate the stress to the strain, which in turn is related to the displacements through (3.36). This leads to the three equations for the three displacement components as

$$\rho \frac{\partial^2 u_x}{\partial t^2} = C_{11} \frac{\partial^2 u_x}{\partial x^2} + (C_{12} + C_{44}) \left(\frac{\partial^2 u_y}{\partial x \partial y} + \frac{\partial^2 u_z}{\partial x \partial z} \right)$$

$$+ C_{44} \left(\frac{\partial^2 u_x}{\partial y^2} + \frac{\partial^2 u_x}{\partial z^2} \right)$$

$$\rho \frac{\partial^2 u_y}{\partial t^2} = C_{11} \frac{\partial^2 u_y}{\partial y^2} + (C_{12} + C_{44}) \left(\frac{\partial^2 u_x}{\partial x \partial y} + \frac{\partial^2 u_z}{\partial y \partial z} \right)$$

$$+ C_{44} \left(\frac{\partial^2 u_y}{\partial x^2} + \frac{\partial^2 u_y}{\partial z^2} \right) \tag{3.45}$$

$$\rho \frac{\partial^2 u_z}{\partial t^2} = C_{11} \frac{\partial^2 u_z}{\partial z^2} + (C_{12} + C_{44}) \left(\frac{\partial^2 u_y}{\partial z \partial y} + \frac{\partial^2 u_x}{\partial x \partial z} \right)$$

$$+ C_{44} \left(\frac{\partial^2 u_z}{\partial y^2} + \frac{\partial^2 u_z}{\partial x^2} \right)$$

With these equations, one can now begin to study the various waves that can be used to determine some of the stiffness constants.

3.2.1 (100) Waves

We begin with the waves that are propagating along a principal axis of the crystal, that is along the cube edge, which we take to be the (100), or x, axis. As usual, we seek solutions of the form of (3.2). Then, the three equations (3.45) become

$$\rho \omega^2 u_x = C_{11} q^2 u_x$$
$$\rho \omega^2 u_{y,z} = C_{44} q^2 u_{y,z} \tag{3.46}$$

Thus, there is a longitudinal wave with displacement u_x and a group velocity

$$v_s = \frac{\partial \omega}{\partial q} = \sqrt{\frac{C_{11}}{\rho}}, \tag{3.47}$$

and a pair of transverse waves with displacements u_y and u_z. The two transverse waves both have a group velocity given by

$$v_t = \frac{\partial \omega}{\partial q} = \sqrt{\frac{C_{44}}{\rho}}. \tag{3.48}$$

The longitudinal wave is a compressional wave, while the two transverse waves are shear waves. The measurements of these waves now determine two of the three independent stiffness constants.

3.2.2 (110) Waves

We now consider waves which propagate in the x-y plane, with $q_x = q_y = q/\sqrt{2}$. Again, we will find that there is a single longitudinal mode and two transverse modes. One of the transverse modes has displacement in the z-direction, which is out of the propagation plane. This mode will be no different than the z-displacement mode in the previous case and brings no new information. This is because of the manner in which the crystal is symmetric for rotations around the z-axis. We can therefore focus on the two modes which have their displacements lying in the plane. For this, we take the wave to be propagating as

$$u \sim e^{i(q_x x + q_y y - \omega t)}. \tag{3.49}$$

Equation (3.45) may now be written for the two waves in the plane of propagation as

$$\rho \omega^2 u_x = \frac{q^2}{2}(C_{11} + C_{44})u_x + \frac{q^2}{2}(C_{12} + C_{44})u_y$$
$$\rho \omega^2 u_y = \frac{q^2}{2}(C_{11} + C_{44})u_y + \frac{q^2}{2}(C_{12} + C_{44})u_x \tag{3.50}$$

These two waves are coupled, and one must solve the determinant

$$\begin{vmatrix} \left(\rho\omega^2 - \frac{q^2}{2}(C_{11} + C_{44})\right) & -\frac{q^2}{2}(C_{12} + C_{44}) \\ -\frac{q^2}{2}(C_{12} + C_{44}) & \left(\rho\omega^2 - \frac{q^2}{2}(C_{11} + C_{44})\right) \end{vmatrix} = 0. \tag{3.51}$$

One finds the two roots of the expansion of this determinant and then uses these to find the relationship between u_x and u_y for each root. This allows us to identify the longitudinal mode, where the two displacements are in phase, and the transverse mode, where the displacements are out of phase. These two modes then have the velocities

$$v_l = \sqrt{\frac{C_{11} + C_{12} + 2C_{44}}{2\rho}}$$

$$v_t = \sqrt{\frac{C_{11} - C_{12}}{2\rho}}$$

(3.52)

Obviously, we can now determine the additional stiffness constant by measuring either in-plane velocity. While one can determine all of the constants by merely measuring the three (110) velocities, it is better to check this with measurements of the (100) modes as well. The stiffness constants have been measured for a great many materials, and collections of these can be found, for example, in [2].

3.3 Models for calculating phonon dynamics

The approach we have followed so far is fairly simple and based upon the sole idea of the forces between the atoms in a one-dimensional chain. These can be viewed to be simple *force constant* models. But, these remain too simple to calculate the full three-dimensional phonon dynamics throughout the Brillouin zone. As a result, more extensive approaches have appeared through the years that attempt to solve this problem, with greater or lesser degrees of success. In some cases, these models are merely extensions of the force constant model to include more force constants and interactions, in order to give more adjustable parameters that allow better fits to the measured phonon spectra. Experimentally, the dispersion curves for the lattice vibrations are often measured by neutron scattering [3]. These results, of course, give the information by which the available parameters of any model can be adjusted to fit the observed curves. In essence, these models are empirical in nature, just as empirical approaches were used in the last chapter to fit the observed electron band structure. In this section, we will examine a few of these models and discuss their applications.

3.3.1 Shell models

One of the most common models is the shell model proposed by Cochran [4]. In this approach, the nuclei and core–shell electrons are considered to be a rigid non-deformable body, while the bonding electrons compose a rigid shell surrounding this body. However, the shell and the nucleus are free to vibrate around each other. The presence of the directed covalent bonds (toward the four nearest neighbors) are treated by having the force constants as general as possible within the symmetry constraints of the crystal. However, the mass of the electron shells, relative to the cores, is considered to be negligible. Hence, the essence of the approach extends the force summation of e.g. (3.4) to four terms instead of the two nearest neighbor terms, and there will be four equations instead of two. In the case of Ge, the two atoms are the same, so that the model has only 5 parameters. First, the forces between the core and the shell are taken to be C_1 and C_2 for the two atoms. While the two atoms are the same, the vibrations of them are not the same, as was shown in section 3.1. Hence, these two forces are treated as different. Then, there is the forces between the two nuclei and the two shells. Here, the forces are described by C_{12} and C'_{12} for the nuclei and the shells, respectively. When the shell is displaced from the nuclei, a local dipole is created. Hence, there will be a dipole–dipole interaction between the two atoms, and this is the fifth force. Cochran was able to adequately fit the measured phonon dispersion for Ge with just these five forces [4].

The five parameter model, however, is not found to work very well for Si. For this purpose, the model can be extended. In the above model, the force between one core and the shell from the second atom is Coulombic in nature. If an elastic interaction is added, this brings two new parameters into the model. In addition, the dipoles can be taken to be different on the two atoms, as well as taking the spring constants to be different for the core–shell interaction of each atom. This brings us to nine parameters, which works reasonably well for Si and the III-Vs. Still more complicated models can be achieved by adding terms to the forces, and 11 and 14 parameter models have shown excellent agreement with experiment [5]. The 11 parameter model is adequate for Si, while the 14 parameter model is used for the zinc-blende materials.

We can write the equations for the shell models quite generally in terms of the Hamiltonian that we have expressed previously. The first step is to extend the potential term in (3.20) as

$$\frac{1}{2}\sum_{i \neq j,j} \frac{\partial^2 V}{\partial r_j \partial r_i} x_j x_i \rightarrow \frac{1}{2}\sum_{\lambda,\lambda'} u(\lambda) \cdot \frac{\partial^2 V}{\partial r_\lambda \partial r_{\lambda'}} \cdot u(\lambda'), \tag{3.53}$$

where the second derivative of the potential is a second-rank tensor. Each subscript corresponds to a pair of indices denoting the unit cell and the particular atom within the unit cell. After suitable Fourier transformation, then (3.1) becomes

$$-\omega^2 M \cdot U(q) = D(q) \cdot U(q). \tag{3.54}$$

Here M is a 6×6 diagonal mass tensor for the zinc-blende or diamond lattice, whose elements are the mass of the (two) atoms per unit cell, and D is the 6×6 second-rank tensor of the force constants. The vector U holds the three displacements of the two atoms, in the three directions. We need to supplement this with the displacements of the valence electron shells, which we denote by W. The motion between the atoms and the shells leads to the dipoles discussed above, which can be deformable, but the net forces between the atoms and the shells are depicted by the second-rank tensor P. Thus, we must add this interaction term to (3.54) to give

$$-\omega^2 M \cdot U(q) = D(q) \cdot U(q) + P(q) \cdot W(q). \tag{3.55}$$

The dimensions of the vector and the polarization tensor are the same as those of the atom vector and the force constant tensor. To this equation, we now add an equation for the motion of the shells, which is

$$0 = P^\dagger(q) \cdot U(q) + V_{ee}^\dagger \cdot W(q). \tag{3.56}$$

The tensor V_{ee} represents the dipole charge vector due to the Coulomb interactions between the shells.

All of the various tensors included in the above equations consist of a short-range elastic (spring type) force term and a Coulombic long-range term. Each of these may be written in the form

$$D(q) = D_{SR}(q) + D_C(q). \tag{3.57}$$

One advantage of this version of the shell model is that the electrostatic interactions can incorporate a frequency and momentum dependent dielectric function, which can account for the separation of the LO and TO modes at the zone center due to the polar nature of the atoms, that we will discuss in section 3.5 below. It is this latter effect that increases the number of parameters in a zinc-blende lattice over the diamond counterpart [6]. The shell model is best in semiconductors, with their directional tertrahedral bonds, when extended to the valence force field model, discussed next.

3.3.2 Valence force field models

One of the features of covalent semiconductors is the fact that the bonds are highly directional, and point toward the nearest neighbor atoms, with the bond charge itself situated near the midpoint of these bonds. These bonds are very important in understanding the cohesion of the crystal and the nature of the bands. For the present purpose, though, it is equally important to understand that these bonds tend to try to resist motion of the atoms that would vary the angle between the bonds, and these bond bending forces can be added to the natural elastic and Coulombic forces acting upon the atoms. One example of the nature of the potential energy term in (3.53) may be written in the form of the previous section, and the resulting valence force field model is [7]

$$\frac{1}{2}\sum_i\left[\sum_j x_j \cdot D \cdot x_i + \sum_{jk}(B_{ABA} + B_{BAB})\right], \tag{3.58}$$

where the A and B in the subscripts of the last term refer to the A and B atoms in the unit cell. The first term is the normal elastic and Coulombic forces between nearest neighbor atoms, as given in the previous section, although it is common to add an equivalent term for second-neighbor interactions. The last summation is the bond-bending terms that are typically of the type

$$B_{ABA} = K_{AB}x_i^2(\delta\vartheta_{ijk})^2 + K_{AB}'(\delta\vartheta_{ijk})x_i \cdot (x_j - x_k), \tag{3.59}$$

where the first term accounts for separation of two neighboring B atoms that stretches the angle between the two bonds connecting them to the A atom. The second term arises from the equivalent effects, but for the case where only one of the B atoms is moving. A similar term arises for rotations and forces of two A atoms around the B atom, which is the second contribution to the second term of (3.58).

Musgrave and Pople [8] first applied this approach to consider the lattice dynamics of diamond with five parameters, but the results were not particularly good. Nusimovici and Birman [9] increased the number of parameters to eight in order to treat a wurtzite crystal, with somewhat better success. Surprisingly, Keating

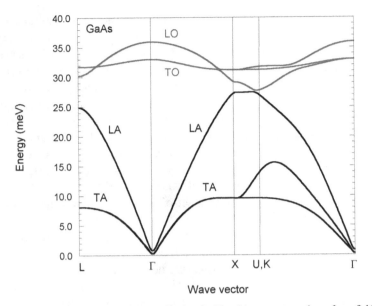

Figure 3.5. The dispersion of phonons in GaAs, calculated with a 14 parameter valence force field model [11]. (Figure reproduced with permission from Ayubi-Moak.)

[10] used a simplified model, with only two parameters to get good success for diamond. His parameters were α, which he called the central first-neighbor constant, and β, which he called the non-central second-neighbor constant. Nevertheless, he achieved good results for diamond, Si, and Ge, as well as some zinc-blende materials. In figure 3.5, the results of a 14 parameter calculation of the phonon spectra for GaAs are shown [11].

3.3.3 Bond-charge models

In some respects, there is a commonality between the inter-atomic forces of the various tetrahedrally coordinated semiconductors and with metals. When we look at the size of the gap versus the width (in energy) of the valence band, we note that it is relatively small whereas the relative dielectric constant is relatively large, being of the order of 10 or more. As a result, the bare atomic potentials of the atoms are screened by essentially the same type of strong Thomas-Fermi screening factor. This screening accounts for most, but not all, of the screening that occurs. It does not account for all of the screening, because the electronic charge is not totally accounted for by this approach. In the covalently bonded materials, a significant amount of charge is localized on the bonds themselves, situated midway between the nearest neighbor atoms. This charge is not incorporated into the Thomas-Fermi screening approach and is treated separately. This localized charge forming the bond is called the *bond charge,* and we discussed it somewhat in the previous section.

While the bond charge was mentioned above, we have not explicitly included it in the models that have been described so far. But these charges can be added to the dynamical matrices in a straight-forward manner. The interactions between the bond charges and the atoms, and those between the bond charges themselves, contribute forces that replicate the contributions of off-diagonal terms (band bending terms) in the dielectric function. The diagonal terms, which are the ones considered so far, lead to short-range, two-body forces between the bond charges and the atoms and between the atoms themselves. The interactions among the bond charges lead to non-central forces which are necessary to stabilize the crystal. Since there are two atoms and four bond charges per primitive, or unit, cell, we may expect that, on average, each atom has twice the charge of each bond charge. As there are two electrons (on average) per bond, we thus expect that the bond charge has a value of something like $-2e/\varepsilon_{r\infty}$, or about $-0.2e$ for most semiconductors.

An early view of a bond-charge model was advanced by Martin [12]. He actually calculated the interatomic forces with a set of parameters determined from pseudo-potential calculations (determining the atomic forces from pseudo-potential approaches is discussed in the next section). These potentials were screened by the diagonal part of the dielectric function. To these he added a Coulomb force between the atoms and the bond charges. However, he kept the bond charges fixed at the midpoint between the two atoms. Weber [13] modified this approach to allow the bond charges to move away from the centroid that is their equilibrium position, and it became clear that these forces among the bond charges, and the bond bending that results, is important in the flattening of the TA mode near the zone boundary. However, he also regressed to treating the elastic parameters as adjustable constants rather than computing them from an electronic structure calculation. Thus, this latter approach returns to an empirical basis.

In figure 2.11, the crystal structure of the diamond, and zinc-blende, structure was shown. The unit cell contains the two atoms of the basis, which are conveniently located at (0,0,0) and at (1/4,1/4,1/4) of the edge of the fcc cube. For this positioning of the two atoms, the corresponding bond charges are located at the points

$$R_3 = \frac{a}{8}(1,\,1,\,1) \qquad R_4 = \frac{a}{8}(-1,\,1,\,-1)$$
$$R_5 = \frac{a}{8}(1,\,-1,\,-1) \quad R_6 = \frac{a}{8}(-1,\,-1,\,1) \tag{3.60}$$

These positions are relative to the atom at the origin of the coordinate system (lower left corner in figure 2.11), and are located at the midpoint of the vectors to the four nearest neighbors of this atom. R_1 and R_2 are the vectors to the two atoms of the unit cell. All of these vectors are the equilibrium positions of the atoms, and not their dynamical deviations from these positions. In the harmonic approximation used previously, the Fourier transformed equations of motion for the motion of the atoms and the bond charges are given by

$$-\omega^2 M \cdot U(q) = D(q) \cdot U(q) + T(q) \cdot B(q). \tag{3.61}$$

The left-hand term and the first term on the right are the same as in (3.54) and (3.55), where U is the displacement vector for the two atoms. The new terms are those of the connection T between the bond charges B and the atoms. The tensor T is a

6×12 matrix, and the four bond charges lead to **B** being a 12×1 vector. As before, the dynamical matrix **D** and **T** have a short-range elastic (spring type) term and a Coulombic term as

$$D(q) = D_{SR}(q) + D_C(q) = D_{SR}(q) + \frac{4Z^2 e^2}{4\pi\epsilon_\infty \Omega} C_G$$

$$T(q) = T_{SR}(q) - \frac{2Z^2 e^2}{4\pi\epsilon_\infty \Omega} C_R \tag{3.62}$$

where Ω is the volume of the unit cell and the **C** matrices describe the Coulomb force directions between the atoms or between the bond charges, as appropriate. A second equation of motion is necessary, but it is assumed that the bond charges have zero mass, as previously, and

$$0 = S(q) \cdot B(q) + T^\dagger(q) \cdot U(q), \tag{3.63}$$

where

$$S(q) = S_{SR}(q) + \frac{Z^2 e^2}{4\pi\epsilon_\infty \Omega} C_S. \tag{3.64}$$

Now, the bond charge variables can be eliminated to yield the reduced equation for the atomic motion as

$$-\omega^2 M \cdot U(q) = [D(q) - T(q)S^{-1}(q)T^\dagger(q)] \cdot U(q). \tag{3.65}$$

The model basically has only three adjustable constants. These are the generalized force constant arising from the second derivative of the interatomic potential, the central force constant for the non-Coulomb interaction between the bond charges and the atoms, and a non-central force constant describing interactions among the bond charges. The Coulomb forces introduce no new constants, but are evaluated for the long range of the Coulomb potential that extends over a great many unit cells. The calculations are carried out in a single small unit cell, but the replication of this unit cell into the entire crystal can be carried out for the long range Coulomb interactions by a summation technique known as the Ewald sum. This adds extra terms to the Coulomb terms in the unit cell to account for the extended crystal.

In figure 3.6, the phonon dispersion relation for Si, calculated by Valentin *et al* is shown [14], and compared with experimental data. Also shown on the right-hand side of the figure is the computed density of phonon states (shown in arbitrary units). The approach can readily extended to more complicated structures. To illustrate this, we show in figure 3.7 the results of the phonons on the (110) surface of GaAs [15].

3.3.4 First principles approaches

One of the most useful results that has been obtained in condensed matter theory is the fact that the force constants within a crystal are directly determined by the static electronic response of the crystal [16, 17]. That is, within the adiabatic

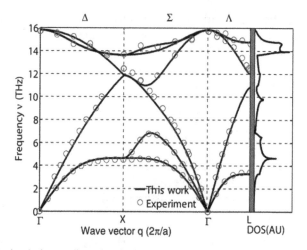

Figure 3.6. The calculated phonon dispersion for bulk silicon. The circles are the experimental data. On the right is the density of states (arbitrary units). Reprinted with permission from Valentine *et al* [14], copyright 2008 by IOP Publishing.

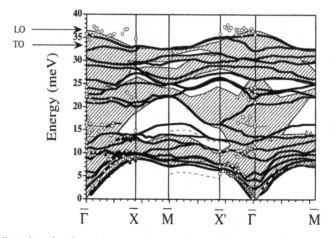

Figure 3.7. The dispersion of surface phonon modes on the GaAs (110) surface. The calculated results are the thick solid curves, while an ideal terminated surface is shown by the dashed lines. Modes with complex displacement patterns that have large amplitude at the outermost atoms are shown by thin curves. The open circles are experimental results. Reprinted with permission from Tütüncü and Srivastava [15], copyright 1996 by IOP Publishing.

approximation we have been using in this book, the lattice distortion associated with a phonon can be seen as a static perturbation acting upon the electrons. Thus, we can use the band structures determined in the last chapter to actually determine what the elastic force constants should be (this was briefly mentioned above for work by Martin [12]). What is needed however is the total energy of the crystal, which is a summation over all the occupied electron states in the valence band. The ease with which these calculations can be carried out via such a direct method has made it possible to determine phonon properties with only local pseudo-potentials and a local density approximation for the exchange and correlation energies [18]. The drawback of the method has been that the calculations for points away from the zone center require much larger calculations due to a need to compute the entire dielectric matrix [19], a formidable task even though only a small part of this matrix is needed for the phonons. However, more recent work has shown that this direct approach can be extended to the entire phonon dispersion curve via a supercell approach [20–22], which is adequate to incorporate all of the Coulomb forces. It is the long-range nature of the Coulomb force that creates some of the problem, and here we discuss the approach of Giannozzi *et al* [23]. While the approach can be extended to nonlocal pseudo-potentials [23], we follow only the local approach here.

To begin, we consider the total energy of the electrons in the fully bonded lattice under consideration. It is assumed that this energy is a continuous function of a set of parameters (which can be, e.g. atomic positions), which are described by the set $\lambda \equiv \{\lambda_i\}$. The Hellmann-Feynman theorem [24, 25] connects a set of forces to the variation of the energy with respect to these parameters [21, 22]. These forces can then be used to study many effects, such as lattice relaxation or surface reconstruction, for example. Here, it is just these forces which will be connected to the phonons. The variation of the energy with one of these parameters may then be expressed as

$$\frac{\partial E(\lambda)}{\partial \lambda_i} = \int n(\lambda, r) \frac{\partial V(\lambda, r)}{\partial \lambda_i} dr. \tag{3.66}$$

Here, $E(\lambda)$ is the ground-state energy relative to a set of given values for the parameters, while n is the corresponding electron density distribution. It turns out that to obtain the variation in the energy to second order, it is only necessary for the right-hand side of (3.66) to be correct to first order. The expansion in which we are interested is that of the density about its 'equilibrium' value, which is taken as n_0. By 'equilibrium' here, we mean that this is the state when the parameters are at their nominal values, such as the atom positions are at their equilibrium values with no displacements, as was the case in the determination of the energy bands. Then, (3.66) can be expanded in the parameters as

$$\frac{\partial E(\lambda)}{\partial \lambda_i} = \int \left\{ n_0(r) \left[\frac{\partial V(\lambda, r)}{\partial \lambda_i} + \sum_j \lambda_j \frac{\partial^2 V(\lambda, r)}{\partial \lambda_i \partial \lambda_j} \right] \right. \\ \left. + \frac{\partial V(\lambda, r)}{\partial \lambda_i} \sum_j \lambda_j \frac{\partial n(\lambda, r)}{\partial \lambda_j} \right\} dr \tag{3.67}$$

All of the derivatives in this last expression are evaluated at the equilibrium condition, which means that if we associate the parameters with the displacements of the atoms, then $\lambda = 0$. Integration of (3.67) with respect to the parameters gives the energy as

$$E(\lambda) = E_0 + \sum_i \lambda_i \int n(\lambda, \, r) \frac{\partial V(\lambda, \, r)}{\partial \lambda_i} dr$$

$$+ \frac{1}{2} \sum_{ij} \lambda_i \lambda_j \int \left[\frac{\partial n(\lambda, \, r)}{\partial \lambda_j} \frac{\partial V(\lambda, \, r)}{\partial \lambda_i} + n_0(r) \frac{\partial^2 V(\lambda, \, r)}{\partial \lambda_i \partial \lambda_j} \right] dr \qquad (3.68)$$

This energy now has the ionic potential plus the electronic energies included, and so represents the eigenvalue of the total Hamiltonian. If we take positional derivatives, where the parameters are the atomic displacements, then these will be derivatives of the potential energy contributions to this Hamiltonian. Making this connection, we then find that the matrix of the force constants is given as

$$C_{\alpha i, \beta j}(r - r') = \frac{\partial^2 E}{\partial u_{\alpha i}(r) \partial u_{\beta j}(r')} = C_{\alpha i, \beta j}^{\text{ion}}(r - r') + C_{\alpha i, \beta j}^{\text{elec}}(r - r'). \qquad (3.69)$$

In this equation, the indices α, β refer to the polarization of the atomic displacement, while i, j refer to the atomic position within the unit cell. The first term is the ion-ion contribution to the energy, which is a long-range Coulomb interaction, for which the energy contribution can be obtained by an Ewald sum [23] as

$$E_{\text{Ewald}} = \frac{Ne^2}{2\Omega} \left[\sum_{G \neq 0} \frac{e^{-G^2/4\xi}}{G^2} \left| \sum_i Z_i e^{iG \cdot r_i} \right| - \frac{1}{4\xi} \left(\sum_i Z_i \right)^2 \right]$$

$$+ \frac{Ne^2}{2} \sum_{ij} \sum_R \frac{Z_i Z_j}{r_i - r_j - R} \left[1 - erf \left(\sqrt{\xi} |r_i - r_j - R| \right) \right]. \qquad (3.70)$$

$$- Ne^2 \sqrt{\frac{2\xi}{\pi}} \sum_i Z_i^2$$

Here, Z_l denotes the bare pseudo-charge on each atom, and ξ is a parameter with an arbitrary size that is adjusted sufficiently large so that the real-space term can be neglected. This form of the Ewald sum is a construction over the reciprocal lattice vectors of the Coulombic interaction in reciprocal space. The Fourier transform of the ion-ion contribution is then found to be

$$C_{\alpha i, \beta j}^{\text{ion}}(q) = \frac{e^2}{\varepsilon_\infty \Omega_G} \left\{ \sum_{q+G \neq 0} \frac{e^{i(q+G)^2/4\xi}}{(q + G)^2} Z_i Z_j e^{i(q+G) \cdot (r_i - r_j)} (q + G)_\alpha (q + G)_\beta \right.$$

$$\left. - \frac{1}{2} \sum_{G \neq 0} \frac{e^{-G^2/4\xi}}{G^2} \left[Z_i \sum_i e^{iG \cdot (r_i - r_j)} G_\alpha G_\beta + c. \, c. \right] \delta_{ij} \right\} \qquad (3.71)$$

The electronic contribution to the elastic matrix elements is given as

$$C_{\alpha i, \beta j}^{\text{elec}}(r - r') = \int \left[\frac{\partial n(\lambda, \, r)}{\partial \lambda_j} \frac{\partial V_{\text{ion}}(\lambda, \, r)}{\partial \lambda_i} + n_0(r) \frac{\partial^2 V_{\text{ion}}(\lambda, \, r)}{\partial \lambda_i \partial \lambda_j} \right] dr, \qquad (3.72)$$

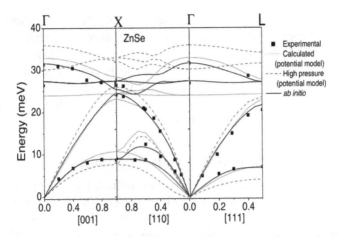

Figure 3.8. The dispersion curves for the phonons in ZnTe. The black solid lines are the results of *ab initio* pseudo-potential calculations for the phonons, while the red curves are the results of a force constant model. The dashed blue curves are the results at high pressure, and the points are experimental data. Reprinted with permission from Basak *et al* [26], copyright 2012 by IOP Publishing.

where $V_{ion}(r)$ is the bare atomic pseudo-potential, which may be expressed as

$$V_{ion}(r) = \sum_{R,i} V_i(r - R - r_i), \tag{3.73}$$

where r_i is the position of the ith atom in the unit cell. Herein lies a problem with use of the empirical pseudo-potentials. The latter are known only at a few reciprocal lattice vector values. But, we need the full real space formulation. With only a few reciprocal values, it is quite difficult to construct the proper real-space (first principles) pseudo-potential, and for this reason, many people begin with first-principles pseudo-potentials, from which they can obtain the Fourier coefficients as a good starting place to compute the band structure. When this is done, the empirical values of the important Fourier coefficients can be used to tweak these starting values when one wants to compute the lattice dynamics. Fortunately, many investigators have published their first-principles pseudo-potentials, and this provides a valuable resource from which to begin. Once these atomic potentials are known, then (3.72) can be constructed and subsequently Fourier transformed to yield

$$C_{\alpha i,\beta j}^{elec}(q) = \int \left[\left(\frac{\partial n(r)}{\partial u_{\alpha i}(q)} \right)^* \frac{\partial V_{ion}(r)}{\partial u_{\beta j}(q)} + \delta_{ij} n_0(r) \frac{\partial^2 V_{ion}(r)}{\partial u_{\alpha i}(0) \partial u_{\beta j}(0)} \right] dr. \tag{3.74}$$

In figure 3.8, we show the results of determining the phonon dispersion for cubic ZnSe [26]. The first principles calculation are done with an *ab initio* pseudo-potential, within the LDA approximation for exchange and correlation, using the readily available package *Quantum Espresso* [27]. These *ab initio* results are shown by the solid black curve. For comparison, results obtained using a shell model are shown by the solid red curves. The results of experiments using inelastic neutron scattering are shown as the solid black circles. While ZnSe crystallizes in the zinc-blende structure normally, it undergoes a phase transition to another structure above 13.7 GPa, and the phonon structure for this high pressure phase is also shown in the figure. An important point in the figure is the difference in the dispersion curves between the *ab initio* calculations and the shell model results. While the results are generally close, there are significant divergences which point to the limitations of the shell model (or perhaps the *ab initio* model).

3.4 Lattice contributions to the dielectric function

In atomic systems which include two (or more) dissimilar atoms per unit cell, such as the zinc-blend materials, the atoms that form the basis have a different number of outer shell electrons which contribute to the bonding. With the tetrahedral bonding each atom will have, on average, four electrons shared with its neighbors to fill the shells. When the dissimilar atoms have different numbers of outer shell electrons, there will be a component of ionic bond in the crystal. For a binary, such as GaAs, the two atoms of the basis will each have a small, but opposite charge, called the *effective charge e**. This leads to a dipole force between the two atoms, or between any one of the two charges and its four neighbors by symmetry. This dipolar field can interact with external electromagnetic waves, which means that it will change the dielectric constant depending upon whether the frequency of the waves lies above or below the characteristic frequency of the optical modes of the lattice vibration. Moreover, it is apparent that the crystal no longer can be inverted through the point between the two atoms, and so the crystal no longer has inversion symmetry. The dipolar force leads to a polarization and it is this polarization that modifies the dielectric constant, through its addition to the electric field effects of the electrons themselves.

Even in the case of SiC, the bonding is not entirely covalent in nature, and an ionic contribution to the bonding leads to the atomic polarization important for the dielectric 'constant.' In fact, even the cubic phase is strongly ionic in nature with an effective charge of nearly $0.9e$ [28, 29]. This leads to a very strong polar interaction which contributes to the dielectric constant. Because the effective charge is so large, it also leads to the crystallization of SiC in many poly-types of wurtzite and hexagonal structures.

Surprisingly, some layered compounds with dissimilar atoms, such as MoS_2, with 3 atoms per unit cell in the layer, do not generate such polar interactions and do not complicate the dielectric function. In particular, MoS_2 and many other TMDC materials, lack this dipolar interaction. In the case of MoS_2, each unit cell contains one transition metal (Mo) and a pair of chalcogenicde atoms (S). The structure exhibits the hexagonal arrangement similar to that of graphene shown in figure 2.8. The transition metal sits at the A atom site, while the two chalcogenide atoms sit above and below each other at the B atom site. The three atoms lead to 9 phonons, three acoustic modes and 6 optical modes. Hence there are 2 LO modes and 2 TO modes. The other two modes are out of plane modes labeled as ZO modes (the z-direction is usually normal to the layer). Careful studies of the phonon dispersion curves show that each LO mode is degenerate with its TO partner at the Γ point [30, 31], which means that an optical signal will not see any polar contribution to the dielectric function, as we show below. Because the maximum of the valence band and minimum of the conduction band lie at the K (and K') point, one might expect some effect, but the vertical transition still involves modes at Γ.

We can demonstrate the role that the polarization plays in the modification of the lattice vibrations by adding the external electric field to the equations of motion for the displacement. The electric field arises from the electromagnetic wave and adds an extra force to the equations of motion. We are interested in the response of the two atoms of the atomic basis pair, so we can use an effective one-dimensional chain which passes through these two atoms. Then, the new equations of motion can be written as

$$
\begin{aligned}
- M_1\omega^2 u_1 &= 2C(u_2 - u_1) + e*E \\
- M_2\omega^2 u_2 &= 2C(u_1 - u_2) - e*E
\end{aligned}
. \tag{3.75}
$$

The electric field E (not to be confused here with the energy) has a different effect on the two atoms because of the opposite effective charges residing on these atoms. It is clear that these equations are in the long wavelength limit, where $q \to 0$. We use this limit as the wave number of the electromagnetic wave is much smaller than any meaningful value of q in the crystal. These two equations can be solved to yield the two displacements as

$$
\begin{aligned}
u_1 &= \frac{e*E/M_1}{\omega_{TO}^2 - \omega^2} \\
u_2 &= - \frac{e*E/M_2}{\omega_{TO}^2 - \omega^2}
\end{aligned}
, \tag{3.76}
$$

where

$$
\omega_{TO}^2 = 2C\left(\frac{M_1 + M_2}{M_1 M_2}\right) \tag{3.77}
$$

is the optical mode calculated in the diatomic lattice as given in (3.14). This is the normal optical mode at $q = 0$, and describes the transverse displacement of this mode. The polarization of the atoms leads to a splitting in the optical mode frequencies of the longitudinal and the transverse modes. This splitting cannot be calculated from the normal approach, as the latter ignores the external electric field. The amount of this splitting depends upon the value of the effective charge. As is apparent in (3.76), both displacements have a singularity at an external frequency equal to this transverse optical mode frequency. There will be a large displacement at this frequency, and we will see that it has a significant effect on the dielectric function, and can lead to absorption at the infrared frequencies of the optical modes. The polarization is defined by the difference in the two displacements and

$$P = \frac{ne_*^2}{2}(u_1 - u_2) = \frac{ne_*^2}{2(\omega_{TO}^2 - \omega^2)}\left(\frac{M_1 + M_2}{M_1 M_2}\right)E. \tag{3.78}$$

In this equation, n is the density of atoms in the crystal, and the number 2 is added to indicate dipole pairs, so that the n/2 gives the number of unit cells per unit volume. The polarization enters the dielectric function through

$$D = \varepsilon(\omega)E = \varepsilon_\infty\left[1 + \frac{S}{\omega_{TO}^2 - \omega^2}\right]E, \tag{3.79}$$

where

$$S = \frac{ne_*^2}{2\varepsilon_\infty}\left(\frac{M_1 + M_2}{M_1 M_2}\right). \tag{3.80}$$

The frequency dependent dielectric function $\varepsilon(\omega)$ has a pole-zero characteristic just above the *transverse* optical mode frequency. The pole is clear from (3.76) and occurs at the transverse optical mode frequency. To find the zero, we set the dielectric function to zero, and find that it occurs at

$$\omega^2 = \omega_{TO}^2 + S \equiv \omega_{LO}^2. \tag{3.81}$$

This defines the *longitudinal* optical mode frequency. Between these two frequencies, the dielectric function is actually negative, which implies an imaginary index of refraction (given by the square root of the dielectric function). In the frequency range between the transverse and the longitudinal modes, electromagnetic waves in the crystal are strongly absorbed, and one finds only evanescent waves. If we let the frequency in (3.79) go to zero, then we find an important relation between the static and optical dielectric constants in terms of these measurable phonon frequencies:

$$\frac{\varepsilon(0)}{\varepsilon_\infty} = 1 + \frac{S}{\omega_{TO}^2} = \frac{\omega_{LO}^2}{\omega_{TO}^2}. \tag{3.82}$$

This last expression is known as the Lyddane–Sachs–Teller relation, which is important as it tells us how the vibrational modes of the lattice affect the electromagnetic wave propagation. Note that we use $\varepsilon(0)$ as the zero frequency dielectric constant rather than ε_0 as this last expression is the free space dielectric function. The high-frequency dielectric function differs from this free space value due to the plasma contribution from the valence electrons, as in the Penn dielectric constant (2.147). At low frequencies, the dielectric function must also account for the polarization of the lattice, while at high (optical) frequencies, this is not the case. Finally, we can rewrite the dielectric function entirely in terms of the two optical mode frequencies as

$$\varepsilon(\omega) = \varepsilon_\infty\left[1 + \frac{\omega_{LO}^2 - \omega_{TO}^2}{\omega_{TO}^2 - \omega^2}\right]. \tag{3.83}$$

It is clear from (3.82) that when the LO and TO mode are degenerate at Γ, there is no contribution or polarization from the dissimilar atoms, as was discussed above for the TMDC materials. In chapter 5, we will talk about the scattering of electrons by the lattice vibrations, and it is clear from this discussion that this scattering will be different for the two types of optical modes. First, the transverse modes produce a normal displacement of the atoms which can scatter the electrons by an effective potential arising from the strain in the crystal due to the atomic displacement. This strain modifies the band structure ever so slightly, and this produces the scattering potential. Since the band structure is already screened by the high density of bonding (valence) electrons, the scattering potential seen by the free electrons is not screened by them, as they are a small number compared to the valence electrons. On the other hand, the longitudinal optical mode may have a large Coulombic polarization, when the two atoms are dissimilar that is given by (3.79) and which creates an electric field that scatters the free electrons (this will dominate the deformation contribution). In this case, the interaction is screened by the free electrons, as their own interaction among themselves is also Coulombic in nature. Thus these two long-range interactions can interfere with each other. We treat this by the screened interaction of the electrons with the polar modes.

3.5 Alloy complications

Ternary semiconductor alloys, such as $Ga_{1-x}Al_xAs$, not only have important device applications, but they also provide a challenge to our understanding of the physics behind such alloys, especially the disorder-induced effects. While the random alloy theory discussed in the last chapter would suggest that this material should have a single LO mode that arises from the average mass contributions from Ga and Al, this is not what is found in the material. It is well known that this material is a two mode alloy (an effect found in most alloys), in which LO modes arise from both GaAs-type phonons and from AlAs-type phonons [32, 33]. In a true random alloy, one would expect a single mode whose frequency varied from the AlAs mode to the GaAs mode smoothly with x. However, what is normally found is that there are these two modes extending from either end of the alloy composition, but moving directly to the other extreme, nor following the linear curve. Moreover, more complicated behavior has been seen in $In_xGa_{1-x}As$, such as one mode following the smooth transition and linear in x, and one at lower frequencies [34]. In general though, the strength of each mode varies with the composition of the alloy.

Consider for example, the alloy $Ga_{1-x}Al_xAs$. The AlAs-type modes for the LO and TO phonons start at the AlAs LO and TO frequencies [35]. Then, as the alloy concentration is reduced they both *decrease* in frequency with decreasing concentration until they merge (become degenerate) at $x = 0$. This merging point lies well above the GaAs LO mode frequency, so they do not merge into one of the pure GaAs modes. On the other hand, the GaAs-type modes begin at $x = 0$ with the pure GaAs LO and TO modes. Then, as the concentration is increased, both modes *decrease* in frequency, rather than approaching the AlAs modes. Again, the two modes merge at $x = 1$, and thus no long contribute polarization. In figure 3.9(a), this

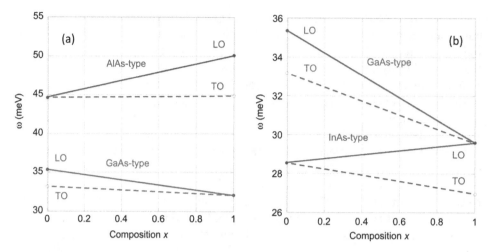

Figure 3.9. Pictorial representation of the two-mode phonon behavior in (a) $Ga_{1-x}Al_xAs$ and (b) $In_xGa_{1-x}As$. Experimental data shows that the modes differ slightly from the linear behavior shown here.

behavior is shown pictorially. Experiments show that the data deviate slightly from this linear behavior.

Somewhat different behavior is seen in the case of $In_xGa_{1-x}As$. Here, the GaAs-type modes also *decrease* with composition just as in $Ga_{1-x}Al_xAs$. But, here, both modes merge at the InAs LO mode. With the InAs-type modes, the LO mode frequency *decreases* as the concentration is reduced, while the TO mode *increases* in frequency until they merge (at $x = 0$) at a frequency well below the GaAs TO mode. As this merge point does not correspond to either of the GaAs mode frequencies, it has been referred to as merging at a 'gap mode' [34], which must be evanescent. In this sense, the gap mode lies between the optical and acoustic modes. But, still this is different than that of the $Ga_{1-x}Al_xAs$, where the AlAs modes merge at a frequency above any GaAs mode. In figure 3.9(b), this behavior is shown pictorially. As in the previous case, experiments show that the data does not exactly follow this linear behavior.

The source of the two mode behavior has been argued for years. One appropriate source that has been suggested is disorder in the alloy [33]. Another is the 'local mode' behavior, that could be likely as the optical modes arise from the two atoms in the unit cell. Hence, the fact that this is a random alloy would suggest that the unit cells with an Al atom (or an In atom) would vibrate differently from those cells with a Ga atom.

When the two-mode behavior is observed, one must then ask the question as to how the dielectric difference between the high-frequency and zero frequency behavior is to be allocated between the two sets of LO and TO modes. Obviously, there is not a single transition between the two values of dielectric function. But, there are a variety of techniques that can be applied. The transition region between the LO and TO frequencies is a region of negative dielectric constant, in the approach of (3.83). In actual fact, the individual modes are damped by anharmonic forces on the atoms; e.g., forces arising from higher order terms in the inter-atomic potentials than the quadratic terms to which we have limited ourselves. These damping forces lead to imaginary parts of the dielectric function which usually keep the latter from going negative. However, this region between the two limiting frequencies has unusual behavior and is often referred to as the 'restrahlen' region (obviously a German phrase). With two-mode behavior, we will have two restrahlen regions.

Let us consider amorphous SiO_2, our common insulator in MOSFETs, which also can be considered as a two-mode system. There is a dominant LO phonon mode at 153 meV, as measured by infrared polarized optical spectroscopy [36]. These results agree well with studies using Raman scattering [37]. These studies also indicated a more complex spectrum with a second important LO mode at about 63 meV [38]. From the known optical and static dielectric constants for SiO_2, one can determine the overall effectiveness of the two modes in giving rise to a polarization. However, if we desire to know the effective coupling of each of the two modes, we need more. The overall coupling strength is known from breakdown studies of Lynch [38]. This can be found from the measured optical and dc relative dielectric constants of 2.5 and 3.9, respectively. From the dielectric analysis, one can then estimate a value for the intermediate region between the two restrahlen regions as about 3.1. Now, with (3.82), we can estimate the frequencies of the two TO modes. Hence, by a variety of measurements, the dielectric properties for two-mode systems can be determined.

However, SiO_2 is not an alloy. In the case of an alloy such as $Ga_{1-x}Al_xAs$, the question arises as to how to assess the effectiveness of each mode as a function of the composition x. Certainly, one can use data from curves such as those of figure 2.9 to estimate how much of the excitation should be given to each mode. But, one can go further. One can use picosecond Raman scattering to characterize quantitatively the effective coupling to each of the two modes as a function of composition [39]. With the pump and probe technique with the picosecond laser, the population of the modes can be driven out of equilibrium. The electron-hole pairs that are generated by the pump provide electrons high in the conduction band when the laser energy is significantly larger than the band gap. These electrons decay to near the bottom of the band by phonon emission at such a rate that these phonons don't have time to relax, and this leads to a non-equilibrium distribution of phonons. The phonon population can be tracked by monitoring both the Stokes and anti-Stokes Raman sidebands. This alloys one to monitor each population separately and thus determine the relative efficiency of either of the two modes. In [39], it was found that the production efficiency of Raman-active LO phonons in the alloy did not differ from pure GaAs, which says that any alloy disorder did not increase the number of modes that could be excited by the electron relaxation. They also showed that the phonon lifetimes did not differ from those in GaAs (see the next section). These measurements show that the fraction of AlAs modes excited for $x < 0.3$ rises linearly with alloy composition, with a fraction estimated to be just less than 0.3 at $x = 0.25$. Studies of the picosecond excitation and Raman scattering with ensemble Monte Carlo techniques (discussed in chapter 7) give roughly identical estimates for the AlAs fraction [40], but that this estimate can have a carrier density dependence, that arises from a mis-match between the Raman scattering wave vector and the dominant wave vector for the non-equilibrium phonons that are produced in the process [41].

3.6 Anharmoic forces and the phonon lifetime

Through the first sections of this chapter, we have treated the forces only through the harmonic expansion; that is, we have kept terms up to the second derivative of the potential. This allowed us to quantize the atomic displacements in order to talk about them in terms of phonons, the excitations of the harmonic oscillator expansions used in the Fourier transform of the atomic motion. But, we have also introduced a lifetime for the phonons in the discussion of sections 3.4 and 3.5. The question arises as to how do non-equilibrium optical phonons relax. The optical branches are generally separated from the acoustic branches by an energy gap. Hence, if the population of non-equilibrium optical phonons is to relax, these optical phonons must disappear in a process that causes the excess energy to be transported to the boundaries of the semiconductor. This cannot occur through just harmonic processes. Instead, we need anharmonic, three phonon processes, which arise only with the third derivative term in the atomic potential expansion (we need three different u_i's). With the consideration of these terms, we can imagine that energy is exchanged between the optical and acoustic phonons. Generally, the dominant method by which energetic electrons lose their energy is by the emission of optical phonons. These must decay into the acoustic modes, which can then carry the heat away to the surface of the semiconductor (and thus to a heat sink). But, if we look carefully at figure 3.4, it is apparent that the optical mode must couple to two or more acoustic modes if we are to conserve energy and momentum in the process. Thus, the leading term in the decay process involves three phonons, and the interaction for this must come from the anharmonic forces of the lattice. To see this, we remind ourselves that the harmonic terms involve only two wave vectors, but we need three to accommodate the three phonons, and that must come from a third derivative of the potential, hence the leading anharmonic term.

3.6.1 Anharmonic terms in the potential

Because of the general flatness of the optical phonon spectrum near the zone center, these phonons do not migrate readily to the surface, as their non-dispersive nature near the zone center gives them a very low group velocity. Rather their energy is dissipated through the anharmonic terms of the lattice potential through three phonon processes. The cubic term that gives rise to this may be written as

$$H_3 = \frac{1}{3!} \sum_{i,j,k} u_i \cdot \left(u_j \cdot \frac{\partial^3 V(r)}{\partial R_i \partial R_j \partial R_k} \cdot u_k \right). \tag{3.84}$$

The third derivative term is a third order tensor, so that there are 27 elements in the force tensor. These are usually reduced to one or a few in order to actually compute the effective perturbation terms to use in the three-phonon interaction. The Fourier representation of this term can be generated in a straight-forward manner, just as has been done previously, and this leads to

$$H_3 = \frac{1}{3!\sqrt{N}} \sum_{qq'q''} u(q) \cdot [u(q') \cdot C_{qq'q''} \cdot u(q'')]. \tag{3.85}$$

Here, the central force constant is a third rank tensor that represents the Fourier transform of the third derivative of the potential in (3.84). This is the 'spring' constant, related to the third-order stiffness constant.

In determining the perturbation potential that induces the emission of an optical phonon and the creation of two acoustic phonons, we do not do the integration over q, as that is the momentum of the initial state of the optical phonon, and we ask about the decay out of that state. As mentioned, we will use only an average value obtained from the large number of elements in the third ranked tensor, and denote this value as $C'_{qq'q''}$, so that we can write the perturbing force term as

$$\delta V(q) = \frac{1}{\sqrt{N}} \sum_{q'q''} \frac{C'_{qq'q''}}{\sqrt{\omega_q \omega_{q'} \omega_{q''}}} a_q a_{q'}^\dagger a_{q''}^\dagger \delta(q + q' + q''). \tag{3.86}$$

Values for the third-order stiffness constants can be found for some materials is collections such as [42].

The corresponding matrix element can be found by utilizing the Fermi golden rule of time-dependent perturbation theory [43]. In this process, the creation and annihilation operators, working on the phonon wave functions, lead to the number operators and hence the populations of the various phonon modes. This process leads to

$$|M(q)|^2 = \frac{\hbar^3}{8NM^3} \sum_{q'} \frac{C'^2_{qq'(q+q')}}{\omega_q \omega_{q'} \omega_{q+q'}} \tag{3.87}$$
$$\times N_q (N_{q'} + 1)(N_{q+q'} + 1) \delta(\omega_q - \omega_{q'} - \omega_{q+q'}).$$

While the energy-conserving delta function in the matrix element has been included here, it is not properly part of this quantity, but arises in the Fermi golden rule. For most of the expressions of interest, the wave vectors in (3.87) belong to different modes of the lattice vibration. One may be an optical phonon and the other two may be acoustic phonons, or all three may be acoustic phonons, for example. But, generally, there are only a few combination of modes that are possible, so that the sums in (3.87) do not lead to a great many terms. Finally, the transition rate for decay of the particular phonon is given by the extra terms that arise in the Fermi golden role, and these lead to

$$\Gamma(q) = \frac{\pi \hbar^2}{4NM^3} \sum_{q'} \frac{C'^2_{qq'(q+q')}}{\omega_q \omega_{q'} \omega_{q+q'}} \tag{3.88}$$
$$\times N_q (N_{q'} + 1)(N_{q+q'} + 1) \delta(\omega_q - \omega_{q'} - \omega_{q+q'})$$

As an example, we consider the decay of the LO mode into two LA modes, in which the latter may have different frequencies. But, we consider the LO mode frequency to be relatively constant due to the non-dispersive nature of this mode, particularly near the zone center, so we will also assume that the LA modes also have well defined frequencies (this negates the need for the summation over the momenta). This leads to a simplified version as

$$\Gamma_{LO}(q) = \frac{\pi\hbar^2}{4NM^3} \sum_{q'} \frac{C'^2}{\omega_{LO}\omega_{LA}\omega_{LO-LA}}$$
$$\times N_{LO}(N_{LA} + 1)(N_{LO-LA} + 1)\delta(\omega_{LO} - \omega_{LA} - \omega_{LO-LA})$$

(3.89)

The last summation is just an integration over the density of modes (density of states, discussed in the next chapter) to which the LO mode can be connected. This density is not large, but because the LO mode is a long-wavelength mode, the summation essentially corresponds to the number of states in a spherical surface of the Brillouin zone. This can be seen by expanding the summation to an integral as

$$\sum_{q'} \delta(\omega_q - \omega_{q'} - \omega_{q+q'}) = \frac{V}{8\pi^3} \int_{S_{q'}} \delta(\omega_q - \omega_{q'} - \omega_{q+q'})q'^2 dq' dS_{q'}.$$

(3.90)

This can be reduced further by integrating out the solid angle and introducing the group velocity as

$$\sum_{q'} \delta(\omega_q - \omega_{q'} - \omega_{q+q'}) = \frac{V}{2\pi^2} \frac{q_{LA}^2}{\hbar v_{LA}}.$$

(3.91)

This finally leads to

$$\Gamma_{LO}(q) = \frac{V\hbar^2}{8\pi NM^3 v_{LA}} \frac{C'^2}{\omega_{LO}\omega_{LA}\omega_{LO-LA}} N_{LO}(N_{LA} + 1)(N_{LO-LA} + 1).$$

(3.92)

The subscripts on the various Bose–Einstein distributions refer to the appropriate phonon energies to be included in evaluating these terms.

3.6.2 Phonon lifetimes

The lifetime of the excess polar optical phonons or even the non-polar optical phonons, that are emitted by the electrons, is related to the rate at which these modes can decay into the acoustic modes. Thus, one can write a continuity equation for the optical phonons as

$$\frac{dN_{LO}}{dt} = G - \Gamma_{LO},$$

(3.93)

where G is the rate at which the optical phonons are generated, either by absorption of the lower energy acoustic modes or by emission from the electrons. If we ignore the generation by the emission of phonons by the electrons for the moment, then we can establish a *lifetime* for the optical modes. Then, the matrix elements for G are the

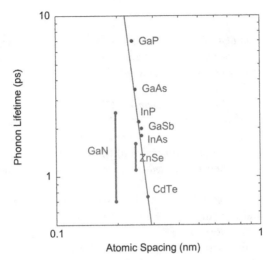

Figure 3.10. The experimentally measured non-equilibrium LO phonon lifetime, measured by sub-picosecond Raman scattering, for a number of zincblende and wurtzite semiconductors. The violet line is a d^{-10} behavior based upon dependence of the lifetime on the bulk modulus. Variation in zinc selenide is a range of measurements, while that for GaN arises from an unexpected dependence on the carrier density.

same as those for the decay process, except that the term in (3.94) goes from N $(N + 1)(N + 1)$ to $(N + 1)NN$. Thus, we can incorporate (3.92) in (3.93) to give the decay terms as

$$\frac{dN_{LO}}{dt} = - P[N_{LO}(N_{LA} + 1)(N_{LO-LA} + 1) - (N_{LO} + 1)N_{LA}N_{LO-LA}]$$
$$= - PN_{LO}\left[1 + N_{LA} + N_{LO-LA} - \frac{N_{LA}N_{LO-LA}}{N_{LO}}\right] \qquad (3.94)$$

The prefactor P is the leading terms in (3.92), and the term in square brackets will vanish in equilibrium. Thus, if we write $N_{LO} = N_{LO} + n_{LO}$, where the last term is the deviation from equilibrium, then the lifetime can be written as

$$\frac{1}{\tau_{LO}} = P(1 + N_{LA} + N_{LO-LA}). \qquad (3.95)$$

In general, the stiffness constant C is thought to scale as the bulk modulus [42]. The latter is thought to scale as d^5 [44], where d is the inter-atomic spacing of the lattice. In figure 3.10, we plot the measured phonon lifetimes for a number of semiconductor crystals at 300 K, and it can be seen that the scaling as $C^2 \sim d^{-10}$ is fit rather well for these materials. The red squares are for the wurtzite phases of some materials, and there is some uncertainty in the value for ZnSe. The GaN results (green) arise from an unexpected dependence upon the carrier density (lower density at the top) [45]. The violet line is the scaling with the bulk modulus.

References

[1] Ferry D K 2001 *Quantum Mechanics* 2nd edn (Bristol: IOP Publishing)

[2] Madelung O (ed) 1996 *Semiconductors–Basic Data* 2nd edn (Berlin: Springer)

[3] Dolling G 1974 *Dynamical Properties of Solids* vol. 1 ed G K Horton and A A Maradudin (Amsterdam: North Holland) ch 10

[4] Cochran W 1959 *Proc. Roy. Soc. (London)* **A253** 260

[5] Dolling G and Cowley R A 1966 *Proc. Phys. Soc. (London)* **88** 463

[6] Bilz H, Gliss B and Hanke W 1974 *Dynamical Properties of Solids* vol 1 ed G K Horton and A A Maradudin (Amsterdam: North Holland) ch 6

[7] Yu P Y and Cardona M 2001 *Fundamentals of Semiconductors* 3rd edn (Berlin: Springer) sec. 3.2.3

[8] Musgrave M J P and Pople J A 1962 *Proc. R. Soc. (London)* **A268** 474

[9] Nusimovici M A and Birman J L 1967 *Phys. Rev.* **156** 925

[10] Keating P N 1966 *Phys. Rev.* **145** 637

[11] Ayubi-Moak J S 2008 dissertation (unpublished), Arizona State University

[12] Martin R M 1969 *Phys. Rev.* **186** 871

[13] Weber W 1977 *Phys. Rev.* **B15** 4789

[14] Valentin A, Sée J, Galdin-Retailleau S and Dollfus P 2008 *J. Phys. Cond. Matt.* **20** 145213

[15] Tütüncü H M and Srivastava G P 1996 *J. Phys. Cond. Matt.* **8** 1345

[16] De Ciccio P D and Johnson F A 1969 *Proc. R. Soc. (London)* **A310** 111

[17] Pick R, Cohen M H and Martin R M 1970 *Phys. Rev.* **B 1** 910

[18] Sham L J and Kohn W 1966 *Phys. Rev.* **145** 561

[19] Van Camp P E, Van Doren V E and Devreese J T 1979 *Phys. Rev. Lett.* **42** 1224

[20] Baroni S, Giannozzi P and Testa A 1861 *Phys. Rev. Lett.* **58**

[21] King-Smith D and Needs R J 1990 *J. Phys. Cond. Matter.* **2** 3431

[22] Kunc K and Martin R M 1982 *Phys. Rev. Lett.* **48** 406

[23] Giannozzi P, de Gironcoli S, Pavone P and Baroni S 1991 *Phys. Rev.* **B 43** 7231

[24] Hellmann H 1937 *Einführung in die Quantenchemie* (Leipzig: Deuticke)

[25] Feynman R P 1939 *Phys. Rev.* **56** 340

[26] Basak T, Rao M N, Gupta M K and Chaplot S L 2012 *J. Phys. Cond. Matt.* **24** 115401

[27] Giannozi P *et al* 2009 *J. Phys. Cond. Matt.* **21** 395502

[28] Birman J L, Lax M and Loudon H 1966 *Phys. Rev.* **145** 620

[29] Patrick L 1966 *J. Appl. Phys.* **37** 4911

[30] Molina-Sanchez A and Wirtz L 2011 *Phys. Rev.* **B 84** 155413

[31] Kaasbjerg K, Thygesen K S and Jacobsen K W 2012 *Phys. Rev.* **B 85** 115317

[32] Lucovsky G, Brodsky M H and Burstein E 1968 *Localized Excitations in Solids* (New York: Plenum Press), p 592

[33] Barker A S Jr. and Sievers A J 1975 *Rev. Mod. Phys.* **47** S1

[34] Brodsky M H and Lucovsky G 1968 *Phys. Rev. Lett.* **21** 990

[35] Jusserand B and Spriel J 1981 *Phys. Rev.* **B 24** 7194

[36] Spitzer W G and Kleinman D A 1961 *Phys. Rev.* **121** 1324

[37] Krishnan R S 1945 *Nature* **155** 452

[38] Lynch W T 1972 *J. Appl. Phys.* **53** 3274

[39] Jash J A, Jha S S and Tsang J C 1987 *Phys. Rev. Lett.* **58** 1869

[40] Shifren L, Ferry D K and Tsen K T 1999 *Physica* B **272** 419

[41] Shifren L, Ferry D K and Tsen K T 2000 *Phys. Rev.* B **62** 15379

[42] Weinrich G 1965 *Solids: Elementary Theory for Advanced Students* (New York: Wiley)

[43] Merzbacher E 1970 *Quantum Mechanics* (New York: Wiley)

[44] Harrison W A 1980 *Electronic Structure and the Properties of Solids* (San Francisco, CA: Freeman)

[45] Tsen K T, Kiang J G, Ferry D K and Morkoç H 2006 *Appl. Phys. Lett.* **89** 112111

IOP Publishing

Semiconductors (Second Edition)
Bonds and bands
David K Ferry

Chapter 4

Semiconductor statistics

The most obvious difference between semiconductors and metals is the presence of an energy gap between the valence band and the conduction band. This gap is of the order of 1 eV. The importance of this lies with the Pauli exclusion principle which says that any quantum state can hold only two electrons, and these have opposite spins. Thus, at low temperature, the electrons will move to the lowest energy levels. Thus, all states in the valence band are occupied at low temperature, and the conduction band states are empty. Hence, the tendency is for the semiconductor to be an insulator. On the other hand, the gap is sufficiently small that electrons can be thermally excited from the valence band to the conduction band at high temperatures. The meaning of 'low' and 'high' depends upon the exact size of the band gap, and therefore differs between semiconductors. In general, one can define an energy that sits between the lowest unfilled state and the highest filled state (which is below the lowest unfilled state). This energy is the Fermi energy, arising from Fermi–Dirac statistics for particles of half-integer spin, such as our electrons. With the pure semiconductors, the Fermi energy will sit near the mid-gap position, or midway between the conduction band and the valence band. But, adding a single impurity with an excess electron or a shortage of an electron changes a number of things, most notably the Fermi energy. For example, the addition of a single donor atom, which possesses an extra valence electron, produces a new state that is introduced near the conduction band edge, and this moves the Fermi energy by almost half the band gap. This is because, at low temperatures, this new state is filled. One single new state moves the Fermi energy almost half the band gap!

How the number of electrons in the conduction band, and the number of holes in the valence band, vary with temperature and the Fermi energy is determined by a combination of the number of states in each band, as a function of energy, and the Fermi–Dirac distribution function, which gives the probability that any particular state is occupied. So, in this chapter, we begin with a determination of the actual density of available states in each band. As one may expect, this will depend upon

doi:10.1088/978-0-7503-2480-9ch4

the shape of the bands and the number of each type of band, which enter the determination via the effective masses we discussed in chapter 2.

Equally important is the quality of the material. When the material is pure, and there are no impurity atoms in it, we call the semiconductor intrinsic. However, we prepare semiconductors to have certain properties, for example when it is used as the channel in a MOSFET. To do this, we incorporate other atoms, known as impurities. These are the atoms mentioned above that have an extra electron, known as donors, or are short by a bonding electron, known as acceptors. The addition of these atoms makes the material extrinsic, and these added atoms move the Fermi energy around in the band to produce differences in the occupation of the conduction and valence bands. Once, we know the various densities of states, we can determine the proper statistics for these extrinsic materials and understand how to control its properties, and this will be discussed later in this chapter.

Then, there are some impurities which produce levels near the center of the gap. And, these can have very different properties. These impurities produce what are known as deep levels, which have very different behavior. Often these deep levels are referred to as traps and in many cases are just generally bad actors which degrade the properties of the semiconductor. Their behavior will also be discussed.

Finally, disorder can dramatically affect the semiconductor. In many cases, the disorder can be so severe that it actually breaks the periodicity of the lattice and produces amorphous semiconductors that have no crystallinity. But, most disorder only disrupts the propagating nature of the waves near the band edges and localizes these states. When this happens, one can usually find a distinct energy level that separates the localized states from the extended conducting states. This energy level is known as the mobility edge, and differs from the energy at the bottom of the conduction band or the energy at the top of the valence band. Fortunately, most semiconductors can tolerate a degree of disorder without disrupting the pristine properties of the two bands. Our impurity donor and acceptor atoms usually lie within this region of tolerance. But, doping levels, that describe the number of impurities added, can become so large that this tolerance is violated. Fortunately, there is a general scaling type of theory that explains the role of disorder, and this will be discussed at the end of the chapter.

4.1 Electron density and the Fermi level

As mentioned, electrons in semiconductors are subject to the Pauli exclusion principle, which means that they are indistinguishable and each available state contains at most 2 electrons of opposite spin angular momentum. These spin-1/2 particles obey Fermi–Dirac statistics for which the probability of a state at energy E being occupied is given by the distribution function

$$P_{FD}(E) = \frac{1}{1 + \exp[(E - E_F)/k_B T]}, \tag{4.1}$$

where E_F is the Fermi energy. By examination of (4.1), we determine that the Fermi energy is the energy at which the probability is exactly 1/2. At the absolute zero of temperature, all electron states have unity probability of being filled for $E \leqslant E_F$. All states for which $E \geqslant E_F$ are completely empty. It is obvious that the energy $E = E_F$ is confused and is defined therefore as having a probability of exactly 1/2, as the probability is 1 below this energy and 0 above it. At higher temperature, there is a gradual spreading of the distribution about the Fermi energy, as shown in figure 4.1. The electrons that would normally exist in region 1 in the figure, at zero temperature, are excited into region 2 *if the levels are in allowed energy states*. It is this latter that becomes important in the case of a semiconductor with a band gap. In this case, electrons that are excited above the Fermi energy are given by the indicated region 2, but *weighted by the appropriate density* of these allowed states. The Fermi energy must adjust so that the number of electrons removed from region 1 is exactly equal to the number of electrons in region 2. Hence, in an intrinsic semiconductor, this means that the Fermi energy will sit near mid-gap, but will move away from it slightly to balance the density of empty states in the valence band with the number of filled states in the conduction band due to differences in the density of states in these two bands, as we will examine in the following discussions.

The total number of electrons in e.g. the conduction band is given by the sum over all of the possible energy states that exist in the band with each state weighted by its probability of being occupied. We write this as

$$n = \frac{1}{V}\sum_i 2P_{\mathrm{FD}}(k_i) \rightarrow \int_{E_c}^{\infty} n(E)P_{\mathrm{FD}}(E)dE, \qquad (4.2)$$

where the factor of 2 accounts for spin degeneracy and the sum is over all of the possible k states in the conduction band (as we know from chapter 2, this number is equal to the number of primitive unit cells in the semiconductor). In the last term, we have gone from a sum over the momentum states to an integral over energy, and the

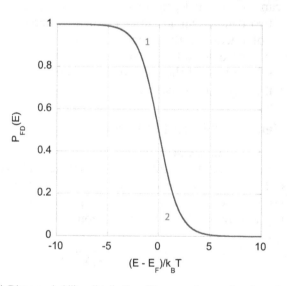

Figure 4.1. The Fermi–Dirac probability distribution (blue curve) as a function of energy. At non-zero temperatures, electrons are excited from region 1 to region 2, and these two regions have equal area.

conversion function $n(E)$ is the appropriate density of states per unit energy per unit volume, a quantity to which we will turn our attention shortly. But, (4.2) tells us that if we know there is density n (per unit volume, while if N is the total number of electrons, $n = N/V$ for a homogeneous semiconductor) of electrons, the Fermi level must adjust itself to satisfy this equation. This is the point of adding donors or acceptors to create a desired density n in the semiconductor.

4.1.1 The density of states

The density of states is one of the most important quantities for determining the electronic properties of semiconductors. Basically, we are asking for the number of states that are available at a given energy in the system. Thus, the density of states is the number of states per unit energy per unit volume, and is a function of the energy itself. The density of states for a semiconductor is easily calculated from the quantized solutions of the Schrödinger equation, that was discussed in section 2.1. But, this depends upon the dimensionality of the system, so that there are variations as we move from bulk to two-dimensional to one-dimensional systems. Moreover, it depends upon the nature of the energy bands, with the traditional behavior arising only for parabolic energy bands.

To understand the properties of the density of states, we expand on section 2.1 to examine the properties of the wave functions themselves and the quantization of the momentum states that can exist in the material. The problem is to solve the Schrödinger equation in the desired material. The most important initial step, and one that really defines the exact density of states that we find, is the assumption of periodic boundary conditions for the wave function, as in (2.3). In fact, the electrons are constrained to stay within the volume of the crystal, so that there are potential barriers that surround the crystal (these are the work function potential required to be overcome to remove an electron from the crystal). In chapter 2, we introduced periodic boundary conditions with a length L in each direction, for reasons that will become clearer below:

$$\psi(x, y, z) = \psi(x + L, y, z)$$
$$\psi(x, y, z) = \psi(x, y + L, z). \tag{4.3}$$
$$\psi(x, y, z) = \psi(x, y, z + L)$$

The use of periodic boundary conditions leads to the correct counting of states; that is, there will be two states (of opposite spin) for each bonding electron in the basic crystal lattice. For example, in Si or GaAs, the crystal lattice is the face-centered cubic lattice which has four lattice sites per unit cell. But, there is a basis of two atoms at each lattice site: either two Si atoms or one Ga and one As atom, so that there are eight atoms per cell. The periodicity required for the atomic lattice thus gives $4a^3$, where a is the edge of the face-centered cubic cell, states, each of which can hold two electrons of opposite spin. Hence, there are $8a^3$ total states. Each atom of the basis has four bonding electrons, so this gives a total of 8 electrons per lattice site, which just exactly fills all the states in the band—the valence band, or bonding band in this case. Using different boundary conditions would not give this experimentally confirmed result.

Because of the atomic periodicity of the crystalline potential, we find that the momentum wave numbers are not continuous, but are discretized due to the boundary conditions. They form a series of discrete points in the three dimensional momentum space. Each point in this space corresponds to a particular momentum state which may be occupied by two electrons of opposite spin, according to the Pauli exclusion principle. The traditional approach is to use a cubic lattice of edge a, appropriate to most semiconductors, so that the length $L = Na$, where N (do not confuse this N with the N above) is the number of atoms in the given direction of the crystal. Then, we find that the quantization of the momentum, leads to

$$
\begin{aligned}
k_x &= \frac{2\pi n_x}{L}, \, n_x = 0, \pm 1, \pm 2, \, \cdots \frac{N}{2}, \\
k_y &= \frac{2\pi n_y}{L}, \, n_y = 0, \pm 1, \pm 2, \, \cdots \frac{N}{2}, \\
k_z &= \frac{2\pi n_z}{L}, \, n_z = 0, \pm 1, \pm 2, \, \cdots \frac{N}{2}.
\end{aligned}
\tag{4.4}
$$

Of course, n_x, n_y, n_z can be larger than the limits indicated above, but these larger values are not independent, and no longer lie in the first Brillouin zone. We assume that our wave functions for these states have the propagating wave

$$
\psi \sim e^{ik \cdot r}.
\tag{4.5}
$$

We see that the phase rolls over (past 2π) when the numbers larger than those of (4.4) are used. So the end result is that the independent values of momentum are precisely equal to the number of unit cells in the lattice (not including the basis).

Three dimensions. There are different ways to determine the density of states. We will show two such methods for three dimensions. For the first one, we assume parabolic bands and will use the band edge mass. In three dimensions, the filling of the lowest energy states at low temperature gives rise to the formation of the Fermi sphere, whose properties may be easily computed. So, this sphere is filled up to the momentum at the Fermi energy, k_F. In an N electron metal, $N/2$ distinct wave vectors are required to hold these electrons. The volume in momentum space for each wave vector is given by our conditions of (4.4) to be $(2\pi/L)^3$. Hence, our $N/2$ states give rise to the requirement that

$$
\frac{N}{2} = \frac{\text{volume of sphere}}{\text{volume per state}} = \frac{4\pi}{3} k_F^2 \frac{L^3}{8\pi^3}.
\tag{4.6}
$$

We can now solve for the Fermi wave number as

$$k_F = \left(3\pi^2 \frac{N}{L^3} \right)^{1/3} = (3\pi^2 n)^{1/3}, \tag{4.7}$$

where n is the number of electrons per unit volume. In a simple metal, these are all free electrons, and they will fill the band up to the half-way point. That is, one half of the available states are filled for this simple metal.

At zero temperature in a semiconductor, all the states are filled up to the Fermi energy. For our case, this value can be found to be

$$E_F = \frac{\hbar^2 k_F^2}{2m_c} = \frac{\hbar^2}{2m_c} (3\pi^2 n)^{2/3}. \tag{4.8}$$

This can be inverted to find the density as a function of the Fermi energy, as

$$n = \frac{1}{3\pi^2} \left(\frac{2m_c E_F}{\hbar^2} \right)^{3/2}. \tag{4.9}$$

This last expression tells us exactly how many electrons, per unit volume, fill all the states up to the Fermi energy. There is no requirement that this be the Fermi energy. If we replace the Fermi energy by the energy itself, then (4.9) tells us how many electrons can be held in the states up to that energy. Thus, we can find the density of states by a simple derivative, as

$$n_{3D}(E) = \frac{dN}{dE} = \frac{1}{2\pi^2} \left(\frac{2m_c}{\hbar^2} \right)^{3/2} E^{1/2}. \tag{4.10}$$

We see that the density of states is proportional to the square root of the energy in three dimensions.

Non-parabolic bands. When we move to semiconductors, there are a few changes. We first consider when the parabolic energy dependence is modified. This parabolic dependence was invoked in the move to (4.8), so is crucial to the development we have used. Now, we want to use the hyperbolic band that was given in (2.107), where we write the energy at momentum k as

$$E = \frac{E_G}{2} \left[\sqrt{1 + \frac{2\hbar^2 k^2}{m_c E_G}} - 1 \right], \tag{4.11}$$

where again m_c is the band-edge mass for the conduction band and E_G is the band gap. To proceed, we will follow a second approach and use (4.2) directly. We begin with the left-hand side of (4.2) and remain in momentum space as

$$n = \frac{2}{V} \sum_k P_{FD}(k) = \frac{2}{V} \frac{V}{8\pi^3} \int P_{FD}(k) d^3 k, \tag{4.12}$$

where the change from discrete summation to continuous integration requires the correction by dividing by the volume occupied by each discrete state (discussed above). The Fermi–Dirac function is a function of energy and not momentum directly, and we even desire the density of states per unit energy per unit volume. Hence, we need to change the integral in (4.12) from one over momentum to one over energy. The three dimensional differential is given by

$$d^3\boldsymbol{k} = k^2\sin\vartheta d\vartheta d\varphi dk, \tag{4.13}$$

with integration over the two angles in polar coordinates and the momentum magnitude. Fortunately, there is no angular variation in the Fermi–Dirac distribution, so the angular integrals yield just 4π. Now, we introduce the energy variables with

$$k = \sqrt{\frac{2m_c E}{\hbar^2}\left(1 + \frac{E}{E_G}\right)}$$
$$kdk = \frac{m_c}{2\hbar^2}\left(1 + \frac{2E}{E_G}\right)dE \tag{4.14}$$

In this last form, we have used the hyperbolic bands from (4.11). Using (4.13) and (4.14) in the right-hand side of (4.12) gives us

$$n = \frac{1}{4\pi^2}\left(\frac{2m_c}{\hbar^2}\right)^{3/2}\int_0^\infty \frac{\sqrt{E\left(1 + \frac{E}{E_G}\right)}\left(1 + \frac{2E}{E_G}\right)dE}{1 + \exp[(E - E_F)/k_B T]}. \tag{4.15}$$

By comparing this result with the right-hand term in (4.2), we can recognize the density of states per unit energy per unit volume to be

$$n_{3D}(E) = \frac{dN}{dE} = \frac{1}{4\pi^2}\left(\frac{2m_c}{\hbar^2}\right)^{3/2}E^{1/2}\xi(E), \tag{4.16}$$

where

$$\xi(E) = \sqrt{\left(1 + \frac{E}{E_G}\right)\left(1 + \frac{2E}{E_G}\right)} \tag{4.17}$$

is the non-parabolic correction factor. Without this factor, we obtain exactly the same results as (4.10). The factor in the square root is a correction to the energy dependence on momentum and the second factor is a correction for the variation of the effective mass with energy.

Silicon. When we deal with Si, we have to make some corrections to the band edge mass that appears in (4.10) and (4.16). In Si, the conduction band has six equivalent minima located about 85% of the way from Γ to X along the (100) directions. A constant energy surface, at energies near the conduction band minimum is an ellipsoid of revolution, which has different masses along the various axes. Along the major axis of the ellipsoid, the band-edge mass is the longitudinal mass m_L, which is about $0.91m_0$ in Si. In the other two directions the band-edge mass is the transverse mass m_T, which is about $0.19m_0$ in Si. When all six ellipsoids are taken together, Si still has the full cubic symmetry. But, now when we determine the density of states, summing over all directions and valleys, we get the density of states mass

$$m_{\Sigma}^{3/2} = 6\sqrt{\left(m_L m_T^2\right)}, \qquad (4.18)$$

from which we obtain the density of states mass

$$m_{d,c} = 6^{2/3}\left(m_L m_T^2\right)^{1/3} \sim 1.06 m_0, \qquad (4.19)$$

where the last value is for Si electrons.

In the Si valence band, we have to count contributions from both the heavy-hole band and the light-hole band. Both of these bands are warped as described in section 2.5.3. So, to begin with, we have to use the average values of the band-edge masses obtained by proper summation over the A, B, and C parameters. From table 2.5, we find these masses are m_{hh}, which is about $0.537 m_0$ and m_{lh}, which is about $0.153 m_0$. These produce a valence band density of states mass

$$m_{d,v}^{3/2} = m_{hh}^{3/2} + m_{lh}^{3/2}, \qquad (4.20)$$

which leads to an average valence band density of states mass $m_{d,v} \sim 0.59 m_0$.

Two dimensions. Exactly the same approach can be used in two dimensions. In this case, an energy describes a circle in the two-dimensional momentum space, and the number of states needed to accommodate $N/2$ distinct wave vectors is given as

$$\frac{N}{2} = \frac{\text{area of circle}}{\text{area per state}} = \pi k^2 \frac{L^2}{4\pi^2}. \qquad (4.21)$$

We can now solve for the wave vector as

$$k = \left(2\pi \frac{N}{L^2}\right)^{1/2} = \sqrt{2\pi n_s}. \qquad (4.22)$$

where n_s is the number of electrons per unit area, or sheet density. Now, the energy is found to be

$$E = \frac{\hbar^2 k^2}{2m_c} = \frac{\hbar^2}{2m_c} 2\pi n_s \rightarrow n_s = \frac{m_c}{\pi \hbar^2} E. \qquad (4.23)$$

From this, we can now find the density of states per unit energy per unit area to be

$$n_{2D}(E) = \frac{dN}{dE} = \frac{m_c}{\pi \hbar^2}. \qquad (4.24)$$

Again, this is for the case of parabolic bands.

One dimension. Once more, the same approach can be used in one dimension. In this case, an energy describes a point in the one-dimensional momentum space, except that this point can be at positive or negative momentum, so we have to consider another factor of 2. The number of states needed to accommodate $N/2$ distinct wave vectors is given as

$$\frac{N}{2} = \frac{\text{length per line}}{\text{length per state}} = 2k\frac{L}{2\pi}. \tag{4.25}$$

We can now solve for the wave vector as

$$k = \frac{\pi}{2}\frac{N}{L} = \frac{\pi n_l}{2}, \tag{4.26}$$

where n_l is the number of electrons per unit length, or line density. Now, the energy is found to be

$$E = \frac{\hbar^2 k^2}{2m_c} = \frac{\hbar^2}{2m_c}\left(\frac{\pi n_l}{2}\right)^2 \rightarrow n_l = \frac{2}{\pi}\sqrt{\frac{2m_c E}{\hbar^2}}. \tag{4.27}$$

From this, we can now find the density of states per unit energy per unit length to be

$$n_{1D}(E) = \frac{dN}{dE} = \frac{1}{\pi}\sqrt{\frac{2m_c}{\hbar^2 E}}. \tag{4.28}$$

4.1.2 Intrinsic material

As we discussed above, intrinsic material has an equal number of electrons and holes, so that the Fermi energy sits near the middle of the energy gap. In this section, we want to determine how many electrons and holes exist at a given temperature, and locate the exact position of the Fermi energy. To proceed, we use (4.2) which we rewrite as

$$n = \int_0^\infty n_{3D}(E)P_{FD}(E)dE. \tag{4.29}$$

Now, if we insert (4.1) and (4.10), using the conduction band density of states mass (4.19), we can rewrite (4.29) as

$$n = \frac{1}{2\pi^2}\left(\frac{2m_{d,c}}{\hbar^2}\right)^{3/2} \int_0^\infty \frac{E^{1/2}dE}{1 + \exp[(E - E_F)/k_B T]}. \tag{4.30}$$

Here, as in (4.29), the zero of energy is taken as the conduction band minimum. In general, the Fermi energy lies below this point, so is normally a negative quantity, but this is not always the case. We discuss this further below. To simplify the equations, we introduce some reduced parameters, as

$$x = \frac{E}{k_B T}$$
$$\eta = \frac{E_F}{k_B T}.$$

(4.31)

Then, we may rewrite (4.30) as

$$n = \frac{1}{2\pi^2}\left(\frac{2m_{d,c}k_B T}{\hbar^2}\right)^{3/2} \int_0^\infty \frac{x^{1/2}dx}{1 + \exp(x - \eta)}.$$

(4.32)

We want to gather the various coefficients into a quantity we call the effective density of states for the conduction band

$$N_c = \frac{1}{4}\left(\frac{2m_{d,c}k_B T}{\pi\hbar^2}\right)^{3/2},$$

(4.33)

so that (4.32) can be written as

$$n = N_c \cdot \frac{2}{\sqrt{\pi}} \int_0^\infty \frac{x^{1/2}dx}{1 + \exp(x - \eta)}.$$

(4.34)

Everything on the right-hand side of (4.34), except for N_c is known as a Fermi–Dirac integral, written in this case as $F_{1/2}(\eta)$. The subscript of the Fermi–Dirac integral corresponds to the power of x in the integrand. These integrals cannot usually be integrated in closed form. Tabulated values of the Fermi–Dirac integral can be found on the web or in an excellent older book by Blakemore [1].

The size of the density n relative to the effective density of states N_c, is an important guideline in evaluating the Fermi–Dirac integral. If $n > N_c$, the full integral must be kept and the material is said to be *degenerate*. In the case where η is large, the integral approaches a limit given as

$$F_{1/2}(\eta) \rightarrow \frac{4}{3\sqrt{\pi}}\eta^{3/2}.$$

(4.35)

Hence for strongly degenerate systems, the density increases slightly faster than linear in the Fermi energy.

On the other hand, when η is negative with a magnitude larger than $3k_B T$, the exponential term in the denominator is larger than 1, and we can approximate the integral as

$$F_{\frac{1}{2}}(\eta) \approx \frac{2}{\sqrt{\pi}} \int_0^\infty x^{\frac{1}{2}}e^{\eta-x}dx = e^\eta, \quad -\eta > k_B T.$$

(4.36)

When the exponential in η is taken out of the integral, the remaining integral is just Γ (3/2), and this cancels the prefactor. In this situation, we refer to the semiconductor as being non-degenerate, and we find

$$n = N_c e^\eta, \quad -\eta > k_B T.$$

(4.37)

This will certainly be the case for intrinsic material.

Now, let us turn to the holes and the valence band. These relate to the empty states that exist in the valence band when electrons are excited to the conduction band. We know that a band in which all possible states are occupied cannot carry any current, simply for the reason that these electrons cannot gain any energy from an applied field. To gain energy, they have to be able to move to a higher energy state, and since there are none in the full band, this band just doesn't carry any current. But, when empty states exist, these may be regarded as positive electrons. This follows from the current, which is

$$I = \sum_{\text{full states}} ev, \tag{4.38}$$

where the electron charge is negative. To this, we add and subtract current that could be carried by electrons in the empty states, as

$$I = \sum_{\text{full states}} ev + \sum_{\text{empty states}} ev - \sum_{\text{empty states}} ev, \tag{4.39}$$

The first two terms on the right-hand side add up to a full band, so must equal to zero current, which leads us to

$$I = -\sum_{\text{empty states}} ev. \tag{4.40}$$

So, the empty states act as quasi-particles with positive charge. And, with the zero of energy sitting at the conduction band edge, these are negative energy particles.

Thus, we are interested in the empty states at the top of the valence band, which are called *holes*. The probability that a state is empty is given by

$$\begin{aligned} P_h(E) &= 1 - P_{\text{FD}}(E) \\ &= \frac{1}{1 + \exp(\eta - x)}. \end{aligned} \tag{4.41}$$

Notice that this reverses the order of the variables in the argument of the exponential. The energy at the top of the valence band is $-E_G$, referenced to the conduction band edge. With this we can write the number of positive holes in the valence band as

$$p = \frac{1}{2\pi^2} \left(\frac{2m_{d,v}k_{\text{B}}T}{\hbar^2} \right)^{3/2} \int_{-\infty}^{-E_G} \frac{\left(-x - \dfrac{E_G}{k_{\text{B}}T} \right)^{1/2} dx}{1 + \exp(\eta - x)}. \tag{4.42}$$

Here, we have used the valence band density of states mass from (4.20), and care has been taken in the square root as the energy is negative relative to the conduction band edge. We note that the argument of the square root term gives zero when the energy is at the top of the valence band, and becomes more negative as we move lower in this band. Now, we introduce some new variables as

$$\xi = \frac{E_G}{k_B T}$$
$$u = -(x + \xi)$$
$$\mu = -\xi - \eta$$

(4.43)

Since η is going to lie *above* the valence band edge, which sits at $-E_G$, μ will be positive. With these changes, we can now write (4.39) as

$$p = N_v \cdot \frac{2}{\sqrt{\pi}} \int_0^\infty \frac{(u)^{1/2} du}{1 + \exp(u - \mu)},$$

(4.44)

where we have introduced the valence band effective density of states

$$N_v = \frac{1}{4}\left(\frac{2m_{d,v}k_B T}{\pi \hbar^2}\right)^{3/2}.$$

(4.45)

Here, we again arrive at the proper Fermi–Dirac integral (4.36) for the non-degenerate situation. Hence, the number of holes is now

$$p = N_v e^\mu, \quad -\mu < k_B T.$$

(4.46)

Now, we note that μ is measured relative to the valence band maximum, and the negative value denotes that it lies above this valence band maximum.

We can now find the value of the intrinsic concentration. For this, we use the fact that n and p have to be equal, as the electrons came from the empty hole states. We denote this as the intrinsic hole concentration n_i. Because these concentrations are all equal, we can write

$$n_i^2 = np = N_c e^\eta \cdot N_v e^\mu = N_c N_v e^{-\xi}.$$

(4.47)

This intrinsic concentration is independent of the position of the Fermi energy and depends upon the two density of states masses and the energy gap, as well as a slew of constants and the temperature of the semiconductor. Thus, the intrinsic concentration becomes

$$n_i = \sqrt{N_c N_v} \exp\left(-\frac{E_G}{k_B T}\right).$$

(4.48)

Using the equality of the densities of electrons and holes, we can also write

$$p = N_v e^\mu = n = N_c e^\eta.$$

(4.49)

We can now use (4.40) to rearrange this equation to give

$$E_F = -\frac{E_G}{2} + \frac{k_B T}{2}\ln\left(\frac{N_v}{N_c}\right).$$

(4.50)

The only difference between the two effective densities of states is the mass values of the two bands. These mass values are thus measures of the number variation of the occupied states with energy. The larger mass causes the Fermi energy to move away from mid gap and away from the larger mass.

4.1.3 Extrinsic material [1, 2]

If impurities are present in the semiconductor in any appreciable quantity, it is likely that at low to moderate temperatures for the impurities to provide more free carriers than given by the intrinsic concentration. Turning this argument around, we can control the properties of the material by the introduction of a set number of impurities. Generally, the impurity introduces an energy state that lies in the band gap of the material. If, the impurities are donors, this energy level lies close to the conduction band edge, and the separation is known as the ionization energy for the impurity. Obviously, then the energy level is known as a donor level. Similarly, if the impurities are acceptors, the new energy level lies near the valence band and is thus known as an acceptor level. The names are descriptions of the type of impurity that is used. Consider Si as the example material. Then, atoms such as As have five outer shell electrons. Only four of these are required for the tetrahedral bonding of the diamond lattice when the As atom replaces a Si atom. The fifth electron is extra, and may be detached from the As atom with a little heating. The energy required to detach this extra electron is the separation of the As donor level from the conduction band, and As is termed a donor as the extra electron is excited into the conduction band. On the other hand, consider adding a boron atom in place of a Si atom. The boron atom has only three outer shell electrons, so it cannot fulfill the tetrahedral bonds. But, some thermal excitation can induce bonding electrons from other atoms to jump to the empty boron bond, and the vacancy can thus move around easily. This is handled by the assumption that an electron jumps from the full valence band to the acceptor, leaving a hole behind that can contribute to the current.

The presence of the donors and acceptors creates additional electrons and holes in the system. But, these do not upset the total charge neutrality of the entire semiconductor. Thus, we can write the charge in the semiconductor as

$$0 = p + N_{di} - n - N_{ai}, \tag{4.51}$$

which must be globally zero. Here, the additional subscript i refers to the fraction of the donors and acceptors that are ionized. Note that when a donor has lost its extra electron, it is positively charged, and when an acceptor has acquired the valence electron, it is negatively charged, and that provides the signs in (4.51). In the following discussion, we are going to assume that the semiconductor is dominated by donors, which we call n-type semiconductors, as the dominant carrier is the electrons. The treatment of acceptor dominated semiconductors, hence p-type material, is complete analogous. And, we will focus mostly in the examples on non-degenerate material to ease the problem of Fermi–Dirac integrals. The net charge will be, from (4.51),

$$Q = N_{di} - N_{ai} = n - p. \tag{4.52}$$

For non-degenerate material, we can use (4.47) to find the electron density as

$$n = p + Q = Q + \frac{n_i^2}{n}, \tag{4.53}$$

for which

$$n = \frac{Q}{2}\left(1 + \sqrt{1 + 4\frac{n_i^2}{Q^2}}\right)$$
$$p = \frac{Q}{2}\left(\sqrt{1 + 4\frac{n_i^2}{Q^2}} - 1\right) \tag{4.54}$$

These equations basically determine the number of free electrons and holes available at a given temperature. If $Q \gg n_i$, then $n \sim Q$ and $p \sim n_i^2/Q$. Of course, one now needs to know precisely how many *ionized* donors and acceptors are in the material, and this quantity is also temperature dependent. Thus, the number of free electrons and holes is determined by the statistics that govern the excitation of electrons from the donor levels (and holes from the acceptor levels), hence by the ionization of the impurity atoms. If all the donors have given up their electrons to the conduction band, or acceptors have given up all their holes to the valence band, the electron concentration is basically independent of temperature and simply related to the number of donor and acceptor atoms in the material. Here, we wish to examine these statistics.

Impurity statistics. The effect of impurities on the number of free carriers is sometimes complex, due to the possibilities for spin degeneracies (the empty level has no spin, so is singly degenerate, but the filled state can be either spin, so is doubly degenerate), excited states, and multiple ionization states of the impurity atom. Thus, the statistics are examined through the probability of occupation of a given energy level—the Fermi–Dirac distribution applied to the impurity level. We consider that the semiconductor has a density of donor atoms N_d per unit volume. These donors ionize with an energy that provides the donor energy level at $E_d < 0$, when referenced to the conduction band edge (which is taken as the zero of energy, as previously). The number of donors that are ionized by giving up their excess electron is taken to be N_{di}, while those that are still neutral by dint of having the excess electron still bound to the donor are taken to be N_{dn}, while of course one of these three numbers is extraneous.

One deviates from the normal Fermi–Dirac statistics in this application due to the spin degeneracy of the electron in the neutral atom. Classically, the ratio of ionized donors to neutral donors is

$$\frac{N_{di}}{N_{dn}} = \frac{1}{2}\exp\left(\frac{E_d - E_F}{k_B T}\right) \equiv \frac{1}{2}e^{\eta_d - \eta}, \tag{4.55}$$

where we have introduced the reduced donor energy $\eta_d = E_d/k_BT$ and used (4.31) (note that both η_d and η are negative when referenced to the conduction band edge, and the factor of ½ describes the spin degeneracy). When the Fermi energy lies above the donor energy, the donors are mainly neutral, while if the Fermi energy lies below the donor energy, the donors are mainly ionized. Now, we can write

$$N_{dn} = N_d - N_{di} = N_d - \frac{1}{2}N_{dn}e^{\eta_d - \eta}, \tag{4.56}$$

and

$$N_{dn} = \frac{N_d}{1 + \frac{1}{2}e^{\eta_d - \eta}}. \tag{4.57}$$

Hence, we see that a slightly modified form of the Fermi-Dirac distribution is used for the donors.

Single monovalent donors. We now turn to the calculation of the actual electron density and the position of the Fermi energy. To proceed, we assume that there are no acceptors in the material and a single type of donor atom, with a single ionization level. With these assumptions, we can rewrite (4.52) as

$$N_{di} = n - p. \tag{4.58}$$

The temperature range of interest will have $n \gg p$, hence away from the intrinsic region, and we can ignore the hole concentration on the assumption that it will be negligible. Thus, we may use (4.49) for the electron concentration (recall we are treating only the non-degenerate case), and

$$N_c e^\eta = N_d - \frac{N_d}{1 + \frac{1}{2}e^{\eta_d - \eta}} = N_d \frac{1}{1 + 2e^{\eta - \eta_d}}. \tag{4.59}$$

If the donor concentration is not too large and the donor ionization energy is not too small, the conditions will remain non-degenerate over most of the temperature range of interest.

At the lowest temperatures, the donors are largely neutral and the electron concentration is quite small. At higher temperatures, the Fermi level lies below the donor energy and all the donors become ionized. We can see this with a little manipulation of the above equation to give

$$n = \frac{N_d}{1 + 2e^{-\eta_d}\left(\dfrac{n}{N_c}\right)}, \tag{4.60}$$

and this may be solved for n as

$$n = \frac{N_c}{4}e^{\eta_d}\left[\sqrt{1 + 8\left(\frac{N_d}{N_c}\right)e^{-\eta_d}} - 1\right].$$ (4.61)

At very low temperatures, where the magnitude of η_d is large, the second term in the square root dominates and

$$n \cong \left(\frac{N_c N_d}{2}\right)^{1/2} e^{\eta_d/2}.$$ (4.62)

In this region, only a few donors are ionized and the effective energy gap lies between the conduction band and the donor energy level. This separation acts like a mini-gap, with the Fermi energy between the two energies. This region is termed the *reserve* region. Here, we talk about the majority of the donor electrons being frozen out of the conduction band and sitting on the donors.

At higher temperatures, where $N_c \gg N_d$, the the second term in the square root becomes small, and the square root can be expanded. This leads to $n \sim N_d$. In this case, the Fermi energy will lie below the donor energy and all donors are ionized. This region is called the *exhaustion* region.

We can also examine the Fermi energy for these non-degenerate conditions. Equation (4.59) is rewritten as

$$e^{\eta} = \frac{1}{4}e^{\eta_d}\left[\sqrt{1 + 8\left(\frac{N_d}{N_c}\right)e^{-\eta_d}} - 1\right].$$ (4.63)

At low temperatures, where the arguments of the exponentials are large, this result leads to

$$E_F = \frac{E_d}{2} + \frac{k_B T}{2}\ln\left(\frac{N_c}{N_d}\right).$$ (4.64)

As expected, the donor level plays a role analogous to the valence band and the Fermi energy lies between the conduction band and the donor energy. At absolute zero of temperature, the Fermi energy is centered in this gap, and moves away from this point as the temperature is raised due to the competition between the two state densities. At still higher temperatures, in the exhaustion region, the square root in (4.63) can be expanded to show that

$$E_F = -\frac{k_B T}{2}\ln\left(\frac{N_c}{N_d}\right).$$ (4.65)

(Remember that this is referenced to the conduction band edge, which is our zero energy level.) This behavior continues until the Fermi energy nears the mid-gap region where the material becomes intrinsic.

The above behavior for the electron density and the Fermi energy is plotted in figure 4.2 for the reserve and exhaustion regions. We take as an example, the alloy $In_{0.65}Al_{0.35}As$ that was discussed in chapter 2. Here, the band gap is 1.2 eV, the

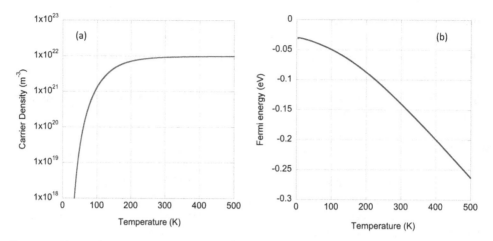

Figure 4.2. The (a) electron density and (b) Fermi energy for a single monovalent donor in $In_{0.65}Al_{0.35}As$. The donor concentration is 10^{22} m^{-3} and donor ionization energy is 60 meV. The other parameters are discussed in the text. The energy in the plot is referenced to the bottom of the conduction band.

electron effective mass is $0.06m_0$, the donor ionization energy is taken to be 60 meV, and a donor concentration of 10^{16} cm^{-3} is assumed.

Partial compensation of donors. While the single impurity is interesting, it is usually the case that other impurities are present, as the growth of bulk material commonly has low concentrations of background impurities. Yet, it is quite normal for one impurity to dominate the behavior, as it has the largest concentration. Nevertheless, if it is a donor, some electrons must go to fill the states of any acceptors that may be present. It is important to treat this case, if for no other reason than to understand how it changes the statistical behavior of the donors. It will be assumed that there is a concentration of donors N_d and, in addition, a lower concentration of acceptors N_a, with $N_a < N_d$. Since the donor level lies above the acceptor level in energy, some donor electrons move to the lower energy offered by the acceptor level. As a result of the donor providing electrons to fill the acceptors, there are only $N_d - N_a$ effective donor atoms to contribute electrons to the conduction band.

At high temperatures, where the position of the Fermi level is below the donor level and moving toward mid-gap as the temperature raised, all of the donors are ionized. The only change here is to replace the donor density with the effective value given above. At low temperatures, however, there is a more substantial change, as we will see below. This arises from the fact that there will be empty donor states at absolute zero temperature, as these electrons have gone to the acceptors. The Fermi level lies at the donor level at absolute zero of temperature. This will also change the nature of the excitation statistics for the donors. Nevertheless, the number of ionized donors is still given by the right-hand side of (4.59), but we have to include the acceptors so that

$$n + N_a = N_{di} = \frac{N_d}{1 + 2e^{-\eta_d}\left(\dfrac{n}{N_c}\right)}. \tag{4.66}$$

We now rearrange the terms as we did above (we keep the intermediate step here, though) to give

$$n\left[1 + 2e^{-\eta_d}\left(\frac{n}{N_c}\right)\right] + 2N_a e^{-\eta_d}\left(\frac{n}{N_c}\right) = N_d - N_a. \tag{4.67}$$

This can be rearranged to give [2]

$$n = \frac{2(N_d - N_a)}{\left[1 + 2\left(\dfrac{N_a}{N_c}\right)e^{-\eta_d}\right] + \sqrt{\left[1 + 2\left(\dfrac{N_a}{N_c}\right)e^{-\eta_d}\right]^2 + 8\dfrac{N_d - N_a}{N_c}e^{-\eta_d}}}. \tag{4.68}$$

For high temperatures the exponential factor is approximately unity (recall that the donor energy is negative) and $N_c \gg N_d, N_a$. Then, we find that

$$n = N_d - N_a. \tag{4.69}$$

This result just tells us that the number of electrons is just given by the number available from the donors. At slightly lower temperatures, where the effective density of states is more comparable to the donor concentration, only the last term in the square root is large, and we find that

$$n \sim \sqrt{\frac{N_c(N_d - N_a)}{2}}\, e^{\eta_d/2}, \tag{4.70}$$

which is a modification of (4.62) to account for the reduced number of available electrons from the donors. At the lowest temperatures, however, the exponentials become large and the other terms become important and we find

$$n \sim \frac{N_c}{2}\left(\frac{N_d}{N_a} - 1\right)e^{\eta_d}. \tag{4.71}$$

Thus, at the lowest temperatures, the slope of the log density as a function of temperature changes by a factor of 2. Yet at somewhat higher temperature, it will change back as given in (4.70). Depending upon the densities of the donors and acceptors, the intermediate range may or may not appear in the data.

We can use the above estimates for the densities in (4.65) to determine the position of the Fermi energies. At the highest temperatures, we then have

$$E_F = -k_B T \ln\left(\frac{N_c}{N_d - N_a}\right). \tag{4.72}$$

As before, this is a simple modification reflecting the presence of the acceptors. At intermediate temperatures, we use (4.70) to give

$$E_F = \frac{E_d}{2} - \frac{k_BT}{2}\ln\left(\frac{2N_c}{N_d - N_a}\right). \tag{4.73}$$

Again, we have to remind ourselves that the energy is referenced to the conduction band edge and the donor energy is therefore negative. Finally, at the very lowest temperatures, we use (4.71) to arrive at

$$E_F = E_d + k_BT\ln\left(\frac{N_d - N_a}{2N_a}\right). \tag{4.74}$$

At these very low temperatures, the Fermi energy tends toward the donor level rather than the mid-point between the donor energy and the conduction-band edge. This is a physical requirement, as mentioned above.

4.2 Deep levels

The impurity levels of interest above were the shallow levels, as they are sometimes called. In these levels, the ionization energies were only a few, or a few tens, of meV. These levels are usually treated within the effective mass approximation as a kind of *hydrogenic impurity*. This means that the Schrödinger equation for the electronic wave function is often treated as that of the hydrogen atom with corrections for the dielectric constant and the free carrier effective mass of the band edge associated with the impurity. This approach introduces little new physics or concepts to the band structure of the semiconductor. In essence, it would be expected that the ionization energy would be that of the hydrogen atom reduced by the dielectric constant and the effective mass, or about two orders of magnitude. However, this is rather inaccurate, because this would be required to move the electron to the vacuum level, whereas we only need to excite it into the nearby band edge, and this may lie near one of the excited states of the equivalent hydrogen atom. This latter would further reduce the required energy from that of the hydrogen atom. The important point of these shallow levels is really that the impurity atom exhibits just a simple Coulombic type potential.

In some defects, however, the impurity creates a significant lattice distortion about the core of the impurity site. The difference between the case above and these so-called *deep levels* lies in the central cell potential—the core potential that interacts on a short range (whereas the Coulombic potential is relatively slowly varying over the unit cell size). When this core potential is sufficiently strong to affect the electrons, this affect is much stronger. For the hydrogenic impurity, the bound electron usually has an orbit that traverses a large number of unit cells. On the other hand, an electron bound to the deep levels is largely localized by the strength of the potential to the range of the local unit cell. Two effects create a sizable contribution to the understanding of these deep levels and their energy levels. These are the Jahn–Teller distortion and the Franck–Condon effect.

The Jahn–Teller theorem states that any electronic system with energy levels that are multiply degenerate may be split when the confining potential is affected by a different symmetry. Thus, if we put a deep level atom into a crystal such that the electronic ground state is multiply degenerate, there can be distortion of the lattice around this atom that raises this degeneracy. For example if an As atom sits on a Ga site (a so called anti-site defect), there are two extra electrons bound to the As atom in the neutral state, and these two atoms have degenerate energy levels. This will lead to a local distortion around the As atom that raises this degeneracy, an effect that has been seen experimentally [3].

There is another distortion, however, that can also arise. The strength of the central cell potential can significantly affect the actual band structure in the neighborhood of the defect atom. The role that this interaction plays is considerably different when the defect is neutral than when it is ionized and becomes charged. Thus, we expect the local band structure and the defect energy levels to change as the defect is ionized by removing an electron. This means that the optical transition energy is different for excitation and for relaxation, and this is the Franck-Condon effect. According to this latter effect, the energy of the localized state will change as the charge level of the defect is modified, and this arises from a local polarization of the lattice. In other words, the atoms in the vicinity of the local defect will relax to a new set of positions, in much the same way in which a surface relaxes. This relaxation is quite local in the lattice and therefore occupies a large part of the Brillouin zone; hence the relaxation is often accompanied by the excitation of a large number of phonons. For this reason, the optical transitions for a deep level are often accompanied by a number of 'phonon sidebands'—transitions that differ from one another by a phonon energy.

The distortion that accompanies the Franck–Condon effect can lead to another interesting phenomenon. When an electron is excited from a localized deep level, the lattice relaxation can lead to a reduction in the conduction band energy that gives a configuration of *lower* energy. The electron can no longer recombine with the charged deep level, since it no longer has sufficient energy to make the transition back; in essence there is an energy barrier preventing the electron from recombining, as it must now not only have the energy to recombine but must also drive the lattice relaxation back to its former state. If this kinetic barrier is sufficiently large, we can have the effect of persistent photoconductivity, which is observed in GaAlAs at low temperatures [4]. In this latter case, it has been hypothesized that this effect is related to a donor-vacancy complex, referred to as the D-X center. Optical absorption excites electrons from the deep levels, which produces a higher conductivity. When the light is removed, the semiconductor remains conducting until the temperature is raised to the order of 200 K or so, where the excited carriers have sufficient thermal energy to surmount the barrier and recombine.

In both physics and chemistry, the details of the Franck-Condon effect and deep impurities is based upon what are called reaction coordinates. One of the earliest papers in this area actually dealt with electron transfer from a donor or acceptor in which there was a lattice relaxation described by a reaction coordinate [5]. As discussed above, the local state of the lattice around the impurity changes with the

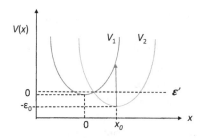

Figure 4.3. A diagram of the reaction coordinate x and the various potentials. V_1 represents the neutral impurity, whose energy level is below those shown here. The potential V_2 represents the shift of the conduction band when the neutral impurity level is taken as the reference point for the energy. The energy $(\varepsilon_0 + \varepsilon')$ represents the energy the electron must gain in order to recombine with the charged impurity.

charge on the impurity, and this is described by the reaction coordinate. This also changes the energy level, so that optical transitions can change depending upon the charge state. In this approach, the interactions can be studied by a spin-boson model that provides a pragmatic, yet realistic formulation for the role of dissipation in the electron transfer [5]. The role of a driving electric field was also considered in the study. Others have also studied electron transfer when coupled to collective boson degree of freedom, including the study of the fluctuations in the system [6]. Again, these latter studies also employed the reaction coordinate for the impurity system.

Let us now describe the approach with the reaction coordinate model. The idea is about the same whether we deal with a molecule or with condensed matter. For an impurity, say a donor, the upper state corresponds in the condensed matter system with the conduction band, while the lower state is the actual impurity level (in the molecule, this might actual be an extended state with dispersion). When the electron is excited from the donor impurity, the local potential changes primarily due to the short range local potential around the impurity. The potential change induces a lattice relaxation around the impurity, and this in turn changes the level of the impurity relative to the conduction band [7], but this is usually treated as a change in the conduction band. There are two contributions in this interaction. First, the Jahn-Teller theorem tells us that any electronic system with multiply degenerate ground states, such as spin degeneracy in the simple donor impurity case, is unstable against a distortion that removes the degeneracy. This leads to the lattice relaxation, which is of course much more prevalent in deep donors as opposed to shallow donors. As a result, the atom is actually displaced in position, and this is the reaction coordinate. But, there is a second distortion, and that is the change in the band structure between the neutral and the ionized states, as mention above, and this is the Franck-Condon effect, and this results in a difference between the optical excitation energy and the thermal excitation energy of the impurity [3, 8]. The overall situation, as mentioned in the previous paragraph, is best described by a reaction coordinate diagram, as shown in figure 4.3. The two potential curves cross at a point x^*, which is the point at which electron transfer occurs. Once the electron is into the second potential V_2, it must gain an energy greater than ε_0 in order to recombine with the donor. Actually, if we consider tunneling, the electron only needs to be excited to positive energy, where it can tunnel to V_1. In semiconductors, this leads to persistent photo-conductivity at low temperatures. Here, the Hamiltonian can be written as [5, 6]

$$H = H_{EL} + H_{RC} + H_B, \tag{4.75}$$

where the bare electronic system is written in terms of a two-state pseudo-spin system as

$$H_{EL} = -\frac{1}{2}[V_1(x) + V_2(x)]\hat{\sigma}_z + \frac{\hbar\Delta}{2}\hat{\sigma}_x. \tag{4.76}$$

The two pseudo-spin operators are given by

$$\begin{aligned}
\hat{\sigma}_z &= |1\rangle\langle1| - |2\rangle\langle2| \\
\hat{\sigma}_x &= |1\rangle\langle2| + |2\rangle\langle1|.
\end{aligned} \tag{4.77}$$

The two states are those of the impurity and band combinations. The coupling term Δ is assumed to be independent of the reaction coordinates. The reaction coordinate term is given by

$$H_{RC} = \left[\frac{p^2}{2m} + V_1(x) + V_2(x)\right]\begin{bmatrix} 1 & 0 \\ 0 & 1 \end{bmatrix}. \tag{4.78}$$

The two harmonic potentials, described in figure 4.3, are given as

$$\begin{aligned}
V_1(x) &= \frac{m\omega_0^2}{2}x^2 \\
V_2(x) &= \frac{m\omega_0^2}{2}(x - x_0)^2 - \varepsilon_0
\end{aligned} \tag{4.79}$$

and the bare term H_B is the energy of the non-interacting particles. Here, the so-called reorganization energy is $V_1(x_0)$, and the excitation energy indicated by the red arrow in figure 4.3 is the sum of this energy and the offset energy ε_0. One notes that the reaction coordinate x is the primary variable in the Hamiltonian and the resulting problem being studied. In both of the approaches in the previous paragraph, the problem is cast in terms of the Zusman equation, which is the equation of motion for the density matrix in the above Hamiltonian. The latter is a 2×2 matrix for the pseudo-spin coordinates. The equation is then transformed into one for the Wigner function, and this is used for studying the electron transfer in the molecular system. The important point is that when the electron is transferred from the 'donor' molecule (state $|1\rangle$) to the 'acceptor' molecule (state $|2\rangle$), the relaxation process makes this a 'one way' reaction, as the barrier ($\varepsilon_0 + \varepsilon'$) to the back reaction is created by the molecular relaxation characterized by the reaction coordinate. This is what leads to the persistent photoconductivity.

4.3 Disorder and localization

Up until now, it has been of primary interest to utilize the Bloch states that exist in a regular crystal lattice that possesses well-ordered parameters. In a sense, these discussions have been predicated upon the fact that the band states have long-range order. But, there is another class of semiconductors, in which only short-range order exists, and these are the amorphous semiconductors. However, the distinction is not all that clear, as significant disorder can occur even in our semiconductors for which we expect long-range order.

The fundamental problem in disordered semiconductors, particularly for those we assumed to have long-range order, is how the spectrum of allowed energy levels, and the corresponding nearest-neighbor coupling energies, are modified in the presence of a significant degree of disorder. Of course, the ultimate aim is to understand how this disorder affects the conductivity of the semiconductor. When disorder is present, the existence of long-range order may or may not be retained, but even if it is retained, this will only be for a reduced set of energy states. Once the long-range order is broken, one can generally think of the disordered material as containing many small regions, with the band properties varying among the regions, but remaining more or less uniform within a single region. Hence, we may have bands in a region, but these bands will have very fuzzy edges and the existence of states within the band gaps. The general experimental fact is that this leads to an excitation gap which is often larger than the expected band gap of the base semiconductor.

Such disorder can also lead to localization of the carriers in these fuzzy edges. In earlier days, it was assumed that this localization, which is often called strong localization in distinction from the weak localization which is a perturbative effect on the conductivity, was caused by this strong disorder. At the same time, it was believed that weak localization was caused by phase interference in the potential landscape created by the weak disorder arising from a large impurity or defect density. In fact, the two are closely related, which should have been obvious from the fact that the same theoretical model is used to simulate both effects! This model is the Anderson model [9], which we will discuss below.

The fundamental problem with disorder is to determine how it affects the allowed energy levels in the material. Normally, we think of (valence or conduction) band semiconductors as having a region of allowed energy states, in which the electron waves are free to propagate, and a band gap, in which the electron waves become localized and non-propagating (evanescent). With disorder in the band semiconductor, the edge between the band and the gap becomes a dispersed gray area (our fuzzy set of states). Usually the width of the band increases allowing states to broaden into the gap area, which is called band tailing. But many of the band states in this tail, as well as somewhat into the normal band region, can become localized. Here, the long range order of the crystal can be compromised, with the ultimate limit producing amorphous material. Because of the extending of the band region and its interaction with localized states, a new parameter arises which plays the role of the normal band edge, and this is the mobility edge. The mobility edge now separates the localized states from the conducting, or long-range ordered, states. The excitation mentioned above then occurs by carriers below the valence mobility edge being excited to states above the conduction mobility edge, and this transition is larger than the normal band gap.

Because of the disorder in the potential, which is assumed to be random, many small Aharonov–Bohm-like loops of various size can exist in the normally conducting states. This can leads to a fluctuation in the conductance as the interferences within a phase coherent area vary with these external variations. In addition, many back-scattering loops will lead to what is known as weak localization. These factors, however, are not the localization which we want to discuss in this section, which is the strong localization and not the phase coherence behavior at low temperatures. In this section, we will begin with the Anderson theory of disorder and how it affects the electronic states and leads to localization. Then, we will illustrate the effect with some simulation and some experimental results in low dimensional semiconductors.

4.3.1 Localization of electronic states

The concept of a rapid transition, at some critical energy, from a set of strongly localized states with only short-range order to a set of extended states with long-range coherence of their wave function, is remarkable. Normally, in band semiconductors, this is just the band edge and we give it no further thought. Perhaps the reason lies in the fact that most people aren't taught about the complex band structure with the continuum of localized states that exist throughout the band gap [10]. But, this idea of such a transition warrants further investigation. Here, we will follow the approach of Anderson [9], using one version of the several he discusses. The model adopts a set of atomic sites in a crystal, in which the atomic energies are randomly, but uniformly, distributed over a fixed energy range, which is commonly denoted as W. This means that the energy is a random function with a probability distribution function given by $1/W$ for energies within a given range, usually denoted as $-W/2 \leqslant E \leqslant W/2$. This will have the effect of *shifting the zero of energy to the center of the band*, whereas we normally associate it with one of the band edges. The reason for this will become clear later. We will find that there is a critical value of the width, such that if W is greater than this critical value, all the states in the band will be localized. The value of W at this critical value is normally known as the Anderson transition.

In this approach, each atomic site (which may not lie on a lattice site in the disorder model) has a 'site' energy, which corresponds to the atomic energy level of that particular atom, and an 'overlap' energy describing the interaction of the wave function of that atom with the wave function of neighboring atoms. This is just the tight-binding model of chapter 2. For example, in chapter 2, our model of graphene assumed the site energy was zero and the overlap energy was the parameter γ_0. There are two general approaches to disorder that have grown up with other approaches to this topic. One is to consider primarily site disorder, as described by the energy distribution discussed above, or some other equivalent form. The second is bond disorder which primarily treats a random variation in the overlap energy. Bond disorder is not typically found in the semiconductors. The Anderson model used here is primarily a site disorder model, but it is mapped into a system in which the lattice is regular with the atoms located at the lattice sites.

In the presence of the other atoms, one can use normal lowest order perturbation theory to write the Schrödinger equation in terms of the wave function and the perturbation of the neighboring sites. We begin by writing the Schrödinger equation as

$$i\hbar\frac{\partial\psi_s}{\partial t} + H\psi_s = E_{0,s}\psi_s + \sum_{s'\neq s} V_{ss'}\psi_{s'}, \tag{4.80}$$

where the on-site energy $E_{0,s}$ is a random variable as described above. The quantity $V_{ss'}$ is the overlap energy between neighboring sites, and the sum runs over only those nearest neighbor sites. This latter energy is essentially constant, as we are considering no bond disorder. The right-hand side of (4.80) is just the first-order perturbation terms, and higher-order terms lead to a series expansion for either the wave function or the energy (normally, one would generate the Hamiltonian energy matrix and diagonalize it, but here we approach the solution with perturbation). In such a series, the random site energy is just the zero-order term, while the actual final energy can be expressed as

$$E_s = E_{0,s} + \sum_{s'\neq s}\frac{V_{ss'}V_{s's}}{E_{0,s} - E_{0,s'}} + \cdots. \tag{4.81}$$

Unless the energy E_s is real, the wave function decays with time since the wave function is normally connected to the energy via a term $\exp(iE_s t/\hbar)$. Thus, the nature of the states will be investigated by examining the convergence properties of the infinite series in (4.81). If this series converges, then the energy is real and the state is localized at that site. On the other hand, if the series does not converge, it must be assumed that the energy lies within an extended band of energies which correspond to wave functions that are extended over the entire crystal. Note that this is opposite to our search for propagating waves with real energy in the band structure for the entire crystal.

The series in (4.81) is a stochastic series since the zero-order site energies are a random variable. Consequently, we can examine the convergence of the series in a statistical sense. Each term in the series contains V^{L+1}, where L is an integer giving the order of the term in the series. In a general lattice, each atom has Z nearest-neighbors, so that there are Z^L contributions to the Lth term, a point we shall use later. The general Lth term contains L product of terms of the form

$$T_{s'} = \frac{V_{ss'}}{E_{0,s} - E_{0,s'}}. \tag{4.82}$$

If the values of the site energies on the primed subscripted sites are statistically uncorrelated, the magnitude of the contribution of a product of such terms can be estimated by taking the average of its logarithm, as

$$\langle ln|T_1\cdots T_L|\rangle \sim L\langle ln|T|\rangle. \tag{4.83}$$

This form of the terms tells us that the series is to be interpreted as a geometric series, which is a special form of a power series that will converge provided that the ratio of subsequent terms approaches a limit which is less than unity. Thus, we require that if the coefficients of the series are A_L, the series will converge if

$$\lim_{L \to \infty} \left| \frac{A_{L+1}}{A_L} \right| x < 1, \tag{4.84}$$

where x is the argument of the series, here the energy. We now apply this convergence criteria to the perturbation series for the energy. This leads us to the conclusion that the series will converge if

$$Z \exp(\langle ln|T| \rangle) < 1. \tag{4.85}$$

We now introduce the probability distribution function discussed above

$$E_{0,s'} = \frac{1}{W} \quad for -\frac{W}{2} \leqslant E_{0,s'} \leqslant \frac{W}{2} \tag{4.86}$$

in order to evaluate the expectation value. This leads to

$$\langle ln|T| \rangle = \frac{1}{W} \int_{-\frac{W}{2}}^{\frac{W}{2}} ln \left| \frac{V_{ss'}}{E_{0,s} - E_{0,s'}} \right| dE_{0,s'}$$

$$= 1 - \frac{1}{2} \left(ln \left| \frac{4E_s^2 - W^2}{4V^2} \right| + \frac{2E_s}{W} ln \left| \frac{2E_s + W}{2E_s - W} \right| \right). \tag{4.87}$$

It is obvious from the above result that the value of W required for localization depends upon the energy in which one is interested. The energy in (4.87) is a smooth variable now. If we take the center of the band, where we have assumed the energy is 0, then only the first term in the larger parentheses survives, and we have the requirement that

$$Z \exp\left(1 - ln\frac{W}{2V} \right) < 1, \tag{4.88}$$

or

$$\frac{W}{2V} > e^1 Z \sim 2.7Z \sim 10.8, \tag{4.89}$$

is required to completely localize the band (we have taken $Z = 4$ for the tetrahedrally coordinated semiconductors). In general, we can say that $W > e^1 \Delta E$, where $\Delta E = 2VZ$, will totally localize the band, so that no extended states exist. At the other extreme, we can take $W = 0$, and we arrive at the equivalent inequality that says states will be localized if $E_s > \Delta E/2$. As ΔE is the width of the normal band, this is just the normal requirement that localized states exist outside the normal band width.

Of interest is to determine also the point at which an energy at the edge of the distribution, $E_s = W/2$ gets localized. We can then combine the two logarithm terms, to find that when $W > e^1 \Delta E/4$, we start to develop localized states at the edge of the band. Hence, there is a critical size of the disorder that must be exceeded before the band states near the edge begin to be localized. These different conditions are sketched in figure 4.4.

4.3.2 Some examples

GaAs/AlGaAs. We can illustrate the localizing effect of disorder and the random potential in another way, and that is to simulate a standard mesoscopic material such as the AlGaAs/GaAs heterostructure. The conducting layer is a quasi-two-dimensional electron (or hole) gas located at the interface between the two materials. The simulation is a study of the conductance of this electron layer. In this approach, we discretize the two dimensional layer with a grid size of 5 nm, and consider a range of widths, in the range of 0.2–0.6 μm, and lengths, in the range of 0.3–0.8 μm. We use the Anderson model to impose a random potential at each grid point according to the uniform distribution discussed above in (4.84). This gives a random potential whose peak-to-peak amplitude is W. While this random potential induces states below the lowest energy of the band (and above the highest energy in the band), it also localizes a fraction of the normally transmitting states. A critical energy separating the localized modes and the propagating modes defines what has been called the 'mobility edge' [10]. Since we are going to vary the amplitude of the random potential, it is useful to see how this localizes the conductance. We consider an electron density of 4×10^{11} cm^{-2}, which gives a Fermi energy about 15 meV in GaAs at 8 mK. Since we have a finite width of sample, the system is quantized in the transverse direction into a set of modes, and the conductance is determined by the number of these modes whose transverse eigen-energies lie below the Fermi energy. Then, we determine the fraction of these modes that actually propagate in the presence of the random potential. This is plotted in figure 4.5, and it may be seen that the behavior suggested by figure 4.4 is certainly followed. The error bars

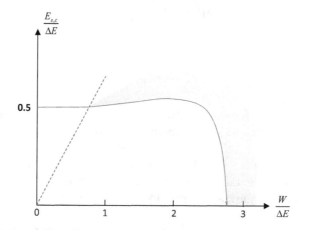

Figure 4.4. The critical energy that differentiates extended from localized states (blue curve), with the localized states outside the curve. The red curve shows the minimum value of the disorder needed to begin to localize states.

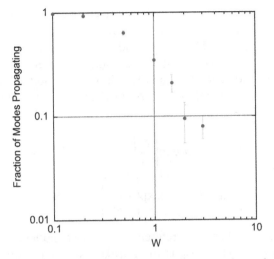

Figure 4.5. The fraction of transverse modes that actually propagate as the random potential of (4.84) is increased in amplitude. This is for a simulation of the two-dimensional electron gas at the interface of a GaAs/GaAlAs hetero-structure at low temperature following the Anderson model.

represent an average over many sizes of sample, all of which have different implementations of the random potential. While there may be 30–60 modes propagating normally, once the disorder gets sufficiently large, only a very few remain propagating. This is a significant point. Nevertheless, the lowest data point in the figure lies at 0.1 eV peak-to-peak disorder potential W. Even though this potential reaches peaks more than twice the Fermi energy, most of the modes remain propagating. So, it takes a significant amount of disorder to really affect the band states.

MoS_2. The transition metal di-chalcogenides (TMDCs) form interesting two dimensional semiconductor layers that have been shown to be useful for electronic applications. However, they currently suffer from the presence of defects that act as both deep levels and random potentials leading to localization. In particular, it has been shown that charge impurities such as S vacancies in MoS_2 and impurities from the dielectric environment not only have significant impact on the mobility of the MoS_2 devices [11], but also can lead to the formation of an impurity band tail within the band gap region as well as localized states above the conduction band edge [12]. The TNDCs tend to have a significant number of chalcogenide vacancies (e.g., missing S in this case). These vacancies lead to both effects. First, the vacancy itself tends to produce a deep level that lies near mid-gap in the material. Secondly, the missing atomic potential produces a modulation of the crystal potential formed from the pseudo-potentials of the atoms themselves. This modulation creates the random potential that leads to the Anderson model and the resulting localization.

To gain a quantitative understanding of the effects of these impurity states, we examine the room temperature device characteristics using a model proposed by Zhu *et al* [12]. From studying extensive capacitance voltage measurements between a back gate on the TMDC and the layer itself (using a SiO_2 insulator grown on the Si back gate), these authors find that there is a mid-gap state that has a population of approximately 10^{12} cm^{-2}. They also find an extremely broad distribution of localized states, as in the Anderson model, and an extremely high mobility edge for the conduction band. They then adapt a simulation model to fit the data. In this model, the impurity band tail is incorporated by the following density of states distribution:

$$D_n(E) = \begin{cases} \alpha D_0 \exp\left[\dfrac{E}{\varphi}\right], & -\dfrac{E_G}{2} < E < 0, \\ \\ D_0 - (1 - \alpha)\exp\left[-\dfrac{E}{\varphi'}\right], & E > 0. \end{cases} \tag{4.90}$$

As before, the energies here are referenced to the conduction band edge, *in the absence of the disorder*. Here, D_0 is the normal two-dimensional density of states given by (4.24), using an effective mass of 0.4 m_0. The $(1 - \alpha)$ term for positive energy accounts for the states that are removed from the conduction band to form the band tail states of the first line. The parameter φ is the characteristic width of the localized states that form the band tail discussed above and is found to be ~100 meV, and φ' is chosen so that the two piece-wise functions have a continuous gradient at $E = 0$. The parameter α is found to be ~0.33. From this model and the measured conductance, the mobility edge is found to lie 10 meV above the conduction band edge.

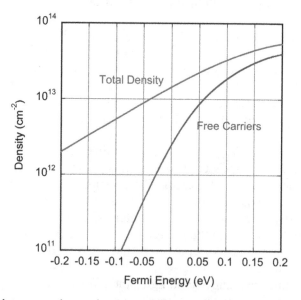

Figure 4.6. Localized states trap electrons into non-mobile states. Here are the total electron density and the free carrier density for a MoS$_2$ monolayer at 300 K. The parameters are discussed in the text.

To calculate the free carrier density, we consider only the extended states above the mobility edge energy E_M with respect to the conduction band edge:

$$n_{\text{free}} = \int_{E_M}^{\infty} D_n(E) P_{\text{FD}}(E) \mathrm{d}E, \qquad (4.91)$$

In different experiments, reported by Xiao *et al* [13], a slightly different value of the band tailing, $\varphi \sim 110$ meV, and a positively charged oxide impurity density of 3.5×10^{12} cm^{-2} were found. As previously, these were found for a layer of MoS_2 placed upon an oxidized Si wafer, so that the Si could be used as a back gate to vary the electron concentration. Using this data and parameters, we plot the total induced electron concentration as well as the free carrier (those above the mobility edge) concentration at room temperature in figure 4.6. The difference in these two densities yields the number of carriers which are trapped in the localized states.

References

[1] Blakemore J S 1962 *Semiconductor Statistics* (New York: Pergamon)

[2] Smith R A 1959 *Semiconductors* (London: Cambridge University Press)

[3] Jaros M 1982 *Deep Levels in Semiconductors* (Bristol: Adam Hilger)

[4] Lang D V and Logan R A 1977 *Phys. Rev. Lett.* **39** 635

[5] Goychuk I, Hartmann L and Hänggi P 2001 *Chem. Phys.* **268** 151

[6] Ankerhold J and Lehle H 2004 *J. Chem. Phys.* **120** 1436

[7] Kittel C 1966 *Introducction to Solid State Physics* 7th edn (New York: Wiley)

[8] Baranowski J M, Grynberg M and Porowski S 1982 *Handbook on Semiconductors* vol 1 ed W Paul (Amsterdam: North-Holland) ch 6

[9] Anderson P W 1958 *Phys. Rev.* **109** 1492

[10] Ferry D K 2000 *Semiconductor Transport* (London: Taylor and Francis) sec 6.2.2

[11] Ma N and Jena D 2014 *Phys. Rev.* X **4** 011043

[12] Zhu W J *et al* 2014 *Nature Commun.* **5** 3087

[13] Xiao Z, Song J, Ferry D K, Ducharme S and Hong X 2017 *Phys. Rev. Lett.* **118** 236801

Chapter 5

Carrier scattering

Scattering of the electrons, or the holes, from one state to another, whether this scattering occurs due to the lattice vibrations or by the Coulomb field of impurities or some other process, is one of the most important processes in the transport of the carriers through the semiconductor. This process is the result of the electron–lattice interaction, discussed in chapter 1, or from the presence of impurities or other random potentials in the material. In one sense it is the scattering that limits the velocity of the charge carriers in any applied fields. If we had no scattering, the carriers will be subject to a uniform increase of the wave vector $\mathbf{k}$ (in an applied dc field) and will cycle continuously through the Brillouin zone and yield a time-average velocity that is zero. It is the scattering that breaks up the correlated, accelerated state and introduces the actual transport process. Transport is seen as a balance between accelerative forces and dissipative forces (the scattering).

The discussion of the adiabatic principle earlier allows separation of the electronic motion from the lattice motion. The former was solved for the static energy bands, while the latter yielded the lattice dynamics—the motion of the atoms—and phonon spectra. There remained still the term that coupled the electronic motion to the lattice motion. This term gives rise to the electron–phonon interactions. There is not a single interaction term. Rather, the electron–phonon interaction can be expanded in a power series in the scattered wave vector $\mathbf{q} = \mathbf{k} - \mathbf{k}'$, and this process gives rise to a number of terms, which correspond to the number of phonon branches and the various types of interaction terms. There can be acoustic phonon interactions with the electrons, and then there are optical interactions with the electrons that can be either through the polar interaction (in compound semiconductors) or through the non-polar interaction. These are just the terms up to the harmonic expansion of the lattice; higher-order terms give rise to higher-order interactions. Random potentials in the semiconductor also create interactions with the electrons. The real part of these interactions produce effects such as the localization discussed in chapter 4. It is the imaginary part of these interactions that lead to decay of the transport. These potentials can arise from impurities, as mentioned, or from surface effects, as another example.

In this chapter, the basic electron–phonon (which may also be hole–phonon) interaction is treated in a general way first. The various interactions that are found to be important in semiconductors are treated to yield scattering rates appropriate to each process. Following this, a summary of the various processes that contribute to the most common semiconductors is presented, followed by a discussion of the non-lattice dynamic scattering processes that arise from other, random potentials. These include ionized impurity scattering, alloy scattering, surface roughness scattering, and deep defect scattering. Throughout the chapter, it mostly will be assumed that the energy bands are parabolic in nature; the extension to non-parabolic bands is a straight-forward expansion, usually through the modification of the density of states, but also at times from the effective mass directly.

5.1 The electron–phonon interaction

The simplest assumption is that vibrations of the lattice cause small shifts in the energy bands. The electronic band structure arises from the atomic potentials of the static crystal lattice. If a wave is present in the lattice, this distorts the lattice and causes a perturbation of the electronic bands. Deviations of the bands from their static lattice value, due to these small shifts from the frozen lattice positions, lead to an additional potential that causes the scattering process. The scattering potential is used in time-dependent, first-order perturbation theory to find a rate at which electrons are scattered out of one state k and into another state k', while either absorbing or emitting a phonon of wave vector q. Each of the different processes, or interactions, leads to a different 'matrix element,' where the matrix element is the key ingredient in the perturbation theory. These matrix elements have a dependence on the three wave vectors and their corresponding energy. These are discussed in the following sections, but here the treatment will retain just the existence of the scattering potential δE which leads to a matrix element

$$M(k, k') = \langle \psi_{k,q} | \delta E | \psi_{k',q} \rangle, \tag{5.1}$$

where the subscripts indicate that the wave function involves both the electronic and the lattice coordinates, and δE is the perturbing potential. This perturbing potential is different for each electron–phonon interaction process. Normally, the electronic wave functions are taken to be Bloch functions that exhibit the periodicity of the lattice. In addition, the matrix element usually contains the momentum conservation condition. Here this conservation condition leads to

$$k - k' \pm q = G, \tag{5.2}$$

where G is a vector of the reciprocal lattice. In essence, the presence of G is a result of the momentum limitation due to the periodic boundary conditions applied to the real space lattice, and the result is that we can only define the unique values of the crystal momentum within a single Brillouin zone. For the upper sign, the final state lies at a higher momentum than the initial state, and therefore also at a higher energy. This upper sign must correspond to the absorption of a phonon by the electron. The lower sign leads to the final state being at a lower energy and momentum, hence corresponds to the emission of a phonon by the electrons.

Straightforward time-dependent, first-order perturbation theory then leads to the equation for the scattering rate, in terms of the *Fermi golden rule* [1]

$$P(k, k') = \frac{2\pi}{\hbar} |M(k, k')|^2 \delta(E_k - E_{k'} \pm \hbar\omega_q), \tag{5.3}$$

where the signs have almost the same meaning as in the preceding paragraph: for example, the upper sign corresponds to the absorption of a phonon. The probability of (5.3) is to be interpreted as the electron in the initial state k is scattered out of this state and into the state k'. This final state lies at a higher energy according to the need for the argument of the delta function to be zero. By the same argument, the lower sign corresponds to the emission of a phonon. A derivation of (5.3) is found in most introductory quantum mechanics texts. Principally, the δ-function limit requires that the collision be fully completed through the invocation of a $t \to \infty$ limit. Moreover, each collision is localized in real space so that use of the well-defined Fourier coefficients k, k', and q is meaningful. The perturbing potential must be small, so that it can be treated as a perturbation of the well-defined energy bands and so that two collisions do not 'overlap' in space or time. On the short time scale, the instantaneous collision indicated in (5.3) can be broadened with a *collision duration*, but this is a complication that arises in quantum transport [2], and will be ignored here.

The scattering rate out of the state defined by the wave vector k and the energy E_k is obtained by integrating (5.3) over all final states. Because of the momentum conservation condition (5.2), the integration can be carried out over either k' or q with the same result (omitting the processes for which the reciprocal lattice vector $G \neq 0$). For the moment, the integration will be carried out over the final state wave vector k', and ($\Gamma = 1/\tau$ is the scattering rate, whose inverse is the scattering time τ)

$$\Gamma(k) = \frac{2\pi}{\hbar} \sum_{k'} |M(k, k')|^2 \delta(E_k - E_{k'} \pm \hbar\omega_q). \tag{5.4}$$

In those cases in which the matrix element M is *independent of the phonon wave vector*, the matrix element can be removed from the summation, and this summation over the delta function just leads to the density of final states

$$P(k, k') = \frac{2\pi}{\hbar} |M(k, k')|^2 n_{3D}(E), \tag{5.5}$$

where the last term is given by (4.10) in three dimensions. This last equation has a very satisfying interpretation: The total scattering rate is just the product of the square of the matrix element connecting the initial state to the final state and the total number of final states (we note, however, that care must be exercised on the evaluation of the density of states: those scattering processes which conserve spin must not include the 'factor of 2 for spin' in the density of states). When (5.5) can be invoked, the scattering angle is a random variable that is uniformly distributed across the energy surface of the final state. Thus any state lying on the final energy surface is equally likely, and the scattering is said to be isotropic.

When there is a dependence of the matrix element on the wave vector of the phonon, the treatment is somewhat more complicated and this dependence must stay inside the summation and be properly treated. For this case, it is slightly easier to carry out the summation over the phonon wave vectors. At the same time, the summation over the wave vectors is changed to an integration and (using spherical coordinates for the description of the final state wave vector, as in section 4.1.1; and shifting the integral to one over q rather than k')

$$\Gamma(k) = \frac{2\pi}{\hbar} \frac{V}{8\pi^3} \int_0^{2\pi} d\varphi \int_0^\pi \sin \vartheta \, d\vartheta \int_0^\infty |M(k, k')|^2 \delta(E_k - E_{k'} \pm \hbar\omega_q) q^2 dq \quad (5.6)$$

where it is assumed the semiconductor is a three-dimensional crystal. There is almost no case where the specific angle of q appears in the matrix element. Rather, it is only the relative angle between q and k that is important, so that it is permissible to consider that the latter vector is aligned in the z-direction, or the polar axis of the spherical coordinates used in (5.6). Since there is no reason not to have azimuthal symmetry in this configuration, there is no reason to have any φ variation and this integral can be done immediately, yielding 2π.

The second angular integral, over the polar angle, involves the delta function, since the latter's argument can be expanded as

$$\begin{aligned} E_k - E_{k'} \pm \hbar\omega_q &= \frac{\hbar^2 k^2}{2m_c} - \frac{\hbar^2 (k \pm q)^2}{2m_c} \pm \hbar\omega_q \\ &= -\frac{\hbar^2 q^2}{2m_c} \mp \frac{\hbar^2}{m_c} kq \cos \vartheta \pm \hbar\omega_q \end{aligned} \quad (5.7)$$

Two effects happen when the integral over the polar angle is performed. The first is that a set of constants (the inverse of the pre-factor in front of the $\cos \theta$) appears due to the functional argument of the delta function, and the second is that finite limits are set on the range of q that can occur. This limit on the integration arises because (5.7) must have a zero that lies within the range of integration of the polar angle. For the case of *absorption* of a phonon (upper sign), this leads to the zero occurring at

$$q = \pm\sqrt{k^2\cos^2 \vartheta + \frac{2m_c\omega_q}{\hbar}} - k \cos \vartheta \quad (5.8)$$

and this leads to the limits

$$\sqrt{k^2 + \frac{2m_c\omega_q}{\hbar}} - k < q < \sqrt{k^2 + \frac{2m_c\omega_q}{\hbar}} + k. \quad (5.9)$$

This can be expressed graphically by recognizing that q now is a scattering from an energy shell of radius E, or k, to an energy shell of $E + \hbar\omega_q$, as shown in figure 5.1.

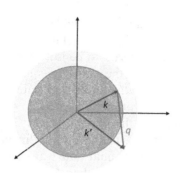

Figure 5.1. The scattering diagram for the absorption of a phonon. The outer shell (yellow) lies at $E + \hbar\omega_q$, while the inner shell lies at E.

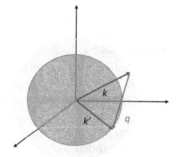
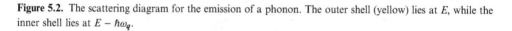

Figure 5.2. The scattering diagram for the emission of a phonon. The outer shell (yellow) lies at E, while the inner shell lies at $E - \hbar\omega_q$.

In the second case, the case for the emission of a phonon (lower sign), the zero occurs at

$$q = k\cos\vartheta \pm \sqrt{k^2\cos^2\vartheta - \frac{2m_c\omega_q}{\hbar}}, \tag{5.10}$$

for which

$$k - \sqrt{k^2 - \frac{2m_c\omega_q}{\hbar}} < q < k + \sqrt{k^2 - \frac{2m_c\omega_q}{\hbar}} \tag{5.11}$$

with the additional requirement that

$$k^2 > \frac{2m_c\omega_q}{\hbar}, \quad E_k > \hbar\omega_q. \tag{5.12}$$

This latter expression simply states that an electron cannot emit a phonon unless it has an energy greater than that of the phonon. Equation (5.11) can be expressed graphically by recognizing that q now is a scattering from an energy shell of radius E, or k, to an energy shell of $E - \hbar\omega_q$, as shown in figure 5.2.

These factors can now be used to evaluate the integral over the polar angle, which finally yields the result

$$\Gamma(k) = \frac{m_c V}{2\pi \hbar^3 k} \int_{q_-}^{q_+} |M(k, q)|^2 q \mathrm{d}q. \tag{5.13}$$

The limits q_+ and q_- are given by (5.9) or (5.11) for phonon absorption or emission, respectively. If the scattering process can scatter into both final spin states, an additional factor of 2 should be added to (5.13). At this point no further progress can be made without specifying the details of the actual matrix element that appears in (5.13).

5.2 Acoustic deformation potential scattering

5.2.1 Spherically symmetric bands

One of the most common phonon scattering processes is the interaction of the electrons (or holes) with the acoustic modes of the lattice through a deformation potential. Here, a long-wavelength acoustic wave moving through the lattice can cause a local strain in the crystal that perturbs the energy bands due to the lattice distortion. This change in the bands produces a weak scattering potential, which leads to a perturbing energy [3]

$$\delta E = \Xi_1 \Delta = \Xi_1 \nabla \cdot \boldsymbol{u}_q, \tag{5.14}$$

where Ξ_1 is the *deformation potential* for a particular band and Δ is the *dilation* of the lattice produced by a wave, whose Fourier coefficient is $\boldsymbol{u}_q$. We note here that any static displacement of the lattice is a displacement of the crystal as a whole and does not contribute, so that it is the wave-like variation of the amplitude within the crystal that produces the local strain in the bands. This variation is represented by the dilation, which is just the desired divergence of the wave. The amplitude $\boldsymbol{u}_q$ is a relatively uniform Fourier coefficient for the overall lattice wave, and may be expressed as [3]

$$\boldsymbol{u}_q = \left(\frac{\hbar}{2\rho_m V \omega_q}\right)^{1/2} \left[a_q e^{i\boldsymbol{q}\cdot\boldsymbol{r}} + a_q^\dagger e^{-i\boldsymbol{q}\cdot\boldsymbol{r}}\right] \boldsymbol{e}_q e^{-i\omega_q t}, \tag{5.15}$$

where ρ_m is the mass density, V is the volume, a_q and a_q^+ are annihilation and creation operators for phonons (as used in chapter 3), $\boldsymbol{e}_q$ is the polarization vector, and the plane-wave factors have been incorporated along with the normalization factor for completeness. This wave amplitude, other than the plane wave factors, is just the displacement of the harmonic oscillator for that mode. Because the divergence operator produces a factor proportional to $\boldsymbol{q}$ in the polarization direction (along the direction of propagation), only the longitudinal acoustic modes couple to the carriers in a spherically symmetry band (the case of ellipsoidal bands will be treated later). The fact that the resulting interaction potential is now proportional to q (i.e., first order in the phonon wave vector) and leads to this term being called a *first-order* interaction.

The matrix element may now be calculated by considering the proper sum over both the lattice and the electronic wave functions. The second term in the square brackets of (5.15) is the term for the emission of a phonon by the carrier. We will consider just this term that leads to the matrix element squared, as

$$|M(k, q)|^2 = \frac{\hbar \Xi_1^2 q^2}{2\rho_m V \omega_q}(N_q + 1)I_{k,q}^2,$$ (5.16)

where N_q is the Bose–Einstein distribution function for the phonons, and

$$I_{k,q} = \int_\Omega u_{k-q}^\dagger u_k d^3 r,$$ (5.17)

is the *overlap* integral between the cell portions of the Bloch waves (unfortunately, similar symbols are used, but the u_k in this equation is the cell periodic part of the Bloch wave and not the phonon amplitude given above) for the initial and final states, and the integral is carried out over the cell volume Ω. For elastic processes, as this will be shown to be, and for both states lying within the same 'valley of the band, this integral is unity. Essentially, exactly the same result (5.16) is obtained for the case of the absorption of phonons by the electrons, with the single exception that $(N_q + 1)$ is replaced by N_q. We will use this last fact below.

One thing that should be recalled is that the acoustic modes have very low energy. If the velocity of sound is 5×10^5 cm s^{-1}, a wave vector corresponding to 25% of the zone edge yields an energy only of the order of 10 meV. This is a very large wave vector, so for most practical cases the acoustic mode energy will be less than a millivolt. This will be important later when this matrix element is introduced into the scattering formulas above. Scattering processes in which the phonon energies are small and may be ignored are termed *elastic* scattering events. Of more interest here is the fact that these energies are much lower than the thermal energy except at the lowest temperatures, and the Bose–Einstein distribution can be expanded under the equipartition approximation as

$$N_q = \frac{1}{\exp\left(\frac{\hbar \omega_q}{k_B T}\right) - 1} \sim \frac{k_B T}{\hbar \omega_q} \gg 1,$$ (5.18)

at all except the lowest temperatures. At low temperatures, where the carrier distribution is degenerate, the transition temperature in (5.18) is known as the Bloch–Gruneisen temperature, given by

$$T_{BG} = \frac{\hbar \omega_q}{k_B} \sim \frac{2\hbar v_s k_F}{k_B},$$ (5.19)

where v_s is the acoustic sound velocity ($= \omega_q/q$) from chapter 3, and $2k_F$ is taken as the maximum value for q (which spans the Fermi surface of the degenerate carrier distribution). If we take the Fermi energy as 15 meV in GaAs, then k_F is 1.6×10^6 cm^{-1}, and $T_{BG} \sim 12$ K. Hence, (5.18) is a good approximation for systems above ~ 35 K.

So, near room temperature we are free to use the assumption of elastic scattering and to use (5.18). In this situation, the emission and absorption terms are about equal, and we can just multiply the matrix element by 2. Now introducing the sound velocity as well, we get

$$|M(k)|^2 \sim \frac{\Xi_1^2 k_B T}{\rho_m v_s^2 V}. \tag{5.20}$$

This final form (5.20) is independent of the wave vector q of the phonons, so that the simple form of (5.5) can be used. For electrons in a simple, spherical energy surface and parabolic bands, this leads to

$$\begin{aligned}
\Gamma(k) &= \frac{2\pi}{\hbar} \frac{\Xi_1^2 k_B T}{\rho_m v_s^2 V} \frac{V}{4\pi^2} \left(\frac{2m_c}{\hbar^2} \right)^{3/2} E^{1/2} \\
&= \frac{\Xi_1^2 k_B T (2m_c)^{3/2}}{2\pi \rho_m v_s^2 \hbar^4} E^{1/2}
\end{aligned} \tag{5.21}$$

It has been assumed that the interaction does not mix spin states, and this factor is accounted for in the density of states. Although most of the parameters may easily be obtained for a particular semiconductor, it is found that the conduction band deformation potential itself is almost universally of the order of 7 to 10 eV for nearly all groups IV and III–V semiconductors (within something of the order of a factor of 2).

In figure 5.3, the acoustic phonon scattering rates are shown for GaN and Si to illustrate the behavior. Both show initial variation as the square root of the energy, according to (5.21), but there is a deviation from this at higher energies due to the non-parabolicity of the bands which has been taken into account in the calculation. While the effective mass in GaN ($0.2m_0$) is smaller than in Si, the deformation potential is larger and other parameters are sufficiently different to lead to the much stronger scattering in this material.

5.2.2 Ellipsoidal bands

In the treatment of spherical energy surfaces above, it was found that the matrix element was independent of the direction in momentum space and was independent of the wave vector (in the equipartition limit). In a many-valley semiconductor, such as the conduction band of silicon or germanium, this is no longer the case. Because the constant energy surfaces are ellipsoidal, shear strains as well as *dilational* strains can produce deformation potentials. The shear strain still leads to a term that depends on the vector direction of $\mathbf{q}$, and it should be expected that band edge shifts will depend on all six components of the shear tensor. Thus there might be as many as six deformation potentials. However, in the semiconductors of interest, such as Si and Ge, the valleys are ellipsoidal and centered on the high symmetry <100> and <111> axes, so that the symmetry properties allow a reduction to just two

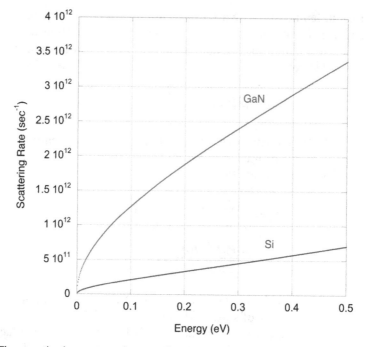

Figure 5.3. The acoustic phonon scattering rates for GaN and Si. A non-parabolic band model has been assumed for the calculation.

independent potentials. These are the *dilational potential* Ξ_d and the *uniaxial shear potential* Ξ_u. In terms of these potentials, the deformation energy is just [4]

$$\delta E = \Xi_d(e_{xx} + e_{yy} + e_{zz}) + \Xi_u e_{zz} \tag{5.22}$$

for an ellipsoid valley whose major axis is aligned with the z-axis. For longitudinal waves in an arbitrary direction q, the factor $\Xi_1^2(e_q \cdot q)^2$ goes over into

$$\Xi_{LA}^2 q^2 = \left(\Xi_d^2 + \Xi_u^2 \cos^2 \vartheta\right) q^2, \tag{5.23}$$

and ϑ is the angle between the z-axis (major axis of the ellipsoid) and the vector q. For transverse waves, only the e_{zz} term couples, and the proper form is just

$$\Xi_{TA}^2 q^2 = \Xi_u^2 \sin^2 \vartheta \cos^2 \vartheta\, q^2. \tag{5.24}$$

It should be remarked that both transverse modes are incorporated here in the general treatment. The differences above lead to different scattering rates for each principal axis within a single ellipsoidal valley. The summation over the multiple valleys (for the current) returns the overall system to cubic symmetry (unless the valleys are taken out of equilibration with each other, for example by hot carriers discussed in chapter 7). To achieve the latter result, each valley must be treated separately in the summation over q, and the various results summed. When numerical evaluations of the angular averages are carried out for Si and Ge, it is found that a fairly good approximation is to use a single energy-dependent scattering rate for the combined longitudinal and transverse acoustic modes. For the case of Ge, for example, it is found that [4]

$$\Xi_l^2 \rightarrow \frac{3}{4}\left(1.31\Xi_d^2 + 1.61\Xi_d\Xi_u + 1.01\Xi_u^2\right) \sim 0.99\Xi_d^2. \tag{5.25}$$

Thus the use of a single deformation potential is not a bad approximation in most cases, especially if the set of ellipsoids remains equivalent under application of electric fields. Values of Ξ_d and Ξ_u that are accepted for Si are -6 and 9 eV, respectively [5].

5.3 Piezoelectric scattering

The piezoelectric effect arises from the polar nature of compound materials, such as GaAs and other III–V compounds. These materials lack a center of inversion symmetry (sitting between the Ga and As atoms, one can understand why there is no inversion symmetry—look one way and you see a Ga atom, look the other and you see an As atom). Strain applied in certain directions in the lattice will produce a built-in electric field, which arises from the distortion of the basic unit cell. This creation of an electric field by the strain is called the piezoelectric effect. In materials with large piezoelectric coefficients, such as quartz, one can use the effect to provide oscillators at precise frequencies. In most semiconductors, the effect is small, but can lead to scattering of the carriers, particularly at low temperatures where other scattering mechanisms are weak. For the purpose here, it is the presence of the acoustic mode that induces a local electric field. The carriers are deflected by this field and are therefore scattered by it. The crystals of interest have a single piezoelectric constant d (in the tensor notation by which stress and strain are discussed in a general cubic material, this is the element d_{14}, and this notation will be used below). By expanding the displacement waves, the polarization components can be found as follows (in Fourier transform form)

$$P_x = i\frac{d_{14}}{\varepsilon_\infty}(e_{q_y}q_z + e_{q_z}q_y)u_q$$

$$P_y = i\frac{d_{14}}{\varepsilon_\infty}(e_{q_z}q_x + e_{q_x}q_z)u_q. \tag{5.26}$$

$$P_z = i\frac{d_{14}}{\varepsilon_\infty}(e_{q_y}q_x + e_{q_x}q_y)u_q$$

Here, ε_∞ is the high frequency dielectric permittivity. The interaction energy shift can be found by

$$\delta E = -\varepsilon_\infty \mathbf{E} \cdot \mathbf{P}, \tag{5.27}$$

where the electric field $\mathbf{E}$ (not to be confused with the energy) arises from the induced potential. The polarization leads to this potential through the piezoelectric inter-action, which couples to form the perturbing energy. For the potential, we shall use a standard screened Coulomb form

$$V(r) = \frac{e}{4\pi\varepsilon_\infty r}e^{-q_d r}, \tag{5.28}$$

where q_d is the reciprocal of the Debye screening length. The exponential factor provides a cut-off in the Coulomb interaction. This potential may be Fourier transformed (in three dimensions) to give

$$V_q = \frac{e}{\varepsilon_\infty}\frac{1}{q^2 + q_d^2}. \tag{5.29}$$

This, in turn, yields the electric field

$$\mathbf{E} = -i\frac{e}{\varepsilon_\infty}\frac{\mathbf{q}}{q^2 + q_d^2}. \tag{5.30}$$

The perturbing potential can now be written, with the definitions above, as

$$\delta E = -i\frac{ed_{14}}{\varepsilon_\infty}\frac{q^2}{q^2 + q_d^2}(\beta\gamma e_{q_z} + \gamma\alpha e_{q_y} + \alpha\beta e_{q_z})u_q, \tag{5.31}$$

where α, β, and γ are the directional cosines between the wave vector $\mathbf{q}$ and the three axes x, y, and z, respectively.

The role of the screening is interesting. In examining (5.31), it is clear that for small $\mathbf{q}$ (large distances), the interaction potential vanishes with q^2. On the other hand, for large $\mathbf{q}$ (small distances), the central q-dependent factor becomes unity. There is a natural cutoff value for q, which is determined by q_d, the reciprocal of the Debye screening length. From this it appears that piezoelectric scattering is a short range effect, much like other Coulomb scatterers to be discussed later. There is a complication in this screening, however. The actual potential, which arises from the electron–phonon interaction, is not a true Coulomb potential because of the harmonic variation at frequency ω_q. With full dynamic screening, if the frequency is sufficiently high, the screening is significantly reduced [6]. This effect strengthens the piezoelectric interaction at longer wave vectors. However, the de-screening is fully effective only when the phonon energy is comparable to the electron energy. Since we are dealing with elastic scattering, this event seldom occurs, and the formulas above may be used freely.

The results of the preceding section now can be used to evaluate the matrix element. Equation (5.31) can be compared directly to (5.20) to yield

$$|M(k)|^2 = \frac{4e^2 d_{14}^2 k_B T}{\varepsilon_\infty^2 \rho_m V\omega_q^2}\left[\frac{q^2}{q^2 + q_d^2}\left(\beta\gamma e_{q_z} + \gamma\alpha e_{q_y} + \alpha\beta e_{q_z}\right)\right]^2, \tag{5.32}$$

in the equipartition limit. The last term can be averaged over the various directions to produce a spherically symmetric average, which gives 12/35 for longitudinal waves and 16/35 for transverse waves [7, 8]. With this in mind, we can introduce average lattice strain constants through

$$c_L = \frac{2}{5}(c_{12} + 2c_{44}) + \frac{3}{5}c_{11},$$

$$c_T = \frac{1}{5}(c_{11} - c_{12}) + \frac{3}{5}c_{44}. \tag{5.33}$$

This allows us to define an effective coupling constant as

$$K^2 = \frac{d_{14}^2}{\varepsilon_\infty}\left(\frac{12}{35c_L} + \frac{16}{35c_T}\right). \tag{5.34}$$

This result may now be used to calculate the scattering rate. However, this scattering is essentially elastic, as it involves the acoustic modes, which have very low energy in comparison with the carriers. Thus the limits can be simplified by ignoring the phonon energy, but the anisotropic approximation (5.6) must be used. With the above considerations, the scattering rate can be written as

$$\Gamma(k) = \frac{2m_c e^2 K^2 k_B T}{\pi \varepsilon_\infty \hbar^3 k} \int_0^{2k} \frac{q^3 dq}{\left(q^2 + q_d^2\right)^2}$$

$$= \frac{m_c e^2 K^2 k_B T}{\pi \varepsilon_\infty \hbar^3 k}\left[\ln\left(1 + \frac{4k^2}{q_d^2}\right) - \frac{4k^2}{4k^2 + q_d^2}\right]. \tag{5.35}$$

This result assumes that the scattering can flip the spin (e.g., it is thought that in piezoelectric scattering, the electron may scatter into either of the final two possible spin states at k'), although this is not well understood and may not be the case. Piezoelectric scattering predominately occurs at relatively low temperatures (where the equipartition approximation may not be valid).

5.4 Optical and intervalley scattering

In the tetrahedrally coordinated semiconductors, there are two atoms per unit cell site and optical mode vibrations are allowed, where the two atoms vibrate relative to each other. These phonons are rather energetic, being of the order of 30 to 50 meV (or more) in energy, and lead to *inelastic scattering processes*, since there is a significant gain or loss of energy by the carrier during the scattering process. The importance of the inelastic scattering processes is quite clear, since the previous processes were essentially elastic. Hence, we need the optical phonons to relax the energy that is obtained from the electric field. We now want to turn to the details of these inelastic processes. Although one normally thinks of scattering occurring just within a single minimum, or valley, of the band, these optical phonons can also cause *inter-valley* or *inter-band* scattering. Such examples are scattering from the light-hole valence band to the heavy-hole valence band by a mid-zone phonon near the Γ point, or a Γ-to-L valley scattering in the conduction band by a zone-edge optical (or high-energy acoustic) phonon.

5.4.1 Zero-order scattering

The scattering due to the optical phonons usually is characterized with a deformable ion model, in which the two sub-lattices are assumed to simply move relative to one another. The potential field of each ion is displaced slightly due to this movement, and this causes a resulting shift in the bond charges, which leaves a small excess positive charge where the ions have moved apart and a slight negative charge in the regions where they are closer together. This produces a macroscopic deformation field D, which is usually given in units of eV cm^{-1}. This scattering is a zero-order process, in that the resulting interaction potential is independent of the wave vector, or

$$\delta E = D u_q, \tag{5.36}$$

so that

$$|M(k, q)|^2 = \frac{\hbar D^2}{2V \rho_m \omega_q} [N_q \delta(E_k - E_{k+q} + \hbar \omega_q) \\ + (N_q + 1)\delta(E_k - E_{k-q} - \hbar \omega_q)] \tag{5.37}$$

The various parameters have their same meanings as for the earlier treatment of acoustic phonon scattering since, other than D, they come from the same source—the lattice wave u_q and the appropriate lattice harmonic oscillator. The two terms in the square brackets are for absorption and emission of a phonon, respectively, and the delta functions are to indicate the energy conservation although they have already been incorporated in the discussion leading to (5.5) and (5.6).

With optical-phonon scattering within a single band (or valley) through the long-wavelength phonons near the zone center, the dispersion relation for the optical modes is quite flat, with very little dependence on the magnitude of the wave vector q. This implies that a reasonable approximation is to take $\omega_q = \omega_0$ with the latter constant in the integrations over the phonon wave vectors. For inter-valley phonon scattering, or for scattering between different valence-band valleys, the dominant part of the phonon wave vector is quite large, so that no significant error is made by continuing to treat the frequency of the optical (or inter-valley) phonon as a constant. Moreover, the scattering is isotropic; that is, there is no q dependence in the matrix element (5.37) once ω_0 is taken as a constant. This means that one can use the density-of-states result (5.5), keeping the emission and absorption terms separate, to find the resulting scattering rate to be

$$\Gamma(k) = \sqrt{\frac{m_c}{2}} \frac{m_c D^2}{\pi \rho_m \omega_0 \hbar^3} [N_q (E + \hbar \omega_0)^{1/2} \\ + (N_q + 1)(E - \hbar \omega_0)^{1/2} u_0 (E - \hbar \omega_0)] \tag{5.38}$$

The last factor in the square brackets is the Heaviside step function $u_0(x)$ that has been added to ensure that the argument of the square root is positive; i.e., that carriers with an energy $E < \hbar \omega_0$ cannot emit a phonon. The optical phonon scattering rate calculated here is the mean free time for collisions, but because this process also relaxes both the energy and the momentum in a very efficient fashion, it is closely related to the relaxation times for the latter quantities. Other than the shift in the onset for the emission term, the energy dependence of the optical scattering is quite similar to that for the acoustic modes, but it is much more temperature dependent because of the complete form of the Bose–Einstein distribution N_q that is retained here.

For inter-valley scattering, (5.38) must still be multiplied by the number of final ellipsoids to which the carrier can scatter. However, this factor should not be included by using density-of-states effective mass m_d instead of the band edge mass that appears in the equation. This is because the density of states mass incorporates e.g. six ellipsoids for Si, whereas electrons in one ellipsoid can only scatter to at most five other ellipsoids. In Si, the intervalley process involves a number of both f and g phonons. The g phonons connect a valley along one axis with its opposite on the same (but negative) axis. The f phonons connect a valley along one axis with its four equivalents along the other two axes. For practical purposes, these can be all combined into a single effective high energy (LO-like) phonon and a single effective low energy (LA-like) phonon, which will be discussed further below. The high energy mode is a normal zero-order coupled phonon, while the low energy is normally forbidden. Hence, this latter phonon leads to a first-order interaction, discussed below. We will return to discussion of this inter-valley scattering in a later section.

5.4.2 First-order scattering

If the zero-order matrix element for the optical or inter-valley interaction vanishes, as is the case, for example, for the *umklapp* phonons (involving a reciprocal lattice vector from (5.2)) via the acoustic modes in Si, it is expected that D is identically equal to zero. However, the general electron–phonon interaction is an expansion in powers of q, and the zero-order interaction is just the q^0 order term. Moreover, the selection rules (to be discussed below) are strictly limiting only upon this zero-order interaction. In first-order interactions, a term arises that is of the form $\Xi_0 q \cdot e_q$. Here, Ξ_0 is the first-order optical coupling constant (in obvious agreement in notation with the acoustic deformation potential in section 5.2). In fact, this approach yields a form exactly like the acoustic deformation potential approach [9, 10], because the latter approach is also a first-order scattering process. It turns out that such an approach can also occur for the optical modes. To proceed, one can use directly (5.16), with the change in notation of the deformation potential and the constant frequency, as

$$|M(\boldsymbol{k}, \boldsymbol{q})|^2 = \frac{\hbar \Xi_0^2 q^2}{2\rho_m V \omega_0}(N_q + 1), \tag{5.39}$$

and an equivalent term for the absorption term. But, here there is a difference from the acoustic mode treatment of section 5.2, and that is the q dependence in the matrix element. Hence, we will have to use (5.6) and we need to generalize the matrix element. Nevertheless, we can write the scattering rate as (5.6), and insert (5.39) to give

$$\Gamma(k) = \frac{m_c \Xi_0^2}{4\pi\rho_m \hbar^2 \omega_0 k} \left\{ (N_q + 1) \int_{q_-^e}^{q_+^e} q^3 dq + N_q \int_{q_-^a}^{q_+^a} q^3 dq \right\}. \tag{5.40}$$

The integrations are straightforward, and the final scattering rate is just

$$\Gamma(k) = \frac{\sqrt{2} (m_c)^{5/2} \Xi_0^2}{\pi\rho_m \hbar^5 \omega_0} \left\{ N_q (2E + \hbar\omega_0) \sqrt{E + \hbar\omega_0} \right.$$
$$\left. + (N_q + 1)(2E - \hbar\omega_0) \sqrt{E - \hbar\omega_0} u_0 (E - \hbar\omega_0) \right\} \tag{5.41}$$

where the Heaviside step function has been added to the emission term to assure that the argument of the square root is positive. The first-order process has a much smaller magnitude at low energies, but has a much stronger energy dependence than the zero-order optical and inter-valley process. Thus, it is much weaker in normal situations, but can become the dominant process for energetic carriers at high electric fields. We will illustrate this in the following section.

5.4.3 Inter-valley scattering

Selection rules. When scattering occurs either within a single valley or band minimum, or between different valleys, whether equivalent or not, it is not always the case that just any old phonon will couple properly to move the carrier from the initial state to the final state. For example, the top of the valence band is predominantly formed from the anion p states, while the bottom of the conduction band at the Γ point is predominantly formed from the cation s states, as discussed in chapter 2. If an electron is going to scatter from the Γ point to the L point in the conduction band, for example, it is necessary that the cation atom be in motion (due to the phonon wave) in order to couple to the electron. We know that the cation motion for the L-point phonon mode is the LO mode if the cation is the lighter of the two atoms and the LA mode if it is the heavier of the two atoms [1]. Although this is a hand-waving argument, it can be placed on quite firm ground through group theory.

Space group selection rules are usually calculated by group-theoretical techniques. It is beyond the scope of this book to go through these techniques, so we merely summarize the macroscopic features of the arguments. If a given set of M physical quantities, such as the matrix elements coupling the carriers in different valleys by the phonons, are to be calculated, the selection rules determine the number of n_M independent matrix elements in the set M. For example, consider the required selection rule for an electron in Si, in which the transition is made from the valley located at $(k_0, 0, 0)$ to the valley at $(-k_0, 0, 0)$, where $k_0 = 0.857\pi/a$ (this is the point at which the minimum of the conduction band appears in figure 2.14). This transition has been termed a g-phonon (the details of the phonon scattering in Si are discussed later), and the selection rule can be written as

Table 5.1. Optical and intervalley selection rules.

Material	Intra-valley	Inter-valley
Si	Forbidden	g: $\Delta_{2'}$(LO)
		f: Σ_1(LA,TO)
Ge	Γ(LO)	X_1(LA,LO)
$A^{III}B^V$	Γ(LO)	$\Gamma \rightarrow L$: L^a
		$\Gamma \rightarrow X$: X^a
		$L \rightarrow L$: X^a

[a] LO if $m_V > m_{III}$, otherwise LA mode

$$\Delta_1(k_0) \otimes \Delta_1(-k_0) = \Delta_1(2k_0), \tag{5.42}$$

where Δ_1 represents the required symmetry for the electron wave function in the appropriate minimum of the conduction band and $\otimes$ represents a group-theoretical convolution operation. In short, the two wave functions on the left can only be coupled by a phonon with the wave function symmetry appearing on the right-hand side. The problem is that the wave vector on the right extends beyond the edge of the Brillouin zone and is therefore termed an *umklapp process*, as the wave vector must be reduced by a reciprocal lattice vector according to (5.2)—but, which reciprocal lattice vector? The point $2k_0$ lies on the prolongation of the (100) direction (Δ direction) beyond the X point into the second Brillouin zone (see figure 2.14). The symmetry Δ_1 passes over into a symmetry function $\Delta_{2'}$ as q passes the X point. Thus the desired phonon must have a wave vector along the (100) axis and have the symmetry $\Delta_{2'}$ for it to couple the two valleys discussed above. If there were no phonons of this symmetry, the transition would be forbidden to zero order, which is the coupling calculated in the previous section. Fortunately, the LO phonon branch has just this symmetry in Si, so that the desired phonon has $q \sim 0.3\pi/a$ and is an LO mode. As a second example, consider the scattering from the central Γ-point minimum in GaAs to the L valleys. The latter valleys lie some 0.29 eV above the central Γ minimum (see figure 2.12), and scattering to these valleys is the process by which inter-valley transfer occurs in this material. An electron that gains sufficient energy in the central valley can be scattered to the satellite valleys, where the mass is heavier and the mobility is much lower. Thus the symmetry operation is given by

$$\Gamma_1(0) \otimes L_1(-k_L) = L_1(k_L), \tag{5.43}$$

where $k_L = (\pi/2a, \pi/2a, \pi/2a)$ is the position of the L point in the Brillouin zone. This value of k_L is now the required phonon wave vector, and the required phonon must have the symmetry given by the right-hand side of (5.43). Not surprisingly, the branch with this symmetry is the LO branch if the cation atom is the lighter atom, and the LA branch if the cation atom is the heavier atom, just as the hand-waving argument above suggested. In table 5.1, the allowed phonons for the materials of interest are delineated, based on the proper group-theoretical calculations [11–13].

In general, study of the electron–phonon scattering process has progressed by using the deformation potential as an adjustable constant. This has been relatively successful, particularly since very few of these have ever been measured (other than for the acoustic modes). Hence, it is perhaps fruitful to review the nature of the understanding of the various scattering processes that are important in typical semiconductors, before turning to methods to actually compute the momentum dependent deformation potentials. First, only Si, Ge, and a few of the group III–V materials are reviewed, primarily because a full understanding of transport in nearly all semiconductors is still lacking. What is presented here is the state of understanding that currently exists, with a few of the speculations that appear in the literature.

Silicon. The conduction band of Si has six equivalent ellipsoids located along the Δ [these are the (100) axes] lines about 85% of the way to the zone edge at X. Scattering within each ellipsoid is limited to acoustic phonons and impurities (to be discussed below), as the intra-valley optical processes are forbidden, as indicated in table 5.1 above. Acoustic mode scattering, by way of the deformation potential, is characterized by two constants Ξ_u and Ξ_d, which are thought to have values of 9 eV and −6 eV, respectively [5]. The effective deformation potential is then the sum of these, or about 3 eV. Non-polar 'optical' scattering occurs for scattering between the equivalent ellipsoids. There are two possible phonons that can be involved in this process. One, referred to as the *g*-phonon, couples the two valleys along opposite ends of the same (100) axis. This is the *umklapp* process discussed above, and has a net phonon wave vector of $0.29\pi/a$. The symmetry allows only the LO mode to contribute to this scattering. At the same time, *f*-phonons couple the (100) valley to the (010) and (001) valleys, and so on. The wave vector has a magnitude of $2^{1/2}(0.85)$ $\pi/a = 1.2\pi/a$, which lies in the square face of the Brillouin zone (figure 2.7) along the extension of the (110) line into the second Brillouin zone. The phonons here are near the X-point phonons in value but have a different symmetry. Nevertheless, table 5.1 illustrates that both the LA and TO modes can contribute to the equivalent inter-valley scattering. Note that the energies of the LO *g*-phonon and the LA and TO *f*-phonons are all nearly the same value, while the low-energy inter-valley phonons are forbidden. Long [14], however, has found from careful analysis of the experimental mobility versus temperature that a weak low-energy inter-valley phonon is required to fit the data. In fact, he treats the allowed high-energy phonons by a single equivalent inter-valley phonon of 64.3 meV, but must introduce a low-energy inter-valley phonon with an energy of 16.4 meV. The presence of the low-energy phonons is also confirmed by studies of magneto-phonon resonance (where the phonon frequency is equal to a multiple of the cyclotron frequency) in Si inversion layers, which indicates that scattering by the low energy phonons is a weak contributor to the transport [15]. The low-energy phonon is certainly forbidden and Long treats it with a very weak coupling constant. Ferry [9] points out that the forbidden low-energy inter-valley phonon must be treated by the first-order interaction and fits the data with a coupling constant of $\Xi_0 = 5.6$ eV, while the allowed transition is treated with a coupling constant of $D = 9 \times 10^8$ eV cm^{-1}. The scattering rate for these effective phonons are shown in figure 5.4. The fit of these

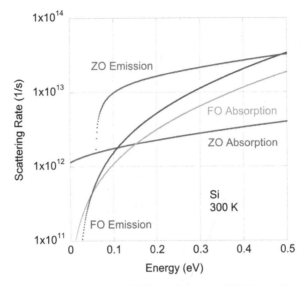

Figure 5.4. Scattering rates for the zero-order (ZO) and first-order (FO) inter-valley phonons in silicon.

scattering rates to the experimentally measured temperature dependence of the Si mobility is shown in figure 5.5. There are few experimental data to confirm these values directly, so they must be taken merely as an indication of the order of magnitude to be expected for these interactions. However, when used in Monte Carlo simulations, they fit quite closely to results computed with a full-band structure used in the calculations (discussed further in chapter 7), although a value of Ξ_0 closer to 6 eV seems to give better behavior. The energy dependence of the first-order phonon leads it to be very important at high electric fields, where the carriers gain significant energy (chapter 7).

The valence band has considerable anisotropy, and the degeneracy of the bands at the zone center can be lifted by strain. Nevertheless, the acoustic deformation potential is thought to have an effective value of about 2.5 eV. Optical modes can couple holes from one valence band to the other, but there is little information on the strength of this coupling.

Germanium. The conduction band of germanium has four equivalent ellipsoids located at the zone edges along the (111) directions—the L points. The acoustic mode is characterized by the two deformation potentials, Ξ_u and Ξ_d, which are thought to have values of about 16 and −9 eV, respectively. These lead to an effective coupling constant of about 9 eV. Optical intra-valley scattering is allowed by the LO mode. Equivalent inter-valley scattering is also allowed by the X-point LA and LO phonons (which are degenerate). The coupling constant for these phonons is fairly well established at 7×10^8 eV cm^{-1} from studies of the transport at both low fields (as a function of temperature) and at high fields [16].

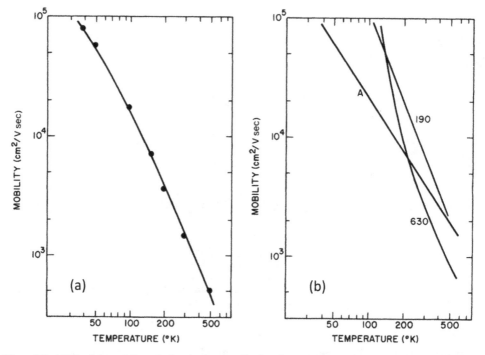

Figure 5.5. (a) Fit of the mobility calculated using the effective phonons, discussed in the text to the mobility of Long [14]. (b) Contribution from the various phonons: 630 refers to the equivalent temperature, $T = \hbar\omega_0/k_B$, of the high energy zero-order coupled phonon, 190 refers to the first-order coupled low energy mode, and A refers to long wavelength acoustic (elastic) phonons. Reprinted with permission from [9], copyright 1976 by the American Physical Society.

The holes in Ge are characterized by the anisotropic valence bands, just as in Si, and the acoustic deformation potentials are very close to the values for Si. Again, little is known about the coupling constants for inter-valence band scattering by optical modes.

Group III–V compounds. In GaAs, InP, and InSb, the acoustic deformation potential is about 7 eV, although many other values have been postulated in the literature. The conduction band is characterized by Γ, L, X ordering of the various minima (figure 2.12). The transport in the Γ valley is dominated by the polar LO mode scattering (discussed below), while at sufficiently high energy, the carriers can scatter to the L and X minima through non-equivalent inter-valley scattering. The deformation fields for these two processes are fairly well established in GaAs to be 7×10^8 and 1×10^9 eV cm^{-1}, respectively [17], through both experimental measurements and theoretical calculations (even though debate has not subsided in the literature). InP is thought to have the same values [18]. The L valleys are similar to those of Ge, so the L-L scattering should be given by the Ge values. L-X scattering is thought to have a deformation field of 5×10^8 eV cm^{-1}, although there is no real experimental evidence to support this. The Γ-L scattering rate in InSb is thought to be somewhat stronger, on the order of 1×10^9 eV cm^{-1} [19, 20].

Again, the holes are characterized by anisotropic valence bands, but in GaAs it is thought that the dominant acoustic deformation potential is about 9 eV, while inter-valence band scattering has been treated through both the polar LO mode and the non-polar TO mode. The latter is thought to have a deformation field of about 1×10^9 eV cm^{-1}, and this value has been used in some discussions of transport [21].

Some comments. While these fitting procedures to available experimental data are useful, they are typically only really reasonable for scattering exactly at the critical Brillouin zone position. If the electrons are away from e.g. the Γ, X, and L points, then the deformation potential will quite likely take on different values. That is, it is quite common for the deformation potential to be momentum dependent through-out the Brillouin zone, and this is not captured in the fitting procedure above. Consequently, one can compute the actual deformation potential using the proce-dures of chapter 3. First, the energy bands are computed using a first principles approach or an empirical pseudo-potential approach, as described in chapter 2. Then, the crystal potential of the deformed crystal, due to a lattice vibration, is computed along with the change in the energy bands. Usually, this is done by a rigid shift according to the rigid-ion model [22]. In this approach it is assumed that the ionic potentials move rigidly with the ions. This affects the pseudo-potential calculation in several ways. First, the positions of the atoms in the unit cell are modified by the shift, and this also changes the form factors. The latter requires knowing the form factors for all Fourier values, not just at the few discussed in chapter 2, which leads to estimates for the actual functions. These are obtained by e.g. spline fits to the 'known' values. This procedure is repeated for a series of displacements around the equilibrium values. At each point the new energy structure is computed, and the set of energies are fit to the displacement as a functional, whose first order coefficients yield the optical deformation potentials.

Modifications to the rigid-ion approach have been made [23], and this has led to deformation potentials somewhat larger than found by other workers. A careful study of the inter-valley scattering deformation potentials using this latter approach has been made by Zollner *et al* [24]. The rigid-ion approach has also been used for quantum wells [25] and for GaN [26]. Exact calculations of the deformation potentials often accompany full-band transport simulations. The full band approaches in semiconductors were first used by Shichijo and Hess [27], as pointed out in chapter 1. It was then developed into a significant simulation package by Fischetti and Laux at IBM [28]. These approaches have a common similarity with the cellular Monte Carlo [29], which utilizes a scattering formulation based upon the initial and final momentum states and can thus take into account this momentum dependent coupling strength to improve the Monte Carlo approach. In the cellular approach, one is concerned with scattering between particular states in the Brillouin zone, and not so much with the energy dependence of the scattering process. Thus, one replaces the integration over the entire Brillouin zone that appears in (5.4) and (5.6) with an integration only over the small cell representing the final state in the discretized Brillouin zone. Hence, a table can be constructed in matrix format, so that any entry Γ_{ij} gives the scattering rate from cell i to cell j.

Finally, it should be remarked that these deformation potentials do not need to be screened by the free carriers. By their nature, arising from the actual band structure, they have already been screened by the bonding (valence) electrons.

5.5 Polar optical phonon scattering

The non-polar optical phonon interactions discussed in the previous sections arose through the deformation of the energy bands. This led to a macroscopic deformation potential or field. In compound semiconductors, the two atoms per unit cell have differing charges, and the optical phonon interaction involving the relative motion of these two atoms has a strong Coulomb potential contribution to the interaction. We encountered this in section 3.4 where this Coulomb interaction led to the TO mode coupling to an electro-magnetic wave and modifies the dielectric function by the dispersion between the long-wavelength LO and TO modes near the zone center. In fact, a split arises between the frequencies of these two modes due to the interaction of the effective charges on the atoms of the lattice in polar semi-conductors. Of interest here, however, is the fact that the LO mode of lattice vibration is a very effective scattering mechanism for electrons in the central valley of the group III-V and II-VI semiconductors. Particularly, in the central Γ conduction band valley, the non-polar interaction is generally weak, and the polar interaction can be dominant. It can also be effective for holes, although the TO non-polar interaction can be quite effective and compete with the polar interaction. In terms of the expansion in orders of q mentioned previously, the polar interaction is q^{-1} which arises from its Coulombic nature.

The polarization of the dipole field, that accompanies the vibration of the polar mode is given essentially by the effective charge times the displacement. The latter is just the phonon mode amplitude u_q, which has been used previously. Hence, we can write the polarization as

$$P_q = \sqrt{\frac{\hbar}{2\gamma V \omega_0}}\, e_q \left(a_q^\dagger e^{-iqr} + a_q e^{iqr} \right), \tag{5.44}$$

where e_q is the polarization unit vector for the mode vibration, a^+ and a are creation and annihilation operators for mode q, and the effective interaction parameter (which is related to the effective charge) is, from section 3.4,

$$\frac{1}{\gamma} = \omega_0^2 \left(\frac{1}{\varepsilon_\infty} - \frac{1}{\varepsilon(0)} \right). \tag{5.45}$$

Here, ε_∞ and $\varepsilon(0)$ are the high-frequency and low-frequency dielectric permittivities, respectively (this difference gives the strength of the polar interaction, and vanishes in non-polar materials where these two values are equal). Comparing (5.44) with (5.37), we see that (5.45) replaces the value D^2/ρ. This polarization leads to a local electric field, which is a longitudinal field in the direction of propagation of the phonon wave, and it is this field that scatters the carriers. The interaction energy arises from this polarization, in a similar manner to the piezoelectric interaction (which is the acoustic mode corresponding to this Coulomb interaction) in terms of the polarization and the interaction field. These lead to a screened version of the polar interaction, in which the perturbing energy is given as

$$\delta E = \sqrt{\frac{\hbar e^2}{2\gamma V \omega_0}} \frac{q}{q^2 + q_d^2} \left(a_q^\dagger e^{-iqr} - a_q e^{iqr}\right) e^{i\omega t}. \tag{5.46}$$

It should be remarked here, in keeping with the use of a simple screening (discussed in the piezoelectric scattering), the harmonic motion of the phonon can lead to a reduction of the screening, so that q_d would be smaller than the Debye screening length. In this case, the phonon energy is often comparable to the electron energy. A good approximation, however, is to ignore this and use the Debye screening value. It should be emphasized that this is a very simple approximation to the full dynamic screening, and its validity has not been tested. Use of (5.46) leads to the matrix element

$$|M(k, q)|^2 = \left(\frac{\hbar e^2}{2\gamma V \omega_0}\right) \frac{q^2}{\left(q^2 + q_d^2\right)^2} [(N_q + 1)\delta(E_k - E_{k-q} - \hbar\omega_0)$$

$$+ N_q \delta(E_k - E_{k+q} + \hbar\omega_0)] \tag{5.47}$$

Again, the delta functions have been included, although they are already taken into account in the derivations of the previous scattering rates in the final integrals rather than in just the matrix elements. This result is now inserted into (5.6) to give the scattering rate as

$$\Gamma(k) = \frac{m_c e^2}{4\pi\gamma\hbar^2 k\omega_0} \left[(N_q + 1)\int_{q_e^-}^{q_e^+} \frac{q^3 dq}{\left(q^2 + q_d^2\right)^2} + N_q \int_{q_a^-}^{q_a^+} \frac{q^3 dq}{\left(q^2 + q_d^2\right)^2}\right]. \tag{5.48}$$

The limits for the emission and absorption terms are given by (5.11) and (5.9), respectively. The final result for the screened interaction is

$$\Gamma(k) = \frac{m_c e^2 \omega_0}{4\pi\hbar^2 k} \left(\frac{1}{\varepsilon_\infty} - \frac{1}{\varepsilon(0)}\right)\left[G(k) - \frac{q_d^2}{2}H(k)\right], \tag{5.49}$$

where

$$G(k) = (N_q + 1)\ln\left[\frac{\left(k + \sqrt{k^2 - q_0^2}\right)^2 + q_d^2}{\left(k - \sqrt{k^2 - q_0^2}\right)^2 + q_d^2}\right]$$

$$+ N_q\ln\left[\frac{\left(\sqrt{k^2 + q_0^2} + k\right)^2 + q_d^2}{\left(\sqrt{k^2 + q_0^2} - k\right)^2 + q_d^2}\right] \quad (5.50a)$$

$$H(k) = (N_q + 1)\frac{4\sqrt{k^2 - q_0^2}}{\left(q^2 + q_d^2\right)^2 + 4k^2 q_d^2} + N_q\frac{4\sqrt{k^2 + q_0^2}}{\left(q^2 + q_d^2\right)^2 + 4k^2 q_d^2}, \quad (5.50b)$$

$$q_0^2 = \frac{2m_c\omega_0}{\hbar}. \quad (5.50c)$$

The emission terms should be multiplied by the Heavyside function to assure that they occur only when the carrier energy is larger than the phonon energy. The value q_0 is the so-called 'dominant phonon' wave vector, and can be used to estimate the reduction in screening that can occur. If this is done, the Debye wave vector q_d is reduced at most by a factor of $2^{1/2}$.

Screening plays a significant role in the scattering of carriers by the polar optical phonon interaction. In both terms, the screening wave vector acts to reduce the amount of scattering that occurs. In the first term, the screening wave vector works to reduce the magnitude of the ratio of terms that occurs inside the logarithm arguments, hence reducing the scattering strength. The second term is negative, which also reduces the strength. In the absence of screening, where $q_d \sim 0$ (which occurs at very low densities of free carriers), the equation above reduces to the more normal form

$$\Gamma(k) = \frac{m_c e^2 \omega_0}{4\pi\hbar^2 k}\left(\frac{1}{\varepsilon_\infty} - \frac{1}{\varepsilon(0)}\right)\left[(N_q + 1)\ln\left(\frac{k + \sqrt{k^2 - q_0^2}}{k - \sqrt{k^2 - q_0^2}}\right)\right.$$

$$\left. + N_q\ln\left(\frac{\sqrt{k^2 + q_0^2} + k}{\sqrt{k^2 + q_0^2} - k}\right)\right]. \quad (5.51)$$

It is assumed here that spin degeneracy of the final states has not been taken into account in the pre-factors, e.g., and that spin is preserved through the scattering process. In figure 5.6, the various scattering rates for electrons in $In_{0.53}Ga_{0.47}As$ grown on InP are shown. It is clear that the polar scattering dominates the acoustic scattering in this, as well as most, III-V compounds.

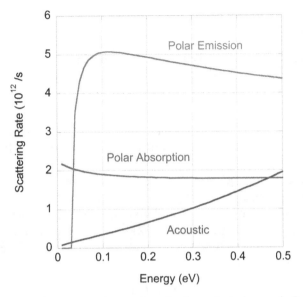

Figure 5.6. Scattering rates for unscreened acoustic and polar optical phonons in $In_{0.53}Ga_{0.47}As$, lattice matched to InP at 300 K.

5.6 Other scattering mechanisms

There are several other scattering processes that can occur in semiconductors that are quite important, but do not involve the phonons from the lattice vibrations. We discuss a number of these important processes in this section. However, we will not discuss electron–electron scattering here, but will deal with it in chapter 9, as we need some details from the dielectric function that will be treated in that chapter as well.

5.6.1 Ionized impurity scattering

In any treatment of electron scattering from the Coulomb potential of an ionized impurity atom, it is necessary to consider the long-range nature of the potential. If the interaction is summed over all space, the integral diverges and a cutoff mechanism must be invoked to limit the integral. One approach is just to cut off the integration at the mean impurity spacing, the so-called Conwell–Weisskopf [30] approach. A second approach is to invoke screening of the Coulomb potential by the free carriers, which was done for piezoelectric scattering. In this case, the potential is induced to fall off much more rapidly than a bare Coulomb interaction, due to the Coulomb forces from the neighboring carriers. The screening is provided by the other free carriers, which provides a background of charge. This is effective over a distance on the order of the Debye screening length in non-degenerate materials. This screening of the repulsive Coulomb potential results in an integral (for the scattering cross section) which converges without further approximations [31].

For spherical symmetry about the scattering center, or ion location, the potential is screened in a manner that gives rise to a screened Coulomb potential, as

$$V(r) = \frac{e}{4\pi\varepsilon_\infty r}e^{-q_d r},$$
(5.52)

where the Debye wave vector q_d is the inverse of the screening length, and is given by

$$q_d^2 = \frac{ne^2}{\varepsilon_\infty k_B T}.$$
(5.53)

Here ε_∞ is the high-frequency permittivity. Generally, if both electrons and holes are present, n is replaced by the summation $n + p$. The above results are for a non-degenerate semiconductor.

In treating the scattering from the screened Coulomb potential, we will take a slightly different approach than merely following the use of the Fermi golden rule. We use a wave scattering approach and compute the *scattering cross section* $\sigma(\vartheta)$, which gives the angular dependence of the scattering. It is assumed the incident wave is a plane wave, and the scattered wave is also a plane wave. Thus the total wave function can be written as

$$\psi(r) = e^{ikz} + U(r)e^{ik' \cdot r}.$$
(5.54)

It is assumed that the incident wave vector k is oriented along the polar (z) axis, and the second term on the right in (5.54) is the scattered wave, whose amplitude U has to be determined. We begin this process by inserting (5.54) into the Schrödinger equation and keep only useful terms up to first order in the scattering vector, so that

$$-\nabla^2 U(r) + k'^2 U(r) = \frac{2m_c}{\hbar^2}\frac{e^2}{4\pi\varepsilon_\infty r}e^{-q_d r}e^{ikz},$$
(5.55)

where an additional factor of e has been added to the Coulomb potential to produce the scattering energy. If the terms on the right-hand side are treated as a charge distribution, the normal results from electromagnetic field theory can be used to write the solution as

$$U(r) = -\frac{m_c e^2}{8\pi^2\varepsilon_\infty\hbar^2}\int\frac{d^3r'}{r'|r - r'|}e^{-q_d r' + ikz'}e^{ik'|r - r'|}.$$
(5.56)

To proceed, it is assumed that $r \gg r'$, and the polar axis in real space is taken to be aligned with r. Further, the scattering wave vector is taken to be $q = k - k'$, so that

$$\int_0^\pi \sin\vartheta\, d\vartheta e^{iq \cdot r'} = \frac{\sin(qr')}{qr'},$$
(5.57)

the φ integration can be done immediately, and the remain integration becomes

$$U(r) = -\frac{m_c e^2}{2\pi\varepsilon_\infty\hbar^2 qr}\int_0^\infty \sin(qr')e^{-q_d r'}dr'$$
$$= \frac{m_c e^2}{2\pi\varepsilon_\infty\hbar^2 qr\left(q^2 + q_d^2\right)}.$$
(5.58)

Now $q = k - k'$, but for elastic scattering $k = k'$, and $q = 2k \sin(\vartheta/2)$, where ϑ is the angle between $\mathbf{k}$ and $\mathbf{k'}$. If we write the scattered wave function $V(\mathbf{r})$ as $f(\vartheta)/r$, then we recognize that the factor $f(\vartheta)$ is the matrix element, and the cross-section is defined as

$$\sigma(\vartheta) = |f(\vartheta)|^2 = \left(\frac{m_c e^2}{8\pi\varepsilon_\infty \hbar^2 k^2} \right)^2 \frac{1}{\left[\sin^2(\vartheta/2) + \left(q_d^2/4k^2 \right) \right]^2}. \tag{5.59}$$

The total scattering cross-section (*for the relaxation time*) is found by integrating over ϑ, weighting each angle by an amount $(1 - \cos\vartheta)$. This last factor accounts for the momentum relaxation effect. The dominance of small-angle scattering prevents each scattering event from relaxing the momentum so this factor is inserted by hand (this factor is not necessary for inelastic processes, and averages to zero for isotropic elastic processes). This is one of the few scattering processes where this factor is included, primarily because each scattering event lasts for a quite long time, and it is necessary to calculate the average momentum loss rate. We note, however, that in transport simulations using the ensemble Monte Carlo process, it is the total scattering rate that is included, not the momentum relaxation rate. Hence, this extra factor is not included in such simulations. Finally, we get the total cross-section as

$$\begin{aligned} \sigma_c &= 2\pi \int_0^\pi \sigma(\vartheta)(1 - \cos\vartheta)d\vartheta \\ &= 16\pi \int_0^\pi \sigma(\vartheta/2)\sin^3(\vartheta/2)d(\sin\vartheta/2) \\ &= \frac{\pi}{2} \left(\frac{m_c e^2}{2\pi\varepsilon_\infty \hbar^2 k^2} \right)^2 \left[\ln\left(\frac{1 + \beta^2}{\beta^2} \right) - \frac{1}{1 + \beta^2} \right] \end{aligned} \tag{5.60}$$

where $\beta = q_d/2k$. The scattering rate is now the product of the cross-section, the number of scatterers, and the velocity of the carrier, or

$$\Gamma(k) = n_{\text{imp}}\sigma_c v = \frac{n_{\text{imp}} m_c e^4}{8\pi\varepsilon_\infty^2 \hbar^3 k^3} \left[\ln\left(\frac{4k^2 + q_d^2}{q_d^2} \right) - \frac{4k^2}{4k^2 + q_d^2} \right]. \tag{5.61}$$

As mentioned above, the actual scattering rate, and not the relaxation rate for momentum, is required in Monte Carlo simulation programs, as developed in chapter 7. In this situation, (5.60) must be modified by removal of the $(1 - \cos\theta)$ term, that produces a $\sin^2(\vartheta/2)$ in the integral in (5.60). This results in the entire set of terms in the square brackets in (5.61) being replaced by the factor

$$\left[\frac{4k^2}{q_d^2 \left(4k^2 + q_d^2 \right)} \right], \tag{5.62}$$

which dramatically changes the energy dependence for small k ($\ll q_D$). The form (5.61) is the one normally found when discussions of mobility and diffusion constants are being evaluated for simple transport in semiconductors (as in chapter 6). However, when weighing various random processes for Monte Carlo approaches, it is the total scattering rate that is important, and this is given by the use of (5.62).

A further important word of caution has to be made about the assumption that $r \gg r'$ just below (5.56). In many semiconductors, this approximation fails, and one must proceed in a different manner. We will discuss in chapter 7 treating the impurity interaction in real space and not via perturbation theory. That is, we take the actual charge of the impurities and the electrons and consider the interaction between (and between the various electrons) in real space using a molecular dynamics approach. The reason for this lies in the fact that studies in GaAs using this approach have shown that the impurity scattering is as much as 50% higher than the above formulas would predict in GaAs at 77 K, for electron densities in the 10^{16}–10^{18} cm^{-3} range [32]. This is because the typical electron violates the assumption mentioned and may be interacting with several impurities at the same time. This means that an approach considering an electron interacting with one impurity at a time fails.

5.6.2 Coulomb scattering in two dimensions

If the Coulomb scatterer is near an interface, the problem becomes more complicated. This is particularly the case for charged scattering centers near the Si–SiO$_2$ (or any semiconductor–insulator) interface as well as in mesoscopic structures. In general, there are always a large number of Coulomb centers near the interface, due to disorder and defects in the crystalline structure in the neighborhood of the interface. In many cases, these defects are associated with dangling bonds and can lead to charge trapping centers which scatter the free carriers through the Coulomb interaction. The Coulomb scattering of carriers lying in an inversion layer (or a quantized accumulation layer) at the interface differs from the case of bulk impurity scattering, due to the reduced dimensionality of the carriers.

Coulomb scattering of surface quantized carriers was described first by Stern and Howard [33] for electrons in the Si-SiO$_2$ system. Since then, many treatments have appeared in the literature, which differ little from the original approach. In general, the interface is treated as being abrupt and as having an infinite potential discontinuity in the conduction (or valence) band, so that problems with interfacial non-stoichiometry and roughness are neglected (the latter is treated below as an additional scattering center). In treating this scattering, it is most convenient to use the electrostatic Green's function for charges in the presence of a dielectric interface, so that the image potential is properly included in the calculation. The scattering matrix element involves integration over plane-wave states for the motion parallel to the interface, and thus one is led to consider only the two-dimensional transform of the Coulomb potential [33]

$$G(q, z - z') = \begin{cases} \dfrac{1}{2q\varepsilon_s}\left(e^{-q|z-z'|} + \dfrac{\varepsilon_s - \varepsilon_{ox}}{\varepsilon_s + \varepsilon_{ox}}e^{-q|z+z'|}\right), & z > 0, \\[4mm] \dfrac{1}{q(\varepsilon_s + \varepsilon_{ox})}e^{-q|z-z'|}, & z < 0. \end{cases} \tag{5.63}$$

Here, q is the two-dimensional scattering vector and ε_s and ε_{ox} are the high-frequency total permittivities for the semiconductor and the oxide, respectively. Equation (5.63) assumes that the scattering center is located a distance z' from the interface (the semiconductor is located in the space $z > 0$) and has an image charge at $-z'$. If the interface were non-abrupt, the ratio of dielectric constants that appears in the second term on the right of the first line of (5.63) would be a function of q.

The Coulomb potential in (5.63) is still unscreened, and it is necessary to divide this equation by the equivalent factor that appears in (5.58): for example,

$$q \rightarrow \sqrt{q^2 + q_0^2}, \tag{5.64}$$

where q_0 is the appropriate two-dimensional screening vector in the presence of the interface, for example

$$q_0 = \frac{n_s e^2}{2\pi(\varepsilon_s + \varepsilon_{ox})k_B T}, \tag{5.65}$$

for a non-degenerate semiconductor [34]. Here, n_s is the sheet carrier concentration in the inversion (or accumulation) layer. More complicated screening approaches are possible, and were considered by Stern and Howard [33], but these are beyond the introductory scope of the present approach.

We note that the second line of (5.63) allows for the situation of remote dopants and trapped charge within an oxide to cite just two example cases of charge not in the 2D electron gas. Here, for example, we refer to the situation in which a monolayer of graphene or of a transition-metal di-chalcogenide is laid on an oxidized silicon substrate. Trapped charge in the oxide or on the oxide surface, adjacent to the monolayer, acts as a remote impurity and scatters the carriers. In this case, the Coulomb potential term is modified by the set-back distance d (of the charge from the monolayer centroid) with the factor [35, 36]

$$e^{-qd}. \tag{5.66}$$

This can be a dominant scattering process in these monolayer materials [37, 38].

For two-dimensional scattering, the scattering cross-section is determined by the matrix element of the screened Coulomb interaction for a charge located at z', with $q = k - k'$ being the difference in the incident and the scattered wave vector as previously. Now, however, the motion in the direction normal to the interface must also be accounted for, as it is not in the two-dimensional Fourier transform. This leads to

$$\langle k|V(z)|k'\rangle = e^2 \int_0^\infty |\zeta(z')|^2 G(q, z - z')dz', \tag{5.67}$$

where $\zeta(z)$ is the z portion of the wave function; that is, we write this wave function as

$$\psi(x, y, z) = \zeta(z)e^{i(k_x x + k_y y)}. \tag{5.68}$$

It is not a bad assumption to consider only scattering from charges located at the interface itself (i.e., $z' = 0$), since this is the region at which the density of scattering charges is usually large. In this idealization, charges are assumed to be uniformly distributed in the plane $z' = 0$. Then, we need only the second line of (5.63), and the scattering rate is

$$\Gamma(k) = \frac{n_{imp}e^4 m_c}{4\pi\hbar^3(\varepsilon_s + \varepsilon_{ox})^2} \int_0^{2\pi} A^2(\vartheta)\frac{(1 - \cos\vartheta)d\vartheta}{q^2 + q_0^2}, \tag{5.69}$$

where

$$A(\vartheta) = \int_0^\infty |\zeta(z')|^2 e^{-qz}dz. \tag{5.70}$$

Here again, the scattering is elastic and $q = 2k\sin(\vartheta/2)$. At this point it is necessary to say something about the envelope function $\zeta(z)$ in order to proceed. In the lowest sub-band, it is usually acceptable to take the wave function as a variational wave function of the form [33]

$$\zeta(z) = 2b^{3/2}ze^{-bz}, \tag{5.71}$$

which leads to an average thickness of the inversion layer of $3b/2$. With this form, (5.70) can be easily evaluated and (5.69) becomes

$$\Gamma(k) = \frac{8b^3 n_{imp}e^4 m_c}{\pi\hbar^3(\varepsilon_s + \varepsilon_{ox})^2} \int_0^\pi \frac{\sin^2(\vartheta)d\vartheta}{[4k^2\sin^2(\vartheta) + q_0^2][2k\sin\vartheta + 2b]^3}, \tag{5.72}$$

where the substitution $\vartheta/2 \to \vartheta$ has been made. (Although (5.71) is usually applied to the triangular potential (infinite wall at the interface, and linear rising potential inside the semiconductor), it can be applied to any potential shape, as it is a general form used to select b by minimizing the energy eigen-value.) In general, the peak of the wave function lies only a few nanometers from the interface, and then dies off exponentially, so that it represents electrons localized in a plane parallel to the interface. The factor b can be a significant fraction of the Brillouin zone boundary distance. For this reason it can generally be assumed that $b \gg k$, so that $A(q)$ is near unity.

Two limiting cases may be found from (5.72), with the approximation of $A(q)$ near unity. For $q_0 \ll q$, the behavior is essentially unscreened, and the integral yields $\pi/4k^2$. In this case, the scattering rate is inversely proportional to the square of the wave vector, which may be assumed to be near the Fermi wave vector for a degenerate inversion layer. Thus the mobility actually increases as the inversion density increases, since the average energy (and hence the average wave vector) increases faster than linear with the density. At the other extreme, $q_0 \gg q$, the scattering is heavily screened by the charge in the inversion layer. In this case, the wave-vector dependence disappears, and the integral yields only $\pi/2q_0^2$, so that the density dependence also disappears from the equation and thus the scattering rate becomes constant.

What if it is not desired to omit the dependence on b above? How does one determine the value for b? It was remarked above that b is a variational parameter. One problem is the form of the potential (band bending) in the semiconductor. Stern and Howard [33] used a Hartree potential, in which the band bending was determined self-consistently by including the potential of the charge itself through Poisson's equation. This is beyond the level of the approach desired here. Instead, one often just uses the triangular potential to give a field $E = e(N_{\text{dep}} + n_s/2)/\varepsilon_s$, where the first term is the depletion charge at the interface and the second is the inversion density. The procedure is the a standard one in mathematical physics, and proceeds by (1) inserting the assumed wave function into the Schrödinger equation, (2) multiplying by its complex conjugate and integrating over all space, and (3) varying b to minimize the energy. This procedure yields

$$b = \left(\frac{3eEm_c}{2\hbar^2} \right)^{1/3}. \tag{5.73}$$

This relationship then gets the dependence on the inversion density into (5.72) and the resulting scattering rate. This gives a density dependence over and above that from the average value of the wave vector k. Some numbers may give further insight. If we assume an inversion layer in Si, with $n_s = 10^{12}$ cm^{-2}, then the effective field is about 0.15 MV cm^{-1} and $b \sim 4 \times 10^6$ cm^{-1}. On the other hand, k_F is about 2.5×10^6 cm^{-1}. The approximation of assuming a very large value for b may not be appropriate for such situations. However, the situation improves at lower densities.

A slightly different form is found for graphene, with its linear Dirac-like bands (see chapter 2). In this case, the electrons are referred to as mass-less fermions, since the so-called rest mass in the relativistic formulation is zero, but the carriers have a dynamic mass found from the energy

$$E = \hbar v_F k, \tag{5.74}$$

where v_F plays the role of the speed of light. Then, the dynamic effective mass may be found as

$$m_c = \frac{\hbar k}{v_F} = \frac{E}{v_F^2}. \tag{5.75}$$

The peculiarity of the energy bands changes the screening and the final scattering process. The screening wave number (5.65) becomes

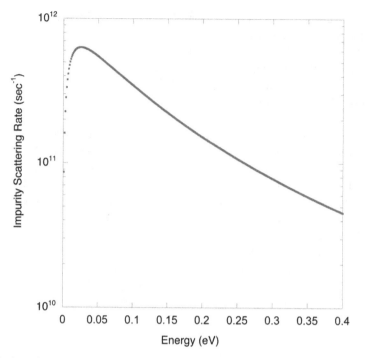

Figure 5.7. The impurity scattering rate in graphene, due to the remote impurities residing on the SiC insulator upon which the graphene sheet was deposited.

$$q_0 = \frac{e^2 k}{2\pi \hbar v_F (\varepsilon_s + \varepsilon_{ox})}. \tag{5.76}$$

Then, the scattering rate due to impurities residing in the underlying oxide will incorporate the set-back term (5.66), and we have [39]

$$\Gamma(k) = \frac{n_{imp} e^4}{4\pi \hbar^2 v_F k (\varepsilon_s + \varepsilon_{ox})^2} \int_0^{\pi/2} d\vartheta \frac{\sin^2(\vartheta)}{[\sin \vartheta + q_0/2k]^2} e^{-4kd \sin \vartheta}. \tag{5.77}$$

As before, the substitution $\vartheta/2 \rightarrow \vartheta$ has been made, and the high b limit has been taken. There are still some subtle differences. The denominator has a somewhat different form, and the $\sin^2(\vartheta)$ term arises from a different source. The normal $1 - \cos \vartheta$ has been left out, but there is an additional $1 + \cos \vartheta$ in graphene which accounts for the forbidden nature of the back-scattering process. In figure 5.7, we show the impurity scattering rate for graphene on SiC, with 2.5×10^{10} impurities.

5.6.3 Surface-roughness scattering

In addition to Coulomb scattering, short-range scattering associated with the interfacial disorder also limits the mobility of quasi-two-dimensional electrons at the interface. High-resolution transmission electron micrographs of the interface between Si and SiO_2 has shown an interface between the two materials that is relatively sharp with a fluctuation on the atomic level [40], as shown in figure 5.8. The fact that the interface is not abrupt on the atomic level, but that variation in the actual position of the interfacial plane can extend over one or two atomic layers along the surface, affects the transport. The local atomic interface actually has a random variation which, coupled with the surface potential, gives rise to fluctuations of the energy levels in any quantum well formed by the potential barrier to the oxide and the band bending in the semiconductor. The randomness induced by the interfacial roughness has some similarity to alloy scattering, treated in the next section, and can lead to limitations on the mobility of the carriers in the inversion layer. At present, a general calculation of the scattering rate based on the microscopic details of the roughness does not exist. Instead, the usual models rely on a semi-classical approach in which a phenomenological surface roughness is parameterized in terms of its height and the correlation length.

In current surface roughness models, displacement of the interface from a perfect plane is assumed to be described by a random function $\Delta(\mathbf{r})$, where $\mathbf{r}$ is a two-dimensional position vector parallel to the interface (the average interface). This model assumes that $\Delta(\mathbf{r})$ varies slowly over atomic dimensions so that the boundary conditions on the wave functions can be treated as abrupt and continuous. This assumption is obviously in error when surface fluctuations occur on the atomic level. However, the model has proven to provide quite good agreement with measured mobility variations in a variety of materials and interface. The scattering potential may be obtained by expanding the surface potential in terms of $\Delta(\mathbf{r})$ as

$$\delta V(\mathbf{r}, z) = V(z + \Delta(\mathbf{r})) - V(z) = eE(z)\Delta(\mathbf{r}) \cdots, \qquad (5.78)$$

where $E(z)$ is the electric field in the inversion layer itself. The scattering rate for the perturbing potential must include the role of the correlation between the scattering centers along the interface. This correlation is described by a Fourier transform $\Delta(\mathbf{q})$, discussed below, where $\mathbf{q}$ is also a two-dimensional vector along the interface. This leads to the scattering matrix element

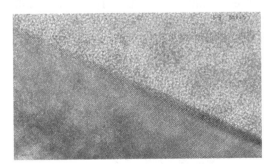

Figure 5.8. A high resolution TEM image of the SiO_2-Si interface. The crystalline Si is to the lower left of the image, and the rows of Si atoms can be distinguished. The image lies in the (111) plane of planar surface of the image. The interface itself is a (100) plane. The techniques used to get the image as well as to determine the roughness are described in [40]. Used with permission from Goodnick.

$$M(k, q) = e\Delta(q) \int_0^{\infty} E(z)|\zeta(z)|^2 dz$$

$$= e^2\Delta(q)\frac{N_{depl} + n_s/2}{\varepsilon_s} \quad , \tag{5.79}$$

where the orthonormality of the wave function has been used and the *average* electric field has been introduced from the discussion just above (5.73). The inversion density appears with the factor of 2 since it is the field at the interface that is of interest. The inversion charge appears almost entirely at the interface, so that it creates a field on each side, of which only one-half of the total field discontinuity appears at the interface. The factor $\Delta(\mathbf{q})$ is the Fourier transform of $\Delta(\mathbf{r})$.

In the matrix element, only the statistical properties of $\Delta(\mathbf{q})$ need be considered. Thus the descriptors discussed earlier may be introduced. There is some debate about the form of the positional auto-correlation function for the interface roughness. In most of the early work, it was assumed to describable by a Gaussian, given by

$$\langle \Delta(\mathbf{r})\Delta(\mathbf{r}' - \mathbf{r}) \rangle = \Delta^2 e^{-(r/L)^2}, \tag{5.80}$$

for which

$$|\Delta(q)|^2 = \pi(\Delta L)^2 \exp\left(-\frac{q^2 L^2}{4}\right). \tag{5.81}$$

The quantity Δ is the rms height of the fluctuation in the interface and L is the correlation length for the fluctuations. In a sense, L is the average distance between 'bumps' in the interface. It must be remembered that the actual interface used for the TEM picture has a finite thickness, and some averaging of the roughness will occur in the image. Nevertheless, typical values obtained from this approach are in the range 0.2 to 0.4 nm for Δ and 1.0 to 3.0 nm for L at the Si–SiO$_2$ interface [40]. Importantly, it was pointed out by these last authors that there was significant evidence that the correlation function was not Gaussian, but had a more exponential character to it. Subsequent measurements by [41] with atomic-force microscopy, and by Feenstra [42] with cross-sectional scanning-tunneling microscopy, have confirmed that the correlation function is an exponential, given by

$$\langle \Delta(\mathbf{r})\Delta(\mathbf{r}' - \mathbf{r}) \rangle = \Delta^2 e^{-\sqrt{2}r/L}, \tag{5.82}$$

for which

$$|\Delta(q)|^2 = \frac{\pi(\Delta L)^2}{\left[1 + \frac{q^2 L^2}{2}\right]^{1/2}}. \tag{5.83}$$

The matrix element is now found by combining the preceding equations to incorporate the correlation function as

$$|M(k, q)|^2 = \pi \left(\frac{\Delta L e^2}{\varepsilon_s} \right)^2 \left(N_{\text{depl}} + \frac{n_s}{2} \right)^2 \frac{1}{\sqrt{1 + \frac{q^2 L^2}{2}}}. \tag{5.84}$$

The actual scattering rate is calculated for two-dimensional scattering as in the preceding section. The scattering is elastic, so that $|k| = |k'|$, and $q = 2k\sin(\theta/2)$ arises from the delta function, which produces energy conservation. Thus one finds that

$$\Gamma(k) = \frac{1}{2\pi\hbar} \int_0^{2\pi} d\vartheta \int_0^\infty q\,dq\, |M(k, q)|^2 \delta(E_k - E_{k+q})$$

$$= \frac{m_c}{2\hbar^3} \int_0^{2\pi} d\vartheta \left(\frac{\Delta L e^2}{\varepsilon_s} \right)^2 \left(N_{\text{depl}} + \frac{n_s}{2} \right)^2 \frac{1}{\sqrt{1 + k^2 L^2 \cos^2 \left(\frac{\vartheta}{2} \right)}}. \tag{5.85}$$

$$= \frac{m_c}{2\hbar^3} \left(\frac{\Delta L e^2}{\varepsilon_s} \right)^2 \left(N_{\text{depl}} + \frac{n_s}{2} \right)^2 \frac{1}{\sqrt{1 + k^2 L^2}} E\left(\frac{kL}{\sqrt{1 + k^2 L^2}} \right)$$

In the last line, $E(x)$ is a complete elliptic integral. The explicit dependence of the scattering rate on the square of the effective field at the interface results in a decreasing mobility with increasing surface field (and increasing inversion density), which agrees with the trends observed in the experimental mobility data of most materials. This decrease in the experimental mobility with surface density qualitatively arises from the increased electric field dispersion around interface discontinuities at higher surface fields, which in turn gives rise to a larger scattering potential. In general, the entire mobility behavior in inversion layers at low temperature is explainable in terms of surface-roughness scattering and Coulomb scattering from interfacial charge, as discussed in the preceding section. This is shown in figure 5.9, in which the mobility in a Si inversion layer is plotted using these two scattering mechanisms [43]. Two different acceptor concentrations are shown in the plot. The downturn at low fields is due to the impurity scattering, while the main behavior of lower mobility at higher fields arises from the surface-roughness scattering. The parameters were adjusted to fit the experimental data of [44].

5.6.4 Alloy scattering

In a semiconductor alloy, the scattering of free carriers due to the deviations from the virtual crystal model has been termed alloy scattering. The virtual crystal concept was introduced in chapter 2 in connection with the alloys of various semiconductor materials. The general treatment of alloy scattering has usually followed an unpublished, but well-known, result due to Brooks, and extended by Makowski and Glicksman [45]. Although this scattering mechanism generally supplements the normal phonon and impurity scattering, it has on occasion been conjectured to be sufficiently strong as to be the dominant scattering mechanism in alloys. The work of Makowski and Glicksman, however, showed that the scattering was in general quite weak. They found that it was probably important only in the InAsP system, although even here it was likely to be much weaker than experimental

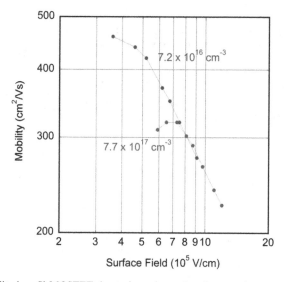

Figure 5.9. The mobility in a Si MOSFET due to impurity and surface-roughness scattering at 300 K. The labels give the acceptor doping in the *p*-type layer. Adapted from Vasileska *et al* [43].

data would suggest (which itself contains additional scattering due to defects in the alloy material, which always seem to be overlooked). These authors utilized a scattering potential given by the difference in the band gaps of the constituent semiconductors. Harrison and Hauser [46] suggested that the scattering potential is related to the differences in the electron affinities. However, as pointed out by Kroemer [47], the electron affinity is a true surface property, but not a qualitatively useful quantity in the bulk and is even a very bad indicator of bulk band offsets. Its use in scattering theories for carrier transport in bulk materials should therefore be treated with a degree of scepticism. A subsequent effort has suggested that the proper estimator for the disorder potential can be deduced from the bowing parameter and should therefore affect the random potential that leads to the scattering. While the alloy scattering process has been asserted to be weak, it has been shown recently that it can be comparable to acoustic phonon scattering, and should not be neglected [48].

The electron scattering rate for alloy scattering is determined directly by the scattering potential δE, which is the topic of the discussion above. The scattering is elastic, and the matrix element can therefore be given simply by

$$M(k, q) = \delta E e^{iqr}, \tag{5.86}$$

which can be used immediately in (5.5) to give

$$\Gamma(k) = \frac{(\delta E)^2}{2\pi\hbar}\left(\frac{2m_c}{\hbar^2}\right)^{3/2}\sqrt{E}. \tag{5.87}$$

Table 5.2. Alloy scattering parameters.

Alloy	δE_C (eV)	δE_{BG} (eV)	Alloy	δE_C (eV)	δE_{BG} (eV)
GaAlAs	0.12	0.7	InGaP	0.56	1.08
GaInAs	0.5	1.07	InAlP	0.54	1.08
InAsP	0.36	1.0	InGaSb	0.44	1.17
GaAsP	0.43	0.83	InPSb	1.32	1.17
InAsSb	0.82	0.83	GaPSb	1.52	1.57
InAlAs	0.47	1.49	InGaAsP	0.29	0.54
AlAsP	0.64	0.27	InGaPSb	0.54	0.56
InAlAsP	0.28	0.58			

This result is for parabolic bands, of course. A factor describing the degree of ordering has been omitted, which assumes that the alloy is a perfect random alloy. The scattering is reduced if there is any ordering in the alloy.

Besides the effect of ordering, the most significant parameter in (5.87) is the scattering potential δE. One could use the difference in the actual values of parameters mentioned above, but this would ignore the effect that the change in the lattice constant in the alloy would have on their general value. The scattering potential that leads to disorder scattering is just the aperiodic contribution to the crystal potential that arises from the disorder introduced into the lattice by the random siting of constituent atoms. In the virtual crystal approximation, the perfect zinc-blende lattice is retained in the solid solution. Thus the bond lengths are equal and the homopolar energy E_h does not make a contribution to the random potential δE. The random potential arises solely from the fluctuations in the band structure. In a ternary solid $A_x B_{1-x} C$, it has been suggested that the form of δE should be [49]

$$\delta E_C \sim s \left(\frac{1}{r_A} - \frac{1}{r_B} \right) \exp\left(-q_{FT} \frac{r_A + r_B}{2} \right), \qquad (5.88)$$

where the r_i are the atomic radii, q_{FT} is the Fermi–Thomas screening vector mentioned earlier (but for the entire set of valence electrons rather than the free carriers), s is a factor ~1.5 to account for the typical over-screening of the Fermi-Thomas approximation, and it has been assumed that the valence of the A and B atoms is the same. This can now be used to calculate the aperiodic potential used for alloy scattering. The results for a number of alloys are presented in table 5.2. Also shown, for comparison, are the equivalent values estimated by the discontinuity in the band gap. There is a weak dependence on the scattering potential δE from the composition as well, but this is small compared to the $x(1 - x)$ term.

As mentioned above, it is generally found that the role of alloy scattering is very weak, although there are often strong assertions from experimentalists that reduced mobility found in alloys must be due to 'alloy scattering.' In fact, only the work of Makowski and Glicksman [43] was sufficiently careful to exclude other mechanisms. In general, little effort is taken to include the proper strength of optical phonon scattering (discussed above), due to the complicated multimode behavior of the dielectric function in the alloy or to include dislocation or cluster scattering that can arise in impure crystals. Nevertheless, as pointed out above, the actual scattering rate for alloy scattering can be of the order of the acoustic phonon scattering [46], and this is shown in figure 5.10, for $In_{0.65}Al_{0.35}As$ strained to the InAs lattice constant.

5.6.5 Defect scattering

The role of defects as short-range scatterers which limit the mobility was suggested quite long ago. Dislocations can scatter either through the distortion of the crystal lattice (and hence the energy bands) [50] or by trapping charge which leads to a Coulombic behavior [51]. In addition, point defects can be misplaced atoms or neutral impurity atoms, and also lead to short range scattering. Generally, in high quality semiconductors, one does not encounter such scattering mechanisms. For example, bulk silicon can now be grown with essentially zero defects; defects only occur now as a result of the various processing (see chapter 1) performed in creating the integrated circuits. However, events in recent years have shown a deviation from this assumption particularly in connection with GaN and graphene. We will discuss these two materials as an example of the ways in which defect scattering can occur.

GaN and similar compounds. GaN is still found to be populated with dislocations, due to the nature of the crystal growth, and this tends to dominate the low-field mobility in the bulk, and is more important in the accumulation layer. In general, the wurtzite lattice is highly dislocated, which leads to hexagonal columns, or 'prisms,' with inserted atomic planes that fill the space between the columns [49]. Grain boundaries between the columns require arrays of dislocations along the interfaces between them [52]. Scattering arises from the charge on these filled

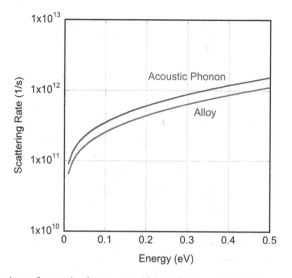

Figure 5.10. A comparison of acoustic phonon scattering and alloy scattering in the ternary $In_{0.65}Al_{0.35}As$. While alloy scattering is relatively small, it cannot be ignored.

dislocation states, and the potential is a modified Coulomb potential in two dimensions (the third dimension is along the dislocation) [53]

$$\delta E = \frac{ef}{2\pi\varepsilon_s c} K_0(q_d r), \tag{5.89}$$

where f is an occupation factor (typically about 70%–80%), c is the basal plane lattice constant of GaN, and K_0 is a modified Bessel function of the first kind. This potential can be Fourier transformed into the scattering vector $\mathbf{q} = \mathbf{k} - \mathbf{k}'$, and the scattering rate can be determined to be [50]

$$\Gamma(k) = \frac{e^4 f^2 m_c N_{\text{disl}}}{\hbar^3 \varepsilon_s^2 c^2 q_d^2} \frac{1}{[1 + (2k/q_d)^2]^{3/2}}. \tag{5.90}$$

Because of the tilt of the dislocation columns relative to the direction of motion of the carrier, the value of k should be taken to include the angle between the dislocation and the trajectory of the carrier.

A slightly different approach also has been taken, in which the dislocation is treated as a line of charge that is fully occupied, but is screened by a degenerate electron gas, which is perhaps more appropriate for carriers at a GaN–AlGaN interface [54]. In this case, Poisson's equation is solved to give the Fourier transform of the two-dimensional scattering potential as

$$V(q) = \frac{2e\rho_L}{2\varepsilon_s q(q + q_{\text{FT}})}. \tag{5.91}$$

Here, ρ_L is the line charge on the dislocation. This can then be used to give the scattering rate as

$$\Gamma(k) = \frac{m_c N_{\text{disl}} e^2 \rho_L^2}{16\pi\hbar^3 k_F^2 \varepsilon_s^2} \int_0^1 \frac{|V(q)|^2 \, du}{\left(u + \frac{q_{\text{FT}}}{2k_F}\right)\sqrt{1 - u^2}}, \tag{5.92}$$

where $u = q/2k_F$.

Ensemble Monte Carlo (EMC) simulation studies for the transport of photo-excited carriers in $\text{In}_x\text{Ga}_{1-x}\text{N}$ have suggested that a defect density of 10^8 cm^{-2} existed in the material; however, it is found that because of the presence of much more efficient inelastic scattering processes, this elastic defect scattering process can affect the low field mobility but is not important for the high field transient experiments [55]. A similar value for the dislocation density was found in studies of the high electric field transport in bulk GaN [56]. In studies of GaN/AlGaN high-electron mobility field-effect transistors, it was found that dislocation densities up to 10^{10} cm^{-2} had little effect on the drain current or the transconductance, but the device performance degraded significantly for larger values [57]. In figure 5.11, we plot the dislocation scattering rate and the acoustic phonon scattering rate for GaN, using a dislocation density of 10^{12} m^{-2}. The scattering from the dislocation falls off very rapidly due to the Coulomb nature of the potential.

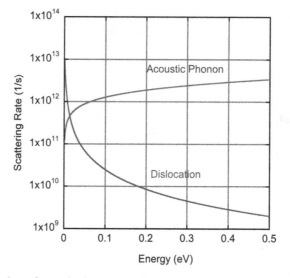

Figure 5.11. A comparison of acoustic phonon scattering and dislocation scattering in GaN, using (5.90) for the latter scattering. The plot is for a dislocation density of 10^{12} m^{-2} at 300 K.

Graphene. The role of defects as short-range scatterers which limit the mobility was suggested for graphene a few years ago [58, 59], and other mechanisms, such as corrugations [60] and steps [61], have been suggested as well. Studies of the defects have shown that atomic-scale lattice defects can lead to appreciable scattering [56, 62]. These are probably all relevant, as experiments on graphene transport usually give much lower mobilities than expected from transport calculations, presumably due to the impact of defect scattering on the transport.

Since the early work of Rutter *et al* [56], it has been known that transport in graphene can be affected by atomic-scale defects, which can mix wave functions of different symmetry and induce both intra-valley and inter-valley transitions. These defects appear to give rise to short-range scatterers [63], which affect both electrons and holes to a comparable extent [64]. At the same time, graphene is known to contain significant numbers of grain boundaries, which seem to exhibit a pentagon-heptagon pairing of adjacent cells [65–69]. However, it is also possible that these can appear as lines of individual defects [70, 71], which also can have the pentagon-heptagon pairing of adjacent cells, a configuration that has also been shown to be stable [72]. It is not clear that dislocations will scatter the same as point defects, as dislocations can be charged, and this leads to Coulomb and/or piezoelectric scattering, or often simpler potential scattering [73]. In graphene, however, it appears that only the potential scattering is seen in transport studies [62, 67, 68]. Hence, it is reasonable then to treat atomic scale defect scattering and dislocation scattering equivalently as a single type of scattering. That is, we can include a local potential scattering center, and discuss the density of such centers, but cannot really separate it as an isolated site or as part of a chain of sites contributing to an extended dislocation. Further, the defects are treated as uncorrelated potentials, which may be a serious error if these defects are part of a chain contributing to a dislocation.

The scattering rate for isolated defects is quite similar to that for the impurity potential. In fact, it has been suggested to just replace the Coulomb potential with a constant potential in the derivation of the scattering rate, but this must be done with care. The Fourier transform of the Coulomb potential has the units of energy-length^2. Hence, this needs to be replaced with a term of the order of $V_0 L^2$, where L is an effective range (L^2 becomes an effective cross-section for scattering) of the potential [74]. When this is done, the scattering rate can be written as

$$\Gamma(k) = \frac{4\alpha^2 E(k)}{3\pi\hbar(\hbar v_F)^2},$$
(5.93)

where

$$\alpha = \sqrt{N_{\text{disl}}}\, V_0 L^2.$$
(5.94)

The energy dependence in (5.93) suggests that we can expect to have a mobility dependence that decreases roughly as the square root of the carrier density. This result is found in some cases, but it is not universal in all studies of graphene.

In studies of the transport in graphene, it is found that values for α range from 0.2 to 0.9 eV-nm [75]. From equation (5.94), one sees that this quantity involves a scattering potential, an effective cross-section L^2, and the density of scattering centers. Measurements of these quantities for individual point defects don't exist, but there are a few values that exist for dislocations, and we can use the results of these measurements to check that the values of α used are reasonable. That is, many groups have studied the structure of the dislocations and point defects via scanning probe techniques as well as more usual analysis approaches, but few have actually measured the scattering potential, or local potential, of the dislocation. Koepke et al [67] have measured this quantity and have estimated the potential at 0.1 eV, with an effective range of about 1 nm on either side of the dislocation. Using these values, and a mid-value of $\alpha \sim 0.5$ eV-nm, we find a value of the needed defect density of some 2.5×10^{14} cm^{-2}, which seems to be an incredibly large value. Even with large numbers of defects strung together into dislocations, this is still a relatively high dislocation density. At the other end, however, Yazyev and Louie [64] have suggested that dislocations could open gaps of perhaps 1 eV and Cervenka and Flipse [68] have suggested a potential range of 4 nm from the dislocation. Using these values reduces the required defect density to just over 10^{12} cm^{-2}, which is not an unreasonable number. Nevertheless, further measurements of defect and dislocation density are required.

Using the defect scattering discussed above, we fit the mobility near to the experimental data [76] at the lowest density, which is 10^{12} cm^{-2} [75]. Here, we use a dislocation density of 10^{12} cm^{-2} and an impurity density of only 2.5×10^{10} cm^{-2}, so that the results are not clouded by the density dependence of the impurity scattering. As mentioned above, it is also assumed that these impurities sit some 4.2 nm away

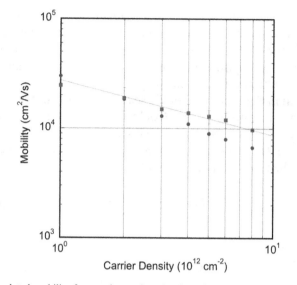

Figure 5.12. The calculated mobility for graphene when dominated by defect scattering (red) at 300 K. The line is a guide to the eye but decreases as the square root of the density according to (5.93). Data from [76] is shown as the blue dots.

from the graphene and below the BN layer. The fit to the data is achieved with a value of $\alpha = 0.4$ eV-nm. We then increase the density and determine the mobility. The results are shown in figure 5.12, with the experimental data from Kim *et al* [76]. The line in the figure is a guide to the eye, but has a slope proportional to the square root of the density. It can be seen that, in this case, the simulation of the mobility has this dependence, but the experiment decreases slightly more rapidly. So, in this case, the overall scattering processes are sufficiently dominated by the defect scattering that the density dependence follows that expected.

So far, we have associated the scattering potential just with point defects and/or dislocations. However, any local potential will lead to such scattering processes. It has been known for some time that mesoscopic conductance effects such as weak localization [77] and conductance fluctuations [78] are seen in graphene at low temperature. These effects are presumably due to a random potential which leads to disorder-induced effects. Martin *et al* [79] used a single-electron transistor to probe the local potential in graphene, and demonstrated the existence of electron and hole 'puddles' near the Dirac point. Subsequently, it was shown that these puddles can be a natural response to the many-electron interactions in the Dirac bands [80], in the presence of any corrugations in the graphene sheet. Thus, while a random distribution of impurities can lead to the disorder potential, this is not necessary and the puddles can conceivably form self-consistently in any graphene sheet. Gibertini *et al* [77] estimate, from their simulations, that the size of the puddles is a few nm. Deshpande *et al* [61] have used scanning tunneling spectroscopy, which shows that the fluctuations of the surface topography shows that puddle-like regions are of the order of 5–7 nm in extent. Even on BN, the size of the puddles is only about 10 nm [81]. Finally, simulations of Rossi and Das Sarma [82] suggest a similar size range of the puddles. Moreover, the peak of the potential could reach as much a 400 meV [83]. Hence, it is conceivable that the random potential, which leads to the puddles around the Dirac point, could in fact create a significant set of scattering centers. If we use a density of 10^{12} cm^{-2} for the scattering centers, an average potential of 0.2 eV, and a range of 5 nm, then we find $\alpha \sim 0.5$ eV-nm, which is exactly within the range needed for the simulations. Hence, the short-range scatterers may well be intrinsic to graphene as a result of the presence of the puddles.

References

[1] Ferry D K 1991 *Semiconductors* (New York: Macmillan)
[2] Ferry D K 2018 *An Introduction to Quantum Transport in Semiconductors* (Singapore: Pan Stanford)
[3] Shockley W and Bardeen J 1950 *Phys. Rev.* **77** 407
   Shockley W and Bardeen J 1950 *Phys. Rev.* **80** 72
[4] Herring C 1955 *Bell Syst. Tech. J.* **34** 237
[5] Ridley B K 1982 *Quantum Processes in Semiconductors* (Oxford: The Clarendon Press)
[6] Ferry D K 2000 *Transport in Semiconductors* (London: Taylor and Francis) ch 7
[7] Hutson A R 1962 *J. Appl. Phys.* **127** 1093
[8] Zook J D 1969 *Phys. Rev.* **A136** 869
[9] Ferry D K 1976 *Phys. Rev.* **B14** 1605
[10] Siegel W, Heinrich A and Ziegler E 1976 *Phys. Stat. Sol. (a)* **35** 269
[11] Birman J L 1962 *Phys. Rev.* **127** 1093
[12] Birman J L and Lax M 1966 *Phys. Rev.* **145** 620
[13] Lax M and Birman J L *Phys. Stat. Sol. (b)* **49** K153
[14] Long D 1960 *Phys. Rev.* **120** 2024
[15] Eaves L, Stradling R A, Tidley R J, Portal J C and Askenazy S 1968 *J. Phys.* C **8** 1975
[16] Paige E G S 1969 *IBM J. Res. Dev.* **13** 562
[17] Kash K, Wolff P A and Bonner W A 1983 *Appl. Phys. Lett.* **42** 173
[18] Shah J, Deveaud B, Damen T C, Tsang W T, Gossard A C and Lugli P 1991 *Phys. Rev. Lett.* **59** 2222
[19] Fawcett W and Ruch J G 1069 *Appl. Phys. Lett.* **15** 368
[20] Curby R C and Ferry D K 1969 *Phys. Stat. Sol. (a)* **20** 569
[21] Osman M A and Ferry D K 1987 *Phys. Rev.* B **36** 6018
[22] Pötz W and Vogl P 1981 *Phys. Rev.* B **24** 2025
[23] Fischetti M V and Higman J M 1991 *Monte Carlo Device Simulations: Full Band and Beyond* ed K Hess (Norwell: Kluwer)
[24] Zollner S, Gopalan S and Cardona M 1990 *J. Appl. Phys.* **68** 1682
   Zollner S, Gopalan S and Cardona M 1991 *Phys. Rev.* B **44** 13446
[25] Lee I, Goodnick S M, Gulia M, Molinari E and Lugli P 1995 *Phys. Rev.* B **51** 7046
[26] Yamakawa S, Akis R, Faralli N, Saraniti M and Goodnick S M 2009 *J. Phys. Cond. Matter.* **21** 1
[27] Shichijo H and Hess K 1981 *Phys. Rev.* B **23** 4197
[28] Fischetti M V and Laux S E 1988 *Phys. Rev.* B **38** 9721
[29] Saraniti M, Zandler G, Formicone G, Wigger S and Goodnick S M 1988 *Semicond. Sci. Technol.* **13** A177

[30] Conwell E M and Weisskopf V 1950 *Phys. Rev.* **77** 388

[31] Brooks H 1955 *Adv. Electron. and Electron Phys.* **8** 85

[32] Joshi R P and Ferry D K 1991 *Phys. Rev.* B **43** 9734

[33] Stern F and Howard W E 1967 *Phys. Rev.* **163** 816

[34] Ferry D K, Goodnick S M and Bird J P 2009 *Transport in Nanostructures* 2nd edn (Cambridge: Cambridge University Press)

[35] Ando T, Fowler A B and Stern F 1982 *Rev. Mod. Phys.* **54** 437

[36] Hamaguchi C 2003 *J. Comp. Electron.* **2** 169

[37] Ferry D K 2012 *IEEE Nano Mag.* **12** 18

[38] Ferry D K 2016 *Semicond. Sci. Technol.* **31** 11LT01

[39] Shishir R S, Chen F, Xia J, Tao N J and Ferry D K 2009 *J. Comp. Electron.* **8** 43

[40] Goodnick S M *et al* 1985 *Phys. Rev.* B **32** 8171

[41] Yoshinobu, Iwamoto A and Iwasaki H 1993 *Proceedings of the 3rd International Conference on Solid State Device Material* (Japan: Makuhari)

[42] Feenstra R M 1994 *Phys. Rev. Lett.* **72** 2749

[43] Vasileska D, Bordone P, Eldridge T and Ferry D K 1996 *Physica* B **227** 333

[44] Takagi S, Toriumi A, Iwase M and Tango H 1994 *IEEE Trans. Electron Dev.* **41** 2357

[45] Makowski L and Glicksman M 1976 *J. Phys. Chem. Sol.* **34** 487

[46] Harrison J W and Hauser J R 1970 *Phys. Rev.* B **1** 3351

[47] Kroemer H 1975 *Crit. Rev. Sol. State Sci.* **5** 555

[48] Welland I and Ferry D K 2019 *Semicond. Sci. Technol.* **34** 064003

[49] Ferry D K 1978 *Phys. Rev.* B **17** 912

[50] Dexter D L and Seitz F 1952 *Phys. Rev.* **86** 964

[51] Read W T 1954 *Phil. Mag.* **45** 775

[52] Weimann N G, Eastman L F, Doppalapudi D, Ng H M and Moustakas T D 1983 *J. Appl. Phys.* **83** 3656

[53] Pödör B 1966 *Phys. Stat. Sol.* **16** K167

[54] Jena D, Gossard A C and Mishra U K 2000 *Appl. Phys. Lett.* **76** 1707

[55] Liang W *et al* 2004 *Semicond. Sci. Technol.* **19** S427

[56] Barker J M, Ferry D K, Koleske D D and Shul R J 2005 *J. Appl. Phys.* **97** 063705

[57] Marino F A *et al* 2010 *IEEE Trans. Electron Dev.* **57** 353

[58] Rutter G M *et al* 2007 *Science* **317** 219

[59] Ni Z H *et al* 2010 *Nano Lett.* **10** 3868

[60] Katsnelson M L and Geim A K 2008 *Phil. Trans. Roy. Soc.* A **366** 195

[61] Low T, Perebeinos J, Tersoff J and Avouris P 2012 *Phys. Rev. Lett.* **108** 096601

[62] Giannazzo F, Sonde S, Lo Negro R, Rimini E and Raineri V 2011 *Nano Lett.* **11** 4612

[63] Deshpande A, Bao W, Miao F, Lau C N and LeRoy B J 2009 *Phys. Rev.* B **79** 205411

[64] Tapasztó L *et al* 2012 *Appl. Phys. Lett.* **100** 053114

[65] Yazyev O V and Louie S G 2010 *Phys. Rev.* B **81** 195420

[66] Yazyev O V and Louie S G 2010 *Nat. Mater.* **9** 806

[67] Huang P Y *et al* 2011 *Nature* **469** 389

[68] Kim K *et al* 2011 *ACS Nano* **3** 2142

[69] Koepke J C *et al* 2012 *ACS Nano* **7** 75

[70] Cervenka J and Flipse C F J 2009 *Phys. Rev.* B **79** 195429

[71] Liu Y and Yakobson B I 2010 *Nano Lett.* **10** 2178

[72] Mesaros A, Papanikolaou S, Flipse C F J, Sadri D and Zaanen J 2010 *Phys. Rev.* B **82** 205119

[73] Jaszek R 2001 *J. Mater. Sci. Mater. Electron.* **12** 1

[74] Harrison W A 1958 *J. Phys. Chem. Sol.* **5** 44

[75] Ferry D K 2013 *J. Comp. Electron.* **12** 76

[76] Kim E, Jain N, Jacobs-Gedrim R, Xu Y and Yu B 2012 *Nanotechnology* **23** 125706

[77] McCann E *et al* 2006 *Phys. Rev. Lett.* **97** 146805

[78] Berger C *et al* 2006 *Science* **312** 1191

[79] Martin J *et al* 2008 *Nat. Phys.* **4** 144

[80] Gibertini M *et al* 2012 *Phys. Rev.* B **85** 201405

[81] Decker R *et al* 2011 *Nano Lett.* **11** 2291

[82] Rossi E and Das Sarma S 2011 *Phys. Rev. Lett.* **107** 155502

[83] Das Sarma S, Adam S, Hwang E H and Rossi E 2011 *Rev. Mod. Phys.* **83** 407

Chapter 6

Carrier transport

Essentially all theoretical treatments of electron and hole transport in semiconductors are based upon a one-electron transport equation, which usually is the Boltzmann transport equation. As with most transport equations, this equation determines the distribution function under the balanced application of the driving and dissipative forces. How do we arrive at a one-electron (or one-hole) transport equation when there are some 10^{15}–10^{20} carriers per cubic centimeter in the device? It turns out that we can achieve this if we ignore correlations between the electrons. There are formal ways of showing this, but they boil down to ignoring two-particle correlations. Boltzmann called this the Stosszahl Ansatz, or molecular chaos, approximation. Even in so doing, the distribution function is not the end product, as transport coefficients arrive from integrals over this distribution. What are these integrals, and how are they determined? Some of them are easy, while others are difficult. And this is the case where the distribution only deviates from equilibrium by a small amount.

In the case of low electric fields, the transport is linear; that is, the current is a linear function of the electric field, with a constant conductivity independent of the field. The approach used is primarily that of the relaxation time approximation, and the distribution function deviates little from that in equilibrium—primarily the Fermi–Dirac distribution or one of its simplifications such as the Maxwellian distribution. In this situation, it must be assumed that the energy gained from the field by the carriers is negligible compared with the mean thermal energy of the carriers.

In this chapter, we begin by discussing a one-electron distribution, and how the Boltzmann equation arrives from those assumptions. The relaxation time approximation is defined, and then used to find approximate solutions for transport in electric and magnetic fields, including transport in a magnetic field. In general, the transport of hot carriers is nonlinear, in that the conductivity is itself a function of the applied electric field. The relationship between the velocity and field is expressed by a mobility, which depends on the average energy of the carriers. In normal linear response theory, a linear conductivity is found by a small deviation from the equilibrium distribution function. This small deviation is linear in the electric field, and the equilibrium distribution function dominates the transport properties. Once the carriers begin gaining significant energy from the field, this is no longer the case, and we will discuss this behavior in the next chapter. The dominant factor in the actual nonlinear transport does not arise directly from higher-order terms in the field, but rather from the implicit field dependence of the non-equilibrium distribution function, such as that of the electron temperature.

6.1 The Boltzmann transport equation

The Boltzmann equation is an equation for a one-electron distribution function that describes the ensemble of carriers. The distribution function gives us a probability for the occupation of a state of given energy and momentum and position. Just what is meant by this distribution function? There are various manners in which this quantity can be defined. For example, it is possible to say that the distribution function $f(\mathbf{v}, \mathbf{x}, t)$ is the probability of finding a particle in the box of volume $\Delta\mathbf{x}$, centered at $\mathbf{x}$, and $\Delta\mathbf{v}$, centered at $\mathbf{v}$, at time t. Here, $\mathbf{v}$ is the particle velocity (momentum is often used instead) and $\mathbf{x}$ is the position, now taken to be vector quantities. In this sense, the distribution function is described in a six-dimensional phase space, and the quantities $\mathbf{x}$ and $\mathbf{v}$ do not refer to any single carrier but to the position in this phase space. This is to be compared with the idea that the N particles are defined in a $6N$-dimensional configuration space, where we have $3N$ velocity variables and $3N$ position variables, all a function of time. It is important to differentiate between these two spaces. The former can be thought of as the one electron phase space while the latter is clearly a many electron configuration space. It is the transition from this configuration space to the physical space that normally yields a hierarchy of distributions: one electron, two electron, etc. And, it is via the molecular chaos approximation that we decouple the one electron distribution from the other distributions. With these ideas, then it is possible to describe one normalization of the distribution function as

$$\int \int d^3x\, d^3v\, f(\mathbf{x}, \mathbf{v}, t) = 1. \tag{6.1}$$

As with all probability functions, the integral over the measure space must sum to unity. However, this is not the only definition that can be made.

An alternative definition is to define the distribution function as the 'average' (the concept of this average will be defined below) number of particles in a phase space box of size $\Delta x \Delta v$ located at the phase space point $(\mathbf{x}, \mathbf{v})$. In this regard, the distribution function then satisfies

$$\int \int d^3x\, d^3v\, f(\mathbf{x}, \mathbf{v}, t) = N(t). \tag{6.2}$$

Here, $N(t)$ is the *total number of particles* in the entire system at time t. At first view it might be supposed that the Fermi–Dirac distribution satisfies (6.2) and in fact defines the Fermi energy level as a function of time. However, this is not correct for two reasons. First, the normalization is wrong; recall that the Fermi–Dirac distribution has a maximum value of unity for energies well below the Fermi energy. Hence the integral in (6.2) must be modified to account for the density of states in the incremental volume. Moreover, one must convert the velocity integration into an energy integration, and this adds additional numerical and variable factors. An additional objection is more serious. The Fermi–Dirac distribution is a point function, and its application to inhomogeneous systems must be handled quite carefully. The Fermi energy is related to the electrochemical potential, which may vary (relative to one of the band edges) with position relative to a point where the band edge is taken to be the reference in energy. Then the Fermi energy is position dependent in this view, especially when the density is position dependent. Yet it is well known from simple theory that the Fermi energy must be position independent if the system is to be in equilibrium (no currents flowing). For this to be the case, the band edges must themselves become position dependent in the inhomogeneous system. Thus, while we can equate (6.2) with the use of the Fermi–Dirac distribution function, this must be done with quite some care in inhomogeneous systems, which can be illustrated by noting that the integration of the Fermi–Dirac distribution over energy yields precisely the Fermi energy (see figure 4.1).

In either of the two above definitions of the distribution function, quantum mechanics further complicates the situation in at least two ways. First, the uncertainty relation requires that $\Delta x \Delta p > \hbar^3/8$, or $\Delta x \Delta v > \hbar^3/8m_c^3$, and the *quantum* distribution function can in fact have negative values for regions of smaller extent than this limit [1]. So, we are constrained over just how finely we can examine the position and momentum coordinates. In addition, the distribution function to be dealt with here is an equivalent one-electron distribution function, so that the many electron aspects, discussed above in terms of the $6N$ dimensional configuration space, are averaged out. In both of these cases, the distribution function is said to be *coarse grained* in phase space, in the first case averaging over small regions in which significant local quantization is significant and in the second case averaging out the many-electron properties that modify the one electron distribution function. This coarse graining in the latter case is the process of the *Stosszahl* ansatz, or molecular chaos, introduced by Boltzmann to justify the use of the one-particle functions or, more exactly, the process by which correlation with early times is forgotten on the scale of the one-particle scattering time τ. The exact manner, by which a multi-electron ensemble is projected onto the one-electron distribution function of (6.1) or (6.2), is best described through the BBGKY Hierarchy. (The letters are taken from the authors Bogoliubov [2], Born and Green [3], Kirkwood [4], and Yvon [5]. The projection approach is described in Ferry [6]).

The variation of the distribution function is governed by an equation of motion, and it is this equation that is of interest here. In equilibrium, no transport takes place since the distribution function is symmetric in **v**-space (more properly, **k**-space). Since the probability of a carrier having the wave vector **k** is the same as for the wave vector −**k** (recall that the Fermi-Dirac distribution depends only on the energy of the carrier, not specifically on its momentum), these balance one another. Since there are equal numbers of carriers with these oppositely directed momenta, the net current is zero. Hence, for transport, the distribution function must be modified by the applied fields (and made asymmetric in phase space). It is this modification that must now be calculated. In fact, the forcing functions, such as the applied field, are themselves reversible quantities and the evolution of the distribution function in phase space is unchanged by these fields. It is only the presence of the scattering processes that can change this evolution, and the classical statement of this fact is (the right-hand side represents changes due to scattering processes)

$$\frac{df(x, v, t)}{dt} = \frac{df(x, v, t)}{dt}\bigg|_{\text{collisions}}. \tag{6.3}$$

By expanding the left-hand side with the chain rule of differentiation, the Boltzmann transport equation is obtained as

$$\frac{\partial f}{\partial t} + v \cdot \nabla f + \frac{dk}{dt} \cdot \frac{\partial f}{\partial k} = \frac{df(x, v, t)}{dt}\bigg|_{\text{collisions}}. \tag{6.4}$$

The first term is the explicit time variation of the distribution function, while the second term accounts for transport induced by spatial variation of the density and distribution function. The third term describes the field induced transport. These three terms on the left-hand side are collectively known as *streaming* terms. Here, the third term has been written with respect to the momentum wave vector rather than the velocity to account for the role of the former in the crystal momentum. Still, in keeping with the discussion above, the change of the distribution function with position must be sufficiently slow that the variation of the wave function is very small in a single unit cell. This ensures that the band model developed in chapter 2 is valid, and a true statistical distribution can be considered. In addition, the force term must be sufficiently small that it does not introduce any mixing of wave functions from different bands, so that the response can be considered semi-classical within the effective mass approximation. Finally, the time variation must be sufficiently slow that the distribution evolves slowly on the scale of either the mean free time between collisions or on the scale of any hydrodynamic relaxation times (still to be developed, although we have already discussed the momentum relaxation time). The force term, the third term on the left-hand side, is just the Lorentz force in the presence of electric and magnetic fields.

The scattering processes, discussed in the last chapter, are all folded into the term on the right-hand side of (6.4). Any scattering process induces carriers to make a transition from some initial state k into a final state k' with a probability $P(k, k')$. Then, the number of electrons scattered depends on the latter probability as well as on the probabilities of the state k being full [given by $f(k)$] and the state k' being empty [given by $1 - f(k')$]. (If the volume in k-space contains only a single pair of spin degenerate states, the number of carriers in this volume is given by the Fermi–Dirac distribution. If the scattering process can flip the spin then a factor of 2 is also included for this spin degeneracy. It is clear that this can now be seen as mixing the two definitions for the distribution function given above, but it is in fact the latter of the two that is being used.) The rate of scattering *out* of state **k** is then found to be given by putting these three factors together, as

$$P(\mathbf{k}, \mathbf{k}')f(\mathbf{k})[1 - f(\mathbf{k}')].\tag{6.5}$$

But there are also electrons being scattered *into* the state **k** from the state **k**′ with a rate given by

$$P(\mathbf{k}', \mathbf{k})f(\mathbf{k}')[1 - f(\mathbf{k})].\tag{6.6}$$

The latter two equations are the basis for the scattering term on the right-hand side of (6.4), which is finally obtained by summing over all states **k**′ as

$$\left.\frac{df(\mathbf{x}, \mathbf{v}, t)}{dt}\right|_{\text{collisions}} = \sum_{\mathbf{k}'}\{P(\mathbf{k}', \mathbf{k})f(\mathbf{k}')[1 - f(\mathbf{k})] \\ - P(\mathbf{k}, \mathbf{k}')f(\mathbf{k})[1 - f(\mathbf{k}')]\}\tag{6.7}$$

In fact, $P(\mathbf{k}, \mathbf{k}')$ also contains a summation over all possible scattering mechanisms by which electrons (or holes) can move from **k** to **k**′. Hence, the right-hand side of (6.7) is a very complicated function of momentum and/or energy that is solved only in some very simple cases.

In *detailed balance*, the two scattering probabilities differ by, for example, differences in the density of final states (the second argument) in energy space and by the phonon distribution difference between emission and absorption processes. This detailed balance is enforced by the fact that the right-hand side of (6.7) vanishes in equilibrium. In fact, (6.7) encompasses four processes when phonons are involved. Carriers can leave by either emitting a phonon (and going to a state of lower energy) or by absorbing a phonon (and going to a state of higher energy). By the same token, they can scatter into the state of interest either by phonon emission from a state of higher energy, or by phonon absorption from a state of lower energy. In equilibrium, the processes connecting our primary state with each of the two sets of levels (of higher and lower energy) must balance. This balancing in equilibrium is referred to as detailed balance. Under this condition, the right-hand side of (6.4) vanishes in equilibrium.

When the distribution function is driven out of equilibrium by the *streaming* forces on the left-hand side of (6.4), the collision terms work to restore the system to equilibrium. Interactions between the carriers within the distribution work to randomize the energy and momentum of the distribution by redistributing these quantities within the distribution. This is known to lead to a Maxwellian distribution in the non-degenerate case through a process that can be shown to maximize the entropy of the distribution. However, it is the phonon interactions that cause the overall distribution function to come into equilibrium with the lattice. If the lattice itself is in equilibrium, it may be considered as the *bath* and the phonons serve to couple the electron distribution to this thermal bath. Under high electric fields, it is also possible for the phonon distribution to be driven out of equilibrium (as we will see in chapter 7) and this makes for a very complicated set of equations to be solved. In the following paragraphs, solutions of the Boltzmann transport equation (6.4) will be obtained for a simplified case. More complicated solutions will be dealt with later.

6.1.1 The relaxation time approximation

If no external fields are present, the collisions tend to randomize the energy and the momentum of the carriers and return them to their equilibrium state. In linear response, it is often useful to assume that the rate of relaxation is proportional to the deviation from equilibrium and that the distribution function decays to its normal equilibrium value in an exponential manner. For this approximation, a relaxation time τ may be introduced by means of the equation

$$\frac{df(x, v, t)}{dt}\bigg|_{\text{collisions}} = -\frac{f - f_0}{\tau}. \tag{6.8}$$

Here f_0 is the equilibrium distribution function, either a Fermi–Dirac distribution or the Maxwellian approximation to this distribution. This is a fairly common approximation that is easily made if the scattering processes are either elastic, or are isotropic and inelastic. Then, in the homogeneous situation, (6.3) and (6.8) lead to

$$f(t) = f_0 + [f(0) - f_0]e^{-t/\tau}. \tag{6.9}$$

This is the equation for the natural decay of the distribution function due to scattering, so $f(0)$ is the non-equilibrium initial state at the beginning of the decay process. On the other hand, if an external force is applied, the distribution function will approach the non-equilibrium steady state with this same exponential behavior, due to the assumption of linear response.

Even when the relaxation time approximation, as (6.8) is called, holds, it is necessary to be able to calculate τ from the scattering rates. As will be seen later in this chapter, this entails an average over the distribution function. Here, it is desired to consider the case of the elastic scattering process in further detail. If the scattering process is elastic, the states $\mathbf{k}$ and $\mathbf{k}'$ lie on the same energy shell [that is, $E(\mathbf{k}) = E(\mathbf{k}')$], and it is feasible to assume that $P(\mathbf{k}, \mathbf{k}') = P(\mathbf{k}', \mathbf{k})$, so that the relaxation term in (6.7) becomes

$$\frac{df(x, v, t)}{dt}\bigg|_{\text{collisions}} = \sum_{k'} P(\mathbf{k}', \mathbf{k})[f(\mathbf{k}') - f(\mathbf{k})]. \tag{6.10}$$

We will return to this term again shortly.

In the presence of an external electric field, we also include the acceleration term in (6.4), so that the combination of this term and (6.8) leads to the simple form for the distribution function. For a homogeneous semiconductor sample, we can solve for the distribution as

$$f(k) = f_0 - \frac{\tau}{\hbar}\mathbf{E} \cdot \frac{\partial f}{\partial k}$$

$$= f_0 - \tau\mathbf{E} \cdot v\frac{\partial f}{\partial E}.$$ (6.11)

$$\approx f_0 - \tau\mathbf{E} \cdot v\frac{\partial f_0}{\partial E}$$

(Note that there is an obvious conflict between the electric field and the energy notation here. One should take care not to confuse the two.) It has been assumed in the last form of (6.11) that the deviation from equilibrium is sufficiently small that the entire right-hand side can be represented as a functional of f_0. If this form now is introduced into (6.10), we can get a better form for the relaxation time approximation to be

$$\frac{df(x, v, t)}{dt}\bigg|_{\text{collisions}} = \sum_{k'} P(k', k)\tau\mathbf{E} \cdot [v(k') - v(k)]\frac{\partial f_0}{\partial E}$$
$$= \int_{S_k} dS_k P(k', k)\tau\mathbf{E} \cdot [v(k') - v(k)]\frac{\partial f_0}{\partial E}.$$ (6.12)

Now, we reuse (6.11) to remove some of the terms, and

$$\frac{df(x, v, t)}{dt}\bigg|_{\text{collisions}} = -(f - f_0)\int_{S_k} dS_k P(k', k)\left[1 - \frac{v(k') \cdot v(k)}{v^2}\right].$$
$$= -(f - f_0)\int_{S_k} dS_k P(k', k)[1 - \cos\vartheta]$$ (6.13)

In (6.12), the fact that the scattering is elastic has been used to assure that the integration is over a single energy shell, hence the integration is only over the surface corresponding to this energy shell. In (6.13), the angular variation has been utilized to write the integral in terms of the angle between the two velocity vectors, and the angular weighting discussed already in the previous chapter is recovered. In truly elastic, isotropic scattering such as acoustic phonons in spherically symmetric bands, the $\cos(\theta)$ term integrates to zero with the shell integration. However, (6.13) now allows us to compute the *momentum* relaxation time for elastic scattering processes as

$$\frac{1}{\tau} = \int_0^\pi d\vartheta \int_0^{2\pi} d\varphi \, \sin\vartheta P(\vartheta, \varphi)[1 - \cos\vartheta],$$ (6.14)

where $P(\vartheta, \varphi) = k^2 P(\mathbf{k}, \mathbf{k}')$, and the latter quantity is calculated by the Fermi golden rule, as was described in the previous chapter (it is in fact $\Gamma(\mathbf{k})$, with k lying on the elastic energy shell).

It is obvious that it will be difficult to compute a simple relaxation time in the case of inelastic scattering, as the two distribution functions in (6.10) come from different energy shells. Thus, they cannot readily be separated from the scattering integrals in order to identify just the relaxation rate. It is in fact possible to calculate the average momentum relaxation rate if a specific form for the non-equilibrium distribution function is assumed. The latter rate can be extrapolated to the equilibrium situation to give an *effective* relaxation rate $1/\tau$, and thus utilize the relaxation time approximation in subsequent calculations. In general, however, this can be done only in very special cases, and more complicated approaches to computing the distribution function in the presence of forces must be utilized.

6.1.2 Conductivity

When the external force is just an electric field, the distribution function is given by just the field streaming term and the scattering term, and we need only solve (6.11). By knowing this distribution function, the electric current density carried by these carriers (in this case, electrons because of the sign used in the force) can be found by summing over the electron states as

$$
\begin{aligned}
J &= - e \int dE \rho(E) v(E) f(E) \\
&= - e^2 \int dE \rho(E) v(E) \tau \mathbf{E} \cdot v \frac{\partial f_0}{\partial E}
\end{aligned}
\tag{6.15}
$$

The first term on the right-hand side of (6.11) does not contribute to the current, as there is no current in equilibrium, and has been discarded. In general, in our semiconductors, the material is cubic and isotropic in the current. So, in (6.15) the two velocities give the current parallel to the electric field. As the thermal velocity is usually larger than the directed velocity, we make the assumption that $v_j^2 \sim v^2/3$, and we then note that $v^2 = 2E/m_c$. We take the distribution function to be normalized as

$$
n = \int dE \rho(E) f(E),
\tag{6.16}
$$

and we can now write the current density as

$$
J = -\frac{2}{3} \frac{ne^2}{m_c} \frac{\int dE \rho(E) E \tau \frac{\partial f_0}{\partial E}}{\int dE \rho(E) f_0(E)} \mathbf{E}.
\tag{6.17}
$$

Normally, we connect the current and the electric field via the relationship and the conductivity σ as

$$
\begin{aligned}
J &= \sigma E = \frac{ne^2}{m_c} \langle \tau_m \rangle E \\
\sigma &= \frac{ne^2}{m_c} \langle \tau_m \rangle
\end{aligned}
\tag{6.18}
$$

where the last line defines the *conductivity* σ and we have dropped the vector notation due to isotropic nature of the conductivity in these semiconductors, and

$$\mu = \frac{e\langle\tau_m\rangle}{m_c} \qquad (6.19)$$

is the definition of the *mobility*. We have introduced in (6.18) the average momentum relaxation time

$$\langle\tau_m\rangle = -\frac{2}{3}\frac{\int dE\rho(E)E\tau\frac{\partial f_0}{\partial E}}{\int dE\rho(E)f_0(E)}. \qquad (6.20)$$

In three dimensions, the density of states in the two integrals varies as the square root of the energy. We can use this to rewrite the integral in the denominator, integrating by parts, as

$$\int dE\rho(E)f_0(E) \sim \int dE E^{\frac{1}{2}}f_0(E) = -\frac{2}{3}\int dE E^{\frac{3}{2}}\frac{\partial f_0}{\partial E}. \qquad (6.21)$$

Then, the average of the relaxation time is given as

$$\langle\tau_m\rangle = \frac{\int dE E^{\frac{3}{2}}\tau_m(E)\frac{\partial f_0}{\partial E}}{\int dE E^{\frac{3}{2}}\frac{\partial f_0}{\partial E}}. \qquad (6.22)$$

The subscript 'm' on the relaxation time refers to the momentum relaxation, so that the scattering time in the integral should include the $(1 - \cos\vartheta)$ term in its evaluation. The same approach can be used in other than three dimensions. In general, the density of states varies as $E^{(d/2)-1}$. Then, the general steps leading to (6.22) can be followed for an arbitrary dimension. However, note that the prefactor $(2/3)$ also arose from the dimensionality, and the argument leading to it gives $2/d$ as the prefactor.

In writing the limits on the above integrals as infinity, it has been assumed that the conduction-band upper edge is sufficiently far removed from the energy range of interest so that the distribution function is zero at this point. In this case, the upper limit does not affect the final result if the limit is taken as infinity rather than just the upper edge of the band (lower edge if we are dealing with holes). In extremely degenerate cases, the upper limit of the integral may be taken as the Fermi energy, and the relaxation time evaluated at the Fermi energy, as the derivatives of the distribution function are sharply peaked at this energy.

In the last chapter, the scattering rates for a number of processes were computed. In each case, these scattering rates were energy dependent, so that (6.22) leads, of course, to a method of incorporating this energy dependence into the observable mobility. In figure 6.1, the mobility for graphene at 300 K is plotted as a function of the electron concentration for two different substrates. Here, scattering by the acoustic and optical inter-valley phonons (between the K and K′ minima of the conduction band), remote impurities, and the remote substrate optical phonon [7, 8] is included. As the density increases, the Fermi energy also increases, and this changes the average energy of the distribution. The break at higher density

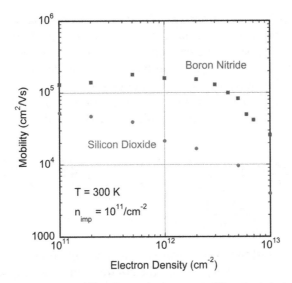

Figure 6.1. The room temperature mobility for graphene on two different substrates, SiO$_2$ and BN. The remote impurity scattering has been kept low so as not to dominate the mobility.

corresponds mainly to the relationship between the Fermi energy and the energy of the remote optical phonon [8].

If the constant-energy surfaces are not spherical, some complication of the problem arises. To begin with, the energy is no longer a function of the single effective mass and becomes expressed as

$$E = \frac{\hbar^2}{2}\left(\frac{k_x^2}{m_x} + \frac{k_y^2}{m_y} + \frac{k_z^2}{m_z}\right) \tag{6.23}$$

for each ellipsoid. To simplify this approach, we introduce the following transformation

$$k_i' = \sqrt{\frac{m_c}{m_i}}\,k_i \tag{6.24}$$

for each direction within a single ellipsoid. This then rescales the energy to be

$$E = \frac{\hbar^2}{2m_c}\left(k_x'^2 + k_y'^2 + k_z'^2\right). \tag{6.25}$$

By introducing the same transformations on the velocity $\mathbf{v}$ ($= \hbar\,\mathbf{k}/m^*$) in each of the ellipsoids, the simple result above is still achieved for the current, but this must be un-transformed to achieve the current in the real coordinates. Carrying out this process for a single ellipsoid yields

$$J_x = \frac{ne^2}{m_x}\langle\tau_m\rangle E_x, \tag{6.26}$$

and so on for each of the other two directions. In most cases of non-spherical energy surfaces, multiple minima are involved, and a summation over these equivalent minima must still be carried through. For example, in silicon with six equivalent ellipsoids in the conduction band, the total conductivity is a sum over the six valleys. However, two of the valleys are oriented in each of the three principal directions and contribute the appropriate amount to each current direction. The total current is then (with 1/6 of the total carrier density in each valley)

$$\frac{J}{E} = \frac{ne^2}{6}\langle\tau_m\rangle\left(\frac{2}{m_1} + \frac{2}{m_2} + \frac{2}{m_3}\right) \to \frac{ne^2}{3}\langle\tau_m\rangle\left(\frac{2m_L + m_T}{m_L m_T}\right)$$
$$= ne^2\langle\tau_m\rangle\frac{2K + 1}{3m_L}.$$

(6.27)

Here, $K = m_L/m_T$, and the masses were discussed in chapters 2 and 4. For silicon, $m_L = 0.91m_0$ and $m_T = 0.19m_0$, so that the conductivity mass $m_c = 3m_L/(2K + 1)$ is about $0.26m_0$. This value is different from either of the two curvature masses and is different from the density-of-states mass. This mass, called the *conductivity mass,* arises from a proper conduction sum over the various ellipsoids. This sum is relatively independent of the actual shape and position of the ellipsoids (the same sum arises in germanium with its four ellipsoids), but arises solely for the sums used in computing the conduction current that is parallel to the electric field. Different sums will arise if a magnetic field is present. Using this multi-valley approach, the mobility for electrons in Si and Ge are shown in figure 6.2 [9].

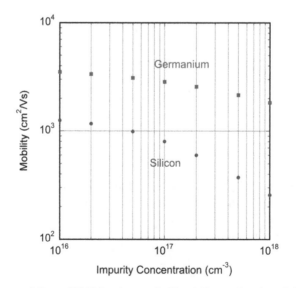

Figure 6.2. Electron mobility at 300 K for electrons in Si and Ge as a function of the donor density.

6.1.3 Diffusion

There are many cases where the distribution function varies with position, either through a change in the normalization as the doping concentration changes, or through the presence of a temperature gradient. Here, we consider the former. Let us consider only the second term on the left-hand side of (6.4), which leads to the expression, in the relaxation time approximation,

$$v \cdot \nabla f = -\frac{f - f_0}{\tau}, \tag{6.28}$$

for which we can write

$$f = f_0 - v \cdot \nabla f \sim f_0 - v \cdot \nabla f_0, \tag{6.29}$$

following the same approximations introduced above. The current is given by the first line of (6.15), and we have

$$J_x = e \int \tau v_x^2 \left(\frac{\partial f_0}{\partial x} \right) \rho(E) dE. \tag{6.30}$$

We can easily bring the spatial derivative outside the integral, as the velocity and scattering time are functions of the energy and not the position. This then gives us

$$\begin{aligned} J_x &= e \frac{\partial}{\partial x} \left(\int \tau v_x^2 f_0 \rho(E) dE \right) \\ &= e \frac{\partial}{\partial x} (D_e n) \end{aligned} \tag{6.31}$$

which allows us to introduce the diffusion 'constant'

$$D_e = \left\langle \frac{\tau v^2}{3} \right\rangle = \frac{\int \tau v^3 f_0 \rho(E) dE}{3 \int f_0 \rho(E) dE}. \tag{6.32}$$

Here, we have used the connection between the component of the velocity along the current to the total velocity that was introduced above (6.16), and the factor of '3' is the dimensionality d of the system. In general, this definition of the diffusion constant is often independent of position due to the normalization with respect to the density. In these cases, it can be brought through the gradient operation to produce the more usual (with the sign for electrons)

$$J = eD\nabla n. \tag{6.33}$$

Clearly, the diffusion coefficient arises from an ensemble average just as the mobility does. However, it should be noted that *the two averages are, in fact, different!* In (6.22), the denominator was integrated by parts in order to get the energy derivative of the distribution function into both the numerator and the denominator. Here, this is not done. While the denominator is technically the same, in fact they differ by the factor $d/2$. One cannot carry out the integration of the numerator by parts to overcome this difference, because we don't know the energy dependence of the momentum relaxation time. Hence, the two averages are quite likely to differ by a numerical factor, except for the case of a Maxwellian distribution where they yield the same amount. In fact, if we assume that $f_0 \sim \exp(-E/k_B T)$, then it is simple to show that there is an Einstein relation

$$D = \mu \frac{k_B T}{e}. \qquad (6.34)$$

This result, of course, holds only for the non-degenerate case for which a Maxwellian is valid. In the degenerate case, (6.34) must be multiplied by the ratio of two Fermi–Dirac integrals, which provides the correction between the average energy and the fluctuation represented by $k_B T$. In figure 6.3, we plot the diffusion coefficient at 300 K for electrons in Si and Ge as a function of the donor concentration.

In high electric fields, however, the connection between mobility and diffusion represented by (6.34) fails. This is because the distribution function is neither a Fermi–Dirac, nor a Maxwellian. Thus, the averages (6.22) and (6.32) have very little in common, and there is no natural way in which to connect them [10]. Worse, it is well known that any estimate of electron temperature in reasonable (or higher) electric fields gives different results along the electric field and perpendicular to the electric field. Evaluations of the diffusivity in these cases also shows such an anisotropy [11], with the anisotropy getting larger as the electric field rises. Thus, it is worse than useless to try to infer an Einstein relationship in any real semiconductor device, where the fields themselves are both high in value and anisotropic and inhomogeneous within the device.

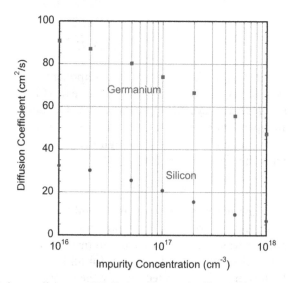

Figure 6.3. The diffusion coefficient at 300 K for electrons in Si and Ge as a function of the donor concentration.

6.1.4 Magnetoconductivity

We now want to introduce the magnetic field into the discussion of the conductivity. This produces what is called a magneto-conductivity. To simplify the notation somewhat, the incremental distribution function is defined through the results found above as $f_1 = f - f_0$, so that the relaxation-time approximation operates only on this incremental quantity. We consider a homogeneous semiconductor, under steady-state conditions, so that the Boltzmann transport equation becomes

$$\frac{e}{\hbar}(\mathbf{E} + \mathbf{v} \times \mathbf{B}) \cdot \frac{\partial f}{\partial \mathbf{k}} = -\frac{f_1}{\tau}. \tag{6.35}$$

It will still be assumed that the incremental distribution function is small compared to the equilibrium one, so that the latter can be used in the gradient (with respect to momentum) term. However, it is known that for the equilibrium distribution function, that the derivative produces a velocity, which yields zero under the dot product with the $\mathbf{v} \times \mathbf{B}$ term $[\mathbf{v} \cdot (\mathbf{v} \times \mathbf{B}) = \mathbf{B} \cdot (\mathbf{v} \times \mathbf{v}) = 0]$. Hence, we must keep the first-order contribution to the distribution function in this term. Then, the Boltzmann equation becomes

$$-e\mathbf{E} \cdot \mathbf{v}\frac{\partial f_0}{\partial E} = -\frac{f_1}{\tau} + \frac{e}{\hbar}(\mathbf{v} \times \mathbf{B}) \cdot \frac{\partial f_1}{\partial \mathbf{k}}. \tag{6.36}$$

In analogy to (6.11), the incremental distribution function is written as

$$f_1 = e\tau_m \mathbf{v} \cdot \mathbf{A}\frac{\partial f_0}{\partial E}, \tag{6.37}$$

where $\mathbf{A}$ plays the role of an equivalent electric field vector that must still be determined. That is, $\mathbf{A}$ is going to contain not only the applied electric field $\mathbf{E}$, but also the induced electric field from the second term in the Lorentz force that is seen in (6.35). If higher-order terms in the distribution function are neglected, as indicated in (6.37), the force functions can be written as

$$-\mathbf{v} \cdot \mathbf{E} = -\mathbf{v} \cdot \mathbf{A} + \frac{e\tau_m}{m_c}(\mathbf{v} \times \mathbf{B}) \cdot \mathbf{A}. \tag{6.38}$$

This equation must hold for any value of the velocity, which is a simple multiplier if the triple product in the last term is rearranged, so we require that

$$\mathbf{E} = \mathbf{A} - \frac{e\tau_m}{m_c}\mathbf{B} \times \mathbf{A}. \tag{6.39}$$

By elementary geometry, it can be shown that the general solution for the vector $\mathbf{A}$ must be (one can back substitute this into the above equation to show that it is the proper solution) [12]

$$A = \frac{\mathbf{E} + \frac{e\tau_m}{m_c}\mathbf{E} \times \mathbf{B} + \left(\frac{e\tau_m}{m_c}\right)^2 \mathbf{B}(\mathbf{B} \cdot \mathbf{E})}{1 + \left(\frac{e\tau_m}{m_c}\right)^2 B^2}. \tag{6.40}$$

This equation can now be used in the distribution function (6.37), and the forms slightly rearranged to give the result

$$f_1 = e\tau_m \mathbf{v} \cdot \mathbf{E} \frac{\mathbf{v} + \frac{e\tau_m}{m_c}\mathbf{v} \times \mathbf{B} + \left(\frac{e\tau_m}{m_c}\right)^2 \mathbf{B}(\mathbf{B} \cdot \mathbf{v})}{1 + \left(\frac{e\tau_m}{m_c}\right)^2 B^2} \frac{\partial f_0}{\partial E}. \tag{6.41}$$

We now consider the case where the magnetic field is perpendicular to the electric field, and to the plane in which the transport is to take place. We take $\mathbf{B} = B\mathbf{a}_z$, and consider the x- and y-directed transport, with the current in the x-direction. The averaging of the distribution function with the current is a straight forward procedure once we have f_1, as given in (6.41). This averaging leads to the equations

$$J_x = \frac{ne^2}{m_c}\left[\left\langle \frac{\tau_m}{1 + \omega_c^2\tau_m^2} \right\rangle E_x - \left\langle \frac{\omega_c\tau_m^2}{1 + \omega_c^2\tau_m^2} \right\rangle E_y\right]$$
$$J_x = \frac{ne^2}{m_c}\left[\left\langle \frac{\omega_c\tau_m^2}{1 + \omega_c^2\tau_m^2} \right\rangle E_x + \left\langle \frac{\tau_m}{1 + \omega_c^2\tau_m^2} \right\rangle E_y\right]. \tag{6.42}$$

In this last form, we have introduced the *cyclotron frequency* $\omega_c = eB/m_c$. Instead of a simple average over the relaxation time, there is now a complicated average that must be carried out. This becomes somewhat simpler if the magnetic field is sufficiently small (i.e., $\omega_c\tau_m \ll 1$). In this case, (6.42) reduces to the more tractable form

$$J_x = \frac{ne^2}{m_c}\left[\langle\tau_m\rangle E_x - \omega_c\langle\tau_m^2\rangle E_y\right]$$
$$J_y = \frac{ne^2}{m_c}\left[\omega_c\langle\tau_m^2\rangle E_x + \langle\tau_m\rangle E_y\right]. \tag{6.43}$$

Now, let us consider a long, filamentary or flat semiconductor—one whose length is much larger than its width (or thickness) so that contact effects at the ends are not important. As discussed above, the current is assumed to be flowing in the x-direction. If we set the y-component of current to zero, a transverse field, the *Hall field*, will develop in the y-direction. This field is related to the longitudinal field as

$$\frac{E_y}{E_x} = -\omega_c\frac{\langle\tau_m^2\rangle}{\langle\tau_m\rangle} = -\omega_c\langle\tau_m\rangle\frac{\langle\tau_m^2\rangle}{\langle\tau_m\rangle^2} = -r\mu B = \tan\vartheta_H. \tag{6.44}$$

The last term defines the *Hall angle* ϑ_H and the *Hall factor*

$$r = \frac{\langle \tau_m^2 \rangle}{\langle \tau_m \rangle^2}, \tag{6.45}$$

while the mobility μ was given in (6.19). If we introduce the transverse field into the x-component of current, the transverse field has little effect on the conductivity if we keep only terms that do not include higher orders of $\omega_c \langle \tau_m \rangle$, and

$$J_x = \frac{ne^2 \langle \tau_m \rangle}{m_c} E_x, \tag{6.46}$$

just as in the absence of the magnetic field. This is merely a consequence of our assumption that the field is small. On the other hand, if the momentum relaxation time is independent of the energy, then the result that the conductivity is unaffected by the magnetic field becomes exact. The Hall scattering factor takes account of the energy spread of the carriers. It is, of course, unity in the case of an energy-independent scattering mechanism or for degenerate semiconductors, where the transport is at the Fermi energy. We will evaluate this for some scattering processes later in this chapter. The *Hall coefficient* R_H is now defined through the relation $E_y = R_H J_x B$, which may be determined by combining the last two equations above, as

$$R_H = \frac{E_y}{J_x B} = -\frac{r}{ne}. \tag{6.47}$$

This is, of course, just the case for electrons (due to the sign assumed on the force earlier). If the Hall factor is not known, it provides a source of error in determining the carrier density from the Hall effect. Worse, in the case of multiple scattering mechanisms, the scattering factor is usually complex and varies in magnitude with both temperature and carrier concentration. However, its value is typically in the range 0.5 to 1.5 so that the absolute measurement of the density is not critically upset by lack of knowledge of this factor.

If both holes and electrons are present, one cannot set the transverse currents to zero separately but must combine the individual particle currents prior to invoking the boundary conditions. The second of equation (6.43) can be rewritten, with both carriers present, in the form

$$J_y = e^2 \left\{ \left[\frac{n\omega_{ce} \langle \tau_{me}^2 \rangle}{m_e} - \frac{p\omega_{ch} \langle \tau_{mh}^2 \rangle}{m_h} \right] E_x + \left[\frac{n\langle \tau_{me} \rangle}{m_e} + \frac{p\langle \tau_{mh} \rangle}{m_h} \right] E_y \right\}, \tag{6.48}$$

so that the Hall angle is given by

$$\frac{E_y}{E_x} = -\frac{n\mu_e^2 Br_e - p\mu_h^2 Br_h}{n\mu_e + p\mu_h} = -\mu_h B \frac{nb^2 r_e - pr_h}{bn + p}, \tag{6.49}$$

and the mobility factor $b = \mu_e / \mu_h$ has been introduced. Since $J_x = (ne\mu_e + pe\mu_h)E_x$, the Hall coefficient now becomes

$$R_H = \frac{1}{e} \frac{nb^2 r_e - p r_h}{(bn + p)^2}. \tag{6.50}$$

It may be ascertained from this equation that the sign of the Hall coefficient identifies the carrier type, but it is important to note that the presence of equal numbers of electrons and holes does not equate to a zero Hall effect. Rather, the difference in the ability of the two types of carriers to move directly affects their lateral motion, and it is a cancellation of the transverse currents, rather than an equality in the carrier concentrations, that is important for a vanishing Hall effect. In fact, it is quite usual to observe a change of sign of the Hall coefficient in p-type semiconductors as they become intrinsic at high temperature due to the fact that the electron mobility is usually greater than the hole mobility.

6.1.5 Transport in high magnetic field

In the above discussion, it was generally assumed that $\omega_c <\tau_\mu> < 1$, so that we did not have to worry about closed orbits around the magnetic field. When the magnetic field is large, however, the electrons can make complete orbits about the magnetic flux lines. In higher magnetic fields, other effects begin to appear, even in non-quantizing magnetic fields. But, the most common is that for quantizing magnetic fields, where $\omega_c <\tau_m> > 1$ and $\hbar\omega_c > k_B T$. It is not usually recognized that both of these conditions are required to see quantization of the magnetic orbits. The first is required so that complete orbits are formed, and the latter is required to have the quantized levels separated by more than the thermal smearing. In this case the orbits behave as harmonic oscillators and the energy of the orbit is quantized [13]. We will assume that the magnetic field is large (e.g., $\omega_c <\tau_\mu> \gg 1$), so that the relaxation effects in can be ignored. Then, in the plane perpendicular to the magnetic field,

$$\frac{dv_x}{dt} = -\frac{e}{m_c} E_x - \omega_c v_y$$
$$\frac{dv_y}{dt} = -\frac{e}{m_c} E_y + \omega_c v_x \tag{6.51}$$

If the derivative with respect to time of the first equation is taken, we can insert the second to yield

$$\frac{d^2 v_x}{dt^2} = \frac{e\omega_c}{m_c} E_y - \omega_c^2 v_x. \tag{6.52}$$

For the present purposes, the electric field will be taken as $E = 0$ without any loss of generality. Then the two velocity components are given by

$$v_x = v_0 \cos(\omega_c t + \varphi)$$
$$v_y = v_0 \sin(\omega_c t + \varphi) \tag{6.53}$$

Here, φ is a reference angle that gives the orientation of the electron at $t = 0$. For the steady-state case of interest here, this quantity is not important and it can be taken as zero without loss of generality. The quantity v_0 is a term that will be equated to the energy, but is the linear velocity of the particle as it describes its azimuthal orbit around the magnetic flux lines.

The position of the particles is found by integrating (6.53). This gives the result in the simple form

$$x = \frac{v_0}{\omega_c}\sin(\omega_c t)$$
$$y = -\frac{v_0}{\omega_c}\cos(\omega_c t)$$

(6.54)

These equations describe a circular orbit with the radius

$$r^2 = x^2 + y^2 = \left(\frac{v_0}{\omega_c}\right)^2.$$

(6.55)

This is the radius of the cyclotron orbit for the electron as it moves around the magnetic field. As mentioned, this motion is that of a harmonic oscillator in two dimensions, and becomes quantized in high magnetic fields when $\hbar\omega_c > k_B T$ and $\omega_c <\tau_m> > 1$ [13]. We introduce the quantization by writing the energy in terms of the Landau energy levels of a harmonic oscillator as

$$E = \frac{1}{2}m_c v_0^2 = \left(n + \frac{1}{2}\right)\hbar\omega_c.$$

(6.56)

Thus the size of the orbit is also quantized. In the lowest level, (6.56) gives the radial velocity and this may be inserted into (6.55) to find the *Larmor radius,* or more commonly called the *magnetic length* l_B,

$$l_B = \sqrt{\frac{\hbar}{eB}}.$$

(6.57)

This is the quantized radius of the lowest energy level of the harmonic oscillator and is the minimum radius, as the higher-energy states involve a larger energy, which converts to a larger radial velocity and then to a larger radius. As the magnetic field is raised, the radius of the harmonic oscillator orbit is reduced, until all the states in that orbit are filled and then jumps to the size of the next orbit. The radial velocity also is increased. In fact, we can define the cyclotron radius at the Fermi surface as

$$r_{c,\max} = k_F r_L^2 = \frac{\hbar k_F}{eB} = \sqrt{2n_{\max} + 1}\, l_B.$$

(6.58)

Here, $n_{\max}$ is the highest occupied Landau level (that in which the Fermi level resides).

The quantized energy levels described by (6.56) are termed *Landau levels,* after the original work on the quantization carried out by Landau [14]. Transport across the magnetic field (e.g., in the plane of the orbital motion) shows interesting oscillations due to this quantization. Let us consider this motion in the two-dimensional plane, such as occurs in an inversion layer at the AlGaAs/GaAs interface, to which the magnetic field is normal. In looking at (6.55), one must consider the fact that the electrons will fill up the energy levels to the Fermi energy level. Thus there are in fact several Landau levels occupied, and therefore the electrons exhibit several distinct values of the orbital velocity v_0. In computing the effective radius, it is then necessary to sum over these levels. If there are n_s electrons per square centimeter, this sum can be written as

$$n_s\langle r^2\rangle = \sum_{i=0}^{i_{max}} r_i^2 = \sum_{i=0}^{i_{max}} \frac{(2i+1)\hbar}{eB},$$

(6.59)

or

$$\langle r^2\rangle = \sum_{i=0}^{i_{max}} (2i+1)\frac{\hbar}{eBn_s}.$$

(6.60)

The number of levels that are filled depends upon the degeneracy that is formed in each Landau level. This depends upon the magnetic field B, and thus connects directly to the radius r. But all of this depends on the areal density n_s as well.

As the magnetic field is raised, each of the Landau levels rises to a higher energy. However, the Fermi energy remains fixed, so at a critical magnetic field, the highest filled Landau level will cross the Fermi energy. At this point, the electrons in this level must drop into the lower levels, which reduces the number of terms in the sum in (6.60). This can only occur due to the increase of the number of states in each Landau level, so a given Landau level can hold more electrons as the magnetic field is raised. As a consequence, the average radius (obtained from the squared average of the radius) is modulated by the magnetic field, going through a maximum each time a Landau level crosses the Fermi level and is emptied. From (6.60), it appears that the radius is periodic in the inverse of the magnetic field. At least in two dimensions, this periodicity is proportional to the areal density of the free carriers and can be used to measure this density. The effect, commonly called the Shubnikov-de Haas effect [15], is normally applied by measuring the conductivity in the plane normal to the field. The Landau level must cross the Fermi energy, as mentioned above, so it can be argued that the magnetic field must move the Landau level this far. Thus the amount the reciprocal field must move is just

$$\Delta\left(\frac{1}{B}\right) = \frac{1}{B}\frac{\hbar\omega_c}{E_F} = \frac{e\hbar}{m_c E_F}.$$

(6.61)

The Fermi energy is related to the carrier density through. The value for the density of states in two dimensions may be taken from (4.23), for which the areal density may be found to be (we assume that the Zeeman effect is sufficiently large to split each Landau level into two spin split levels which reduces the density of states by a factor of two)

$$\Delta\left(\frac{1}{B}\right) = \frac{e}{2\pi\hbar n_s}.$$

(6.62)

If there is spin degeneracy, or a set of multiple valleys that are degenerate, then (6.62) needs to be modified to account for these degeneracies.

When the Fermi level lies in a Landau level, away from the transition regions, there are many states available for the electron to gain small amounts of energy from the applied field and therefore contribute to the conduction process. On the other hand, when the Fermi level is in the transition phase, the upper Landau levels are empty and the lower Landau levels are full. Thus there are no available states for the electron to be accelerated into, and the conductivity drops to zero in two dimensions. In three dimensions it can be scattered into the direction parallel to the field (the z-direction), and this conductivity provides a positive background on which the oscillations ride. In figure 6.4, we plot the longitudinal resistance for a two-dimensional electron gas residing in the inversion layer at the AlGaAs/GaAs hetero-junction interface. We plot the resistance as a function of 1/B in order to illustrate the effects of (6.62), and a clear periodicity in the resistance is seen.

The problem with the foregoing argument is that the transition region between Landau levels occurs over an infinitesimal range of magnetic field. If the conductivity is zero only over this small range, it would be almost undetectable, and the oscillations would be unobservable. In fact, it is the failure of the crystal to be perfect that creates the regions of low conductivity. In nearly all situations in transport in semiconductors, the role of the impurities and defects is quite small and can be treated by perturbation techniques, as with scattering. However, in situations where the transport is sensitive to the position of various defect levels, this is not the case. The latter is the situation here.

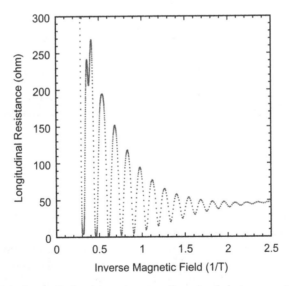

Figure 6.4. A plot of the longitudinal resistance in a two-dimensional electron gas at the interface between AlGaAs and GaAs at 1.5 K. This plot is versus the inverse magnetic field, and the period that is clearly evident determines the sheet density in the inversion layer through (6.62), although spin splitting is only observed in the lowest peak below $0.5/T$.

The end result is that the transition of the Fermi energy between Landau levels is broadened significantly due to the presence of the localized states. Thus, while the Fermi level is passing through the localized levels, the conductivity can drop to zero, since the localized levels also do not contribute to any appreciable conductivity. These levels are essential to the conductivity oscillations but do not contribute to either the periodicity or the conductivity itself. This is an interesting but true enigma of semiconductor physics.

The zeros of the conductivity that occur when the Fermi energy passes from one Landau level to the next-lowest level are quite enigmatic. They carry some interesting by-products. Let us rewrite (6.42) for the case where $\omega_c <\tau_m> > 1$, and the averages for the relaxation time are carried out at the Fermi surface, so $\tau \to \tau_F$. This becomes

$$
\begin{aligned}
J_x &= \frac{n_s e^2}{m_c}\left[\frac{1}{\omega_c^2 \tau_F}E_x - \frac{1}{\omega_c}E_y\right] \\
J_y &= \frac{n_s e^2}{m_c}\left[\frac{1}{\omega_c}E_x + \frac{1}{\omega_c^2 \tau_F}E_y\right]
\end{aligned}
\tag{6.63}
$$

From this equation, the conductivity tensor can be written as

$$
\sigma = \begin{bmatrix} \sigma_{xx} & \sigma_{xy} & 0 \\ -\sigma_{xy} & \sigma_{xx} & 0 \\ 0 & 0 & \sigma_{zz} \end{bmatrix},
\tag{6.64}
$$

where

$$
\begin{aligned}
\sigma_{xx} &= \frac{n_s e^2}{m_c \omega_c^2 \tau_F} \\
\sigma_{xy} &= -\omega_c \tau_F \sigma_{xx}
\end{aligned}
\tag{6.65}
$$

We have also included the z-component for completeness, although it is zero in this type of interface. Inverting this matrix to find the resistivity matrix, the longitudinal resistivity is given by

$$
\rho_{xx} = \frac{\sigma_{xx}}{\sigma_{xx}^2 + \sigma_{xy}^2}.
\tag{6.66}
$$

In the situation where the longitudinal conductivity σ_{xx} goes to zero, we note that the longitudinal resistivity ρ_{xx} also goes to zero. Thus there is no resistance in the longitudinal direction. Is this a superconductor? No! It must be remembered that the conductivity is also zero, so that there is no allowed motion along that direction. The entire electric field must be perpendicular to the current and there is no dissipation since $\mathbf{E} \cdot \mathbf{J} = 0$, but the material is not a superconductor.

The presence of the localized states, and the transition region for the Fermi energy between one Landau level and the next-lower level, leads to another remarkable effect. This is the quantum Hall effect, first discovered in silicon metal-oxide-semiconductor

(MOS) transistors prepared in a special manner so that the transport properties of the electrons in the inversion layer could be studied [16]. Klaus von Klitzing was awarded the Nobel prize for this discovery in 1985. The effect leads to quantized resistance, which can be used to provide a much better measurement of the fine structure constant used in quantum field theory (used in quantum relativity studies), and today provides the standard of resistance in the United States, as well as in many other countries.

A full derivation of the quantum Hall effect is well beyond the level at which we are discussing the topic here. However, we can use a consistency argument to illustrate the quantization exactly, as well as to describe the effect we wish to observe. When the Fermi level is in the localized state region, and lies between the Landau levels, the lower Landau levels are completely full. We may then say that

$$E_F \approx N\hbar\omega_c, \tag{6.67}$$

where N is an integer giving the number of filled (spin-split) levels. At first, one might think that (6.67) would place the Fermi energy in the center of a Landau level, but note that no equality is used. The argument desired is to relate the Fermi level to the number of carriers that must be contained in the filled Landau levels. In fact, the magnetic field is usually so high that spin degeneracy is raised and N measures half-Landau levels rather than full levels. That is, it measures the number of spin-resolved levels. Using (6.67) for the Fermi energy in (4.23), we have

$$n_s = N\frac{eB}{h}. \tag{6.68}$$

The density is taken to be constant in the material, so that the Hall resistivity is given as

$$\rho_{\text{Hall}} = \frac{E_y}{J_x} = -BR_H = \frac{h}{Ne^2}. \tag{6.69}$$

The quantity $h/e^2 \sim 25.81$ kΩ is a ratio of fundamental constants. Thus the conductance (reciprocal of the resistance) increases stepwise as the Fermi level passes from one Landau level to the next-lower level. This is shown in figure 6.5 for the data of figure 6.4. The values of N for this data are shown at the filled levels. Between the Landau levels, when the Fermi energy is in the localized state region, the Hall resistance is constant (to better than 1 part in 10^7) at the quantized value given by (6.69) since the lower Landau levels are completely full. The accuracy with which the Hall resistivity appears in the quantum Hall effect is remarkable. In fact, it is so accurate that it is unlikely to be a result of a few random impurities in the material as was discussed above. The accuracy arises from the fact that the quantum Hall effect is the result of topological stability, and the result is related to a Chern number [17]. The topological nature was first discussed by Laughlin [18]. It is well-known that a periodic lattice in two dimensions can be folded onto the surface of a torus, and that a closed trajectory must make loops around the two dimensions that are rationally related. This leads to allowing only a particular discrete set of magnetic fields [19, 20]. The variation of the wave function with the two angles of

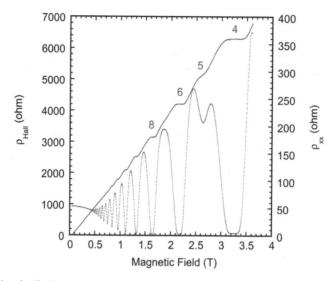

Figure 6.5. The longitudinal and Hall (transverse) resistance for the data of figure 6.4. Plateaus are evident at the highest magnetic fields, and the values of N for these plateaus are indicated. Only the area around $N = 5$ shows spin splitting of the Landau levels.

the torus is known as the adiabatic curvature, and when the integral of this over the surface is an integer, this gives the Chern number. This latter fact stabilizes the structure, and the transport and leads to the stability of the quantum Hall effect.

The magnetic field could, of course, be swept to higher values than shown in these figures, and in high quality material, new features appear. These are not explained by the above arguments. In fact, in high quality samples, once the Fermi energy is in the lowest Landau levels, one begins to see fractional filling and plateaus, in which the resistance differs from h/e^2 (that is, N takes on fractional values which are the ratio of integers) [21]. This *fractional quantum Hall effect* is theorized to arise from the condensation of the interacting electron system into a new many-body state characteristic of an incompressible fluid [22]. Tsui, Störmer, and Laughlin shared the Nobel prize for this discovery. However, the properties of this many-body ground state are clearly beyond the present level, and we leave this discussion of more properties of the quantum Hall effect itself. The interested reader can find a discussion in [23].

6.1.6 Energy dependence of the relaxation time

In the above sections, various averages of the relaxation time τ_m have appeared in which it is necessary to average over the distribution function. These averages give simple relationships that are necessary for computing various transport coefficients. The energy dependence of the scattering rates, for most processes, are quite complicated. In this section, it is desired to examine a general form for the dependence of the momentum relaxation time on the energy, which is taken to be $\tau_m = AE^{-s}$, where A and s are constants that are different for the different scattering mechanisms. The average relaxation time is determined by carrying out the integrations inherent in (6.22) and (6.45) for a specific distribution function. For the latter, we take a Maxwellian so that

$$\frac{\partial f_0(E)}{\partial E} \sim -\frac{1}{k_B T} \exp\left(-\frac{E}{k_B T}\right). \tag{6.70}$$

In addition, reduced units will be defined as $x = E/k_B T$. Then, (6.22) becomes

$$\langle \tau_m \rangle = \frac{A}{(k_B T)^s} \frac{\displaystyle\int_0^\infty x^{(d/2)-s} e^{-x} dx}{\displaystyle\int_0^\infty x^{(d/2)} e^{-x} dx} = \frac{A}{(k_B T)^s} \frac{\Gamma[(d/2) + 1 - s]}{\Gamma[(d/2) + 1]}, \tag{6.71}$$

and the gamma function has been introduced in the last form. Usually, the semiconductor of interest is a bulk material, and therefore a three-dimensional solid. As a consequence, the value $d = 3$ will be used in the remainder of this section. Then, for example, if we consider acoustic phonon scattering, $s = 1/2$, and

$$\langle \tau_m \rangle = \frac{A}{(k_B T)^s} \frac{4}{3\sqrt{\pi}}. \tag{6.72}$$

A second important average of the relaxation time is that arising in the Hall scattering factor, where the average of the square of the relaxation time is required. This can be obtained merely by extending (6.71) to

$$\langle \tau_m^2 \rangle = \frac{A^2}{(k_B T)^{2s}} \frac{\Gamma\left(\frac{5}{2} - 2s\right)}{\Gamma(5/2)}, \tag{6.73}$$

and the Hall scattering factor can readily be determined to be

$$r = \frac{\langle \tau_m^2 \rangle}{\langle \tau_m \rangle^2} \frac{\Gamma\left(\frac{5}{2} - 2s\right)\Gamma(5/2)}{\left[\Gamma\left(\frac{5}{2} - s\right)\right]^2}. \tag{6.74}$$

Another average that is important is that of the diffusion coefficient in (6.32). Here, we can define an average relaxation time for diffusion as

$$\langle \tau_{mD} \rangle = \frac{3m_2}{2\langle E \rangle} \left\langle \frac{\tau_m v^2}{3} \right\rangle = \frac{1}{\langle E \rangle} \langle \tau_m E \rangle = \frac{2A}{3(k_B T)^{s+1}} \frac{\Gamma\left(\frac{7}{2} - s\right)}{\Gamma(5/2)}. \tag{6.75}$$

Naturally, this average is different than that used for the momentum relaxation time in (6.72), but we have achieved the result using the common version of the Einstein relationship between the diffusion coefficient and the mobility.

The problem arises due to the fact that there is usually more than one scattering mechanism present. When this is the case, their effects must be combined prior to computing the averages, which leads to a very complicated average. The most common manner of adding the effect of various scattering mechanisms is introduced by adding the effective resistances of each, which leads to

$$\frac{1}{\tau} = \sum\nolimits_{i=1}^{n} \frac{1}{\tau_i},$$

(6.76)

for which the sum is carried out over the different scattering mechanisms. In a typical semiconductor, impurity scattering, acoustic phonon scattering, and a variety of optical phonon scattering processes will all be involved. The average relaxation time, however, introduces a temperature dependence to the mobility through the temperature term arising in the above equations and from any temperature variation that is in the constant A. In fact, (6.76) is an expression of Mathiesen's rule, in which each scattering process is considered to be independent of all others. This is valid only when there is no correlation between the scattering events, such as may occur between carrier-carrier scattering (screening) and impurity scattering. While one must examine this in each case, it is usually true except in very high carrier density situations. One can infer from the calculations of the various scattering processes in chapter 5, that doing such an average can be quite difficult when several scattering events are present. In this case, one generally accedes to the use of numerical techniques for solving the Boltzmann transport equation (6.4) with (6.8) for a number of scattering processes. Perhaps the best approach for near equilibrium transport is via Rode's technique, discussed in the next section. In the case of transport in far from equilibrium situations, other approaches have been developed, and these are discussed in chapter 7.

6.2 Rode's iterative approach

The problem with the general relaxation-time approximation lies in the presence of inelastic scattering processes. The difficulty lies with the collision integral (6.7). Even when the system is non-degenerate, the collision integral still involves the distribution function at two different energies and momentum. This becomes particularly problematic when the scattering is anisotropic, such as in polar optical phonon scattering. If the scattering is isotropic, however, the problem is somewhat easier to handle. One way to approach this is via the method called Rode's iterative technique [24]. In this approach we assume that the distribution can be split into the equilibrium part and a momentum dependent correction term. This correction term includes the first order Legendre polynomial, which is just a simple cosine term, to account for the angle between the electric field and the momentum. But, the correction term has arbitrary dependence on both the magnitude and the direction of the momentum vector. If we were to use a Legendre expansion, then this correction term would be an isotropic function depending only upon the energy [25], and this is what we want to avoid. Let us reiterate that this approach is for transport that is not far from equilibrium. Then, we write the distribution function as

$$f(\boldsymbol{k}) = f_0(E) + g(\boldsymbol{k})\cos \vartheta,$$

(6.77)

where, as mentioned, the angle ϑ defines the direction of the momentum vector relative to the electric field. In essence, this approach is consistent with the idea of linear response to the electric field. Note here that E is the energy and not the field.

6.2.1 Transport in an electric field

The Boltzmann equation, in the case where the material may in fact be degenerate, is given by the combination of (6.4) and (6.7), although we will deal with the time independent steady-state transport and drop the partial derivative with respect to time that is found in (6.4). Since the first term in (6.77) is the equilibrium distribution function, it will give no contribution to the collision term, as

$$\int \{P(k, k')f_0(E)[1 - f_0(E')] - P(k', k)f_0(E')[1 - f_0(E)]\}d^3k' = 0 \qquad (6.78)$$

in order to satisfy the principle of detailed balance. To make life easier, we will take the electric field and the diffusion gradient in the z direction. As previously in the relaxation-time approximation, we will retain only the equilibrium distribution in the two streaming terms of the Boltzmann equation and keep only terms linear in the small correction term of (6.77) in the collision integral. Then, the Boltzmann equation can be written as (where E is now the electric field)

$$v\frac{\partial f_0}{\partial z} + \frac{eE}{\hbar}\frac{\partial f_0}{\partial k} = -g(k)\int [P(k, k')[1 - f_0(k')] + P(k', k)f_0(k')]d^3k'$$

$$+ \frac{3}{2}\int_{-1}^{1} x \int x'g(k')[P(k', k)(1 - f_0(k)) \qquad , \qquad (6.79)$$

$$+ P(k, k')f_0(k)]d^3k'dx$$

where

$$x = \cos\vartheta_k \qquad (6.80)$$
$$x' = \cos\vartheta_{k'}$$

are the two polar angles between the two momentum vectors and the z-axis. The scattering functions are actually dependent upon the difference between the two momentum vectors rather than the polar angles, and the second polar angle can be expanded into an expression for the first polar angle and the angle between the two vectors by standard geometry. We take this difference angle to be

$$\cos\vartheta_{k'} = \cos\vartheta_k\cos\Theta + \sin\vartheta_k\sin\Theta \rightarrow xX \qquad (6.81)$$

as the sine terms will integrate to zero, and we have introduced the new X for the cosine of the difference angle. Thus, we may write the integration in the second term in (6.79) as

$$\int x'A(X)d^3k' = x\int XA(X)d^3k'. \qquad (6.82)$$

This now allows us to simplify the expression (6.79) to

$$v\frac{\partial f_0}{\partial z} + \frac{eE}{\hbar}\frac{\partial f_0}{\partial k} = \int Xg(k')\{P(k', k)[1 - f_0(k)] + P(k, k')f_0(k)\}d^3k'$$
$$- g(k)\int [P(k, k')[1 - f_0(k')] + P(k', k)f_0(k')]d^3k' \quad . \tag{6.83}$$

This result is exact for small electric fields within the linear response formalism. After the integration over the primed momentum, the result is only a function of the single magnitude k, as there are no angular parts left in the equation and we need consider g only a function of the magnitude of the momentum k. Thus, g is isotropic for isotropic energy bands around the main extrema. The exact form of the Fermi–Dirac or Maxwellian distribution function is maintained by the inelastic processes, as the shape of the distribution is unaffected by elastic scattering. This was evident already for the relaxation-time approximation. Since all the transport coefficients are determined solely by integrations over g, all of these integrals can be performed, even with non-parabolic bands.

Because of these properties, it is desirable to split the scattering functions into the elastic and inelastic parts. The elastic parts are all evaluated on the same energy shell and thus reduce to the single integral

$$g(k)\int (1 - X)P_{el}(k, k')d^3k' \to \frac{g(k)}{\tau_m}. \tag{6.84}$$

In the last term, we have recognized that the left-hand side is precisely the relaxation-time approximation and thus defined the momentum relaxation time appropriately. In order to simplify the remaining terms, we denote them accordingly as those for scattering out of the state k and scattering in to the state k, with

$$S_O = \int [P_{in}(k, k')[1 - f_0(k')] + P_{in}(k', k)f_0(k')]d^3k'$$
$$S_{IN}(g') = S_{IN}[g(k')] \quad . \tag{6.85}$$
$$= \int Xg(k')\{P_{in}(k', k)[1 - f_0(k)] + P_{in}(k, k')f_0(k)\}d^3k'$$

Here, the lower case 'in' refers to inelastic, while the upper case 'IN' refers to the in-scattering term, and one should not confuse these two similar notations. Now, we can rewrite (6.83) as

$$g(k) = \frac{S_{IN}(g') - v\frac{\partial f_0}{\partial z} - \frac{eE}{\hbar}\frac{\partial f_0}{\partial k}}{S_O(k) + 1/\tau_m}. \tag{6.86}$$

The above result is an integral equation for the linear deviation of the distribution function from its equilibrium form. The best method of approaching a solution is by the iterative procedure. An interesting aspect of this is that it is possible to treat the iteration process as a time stepping method to yield the time dependent behavior for the Boltzmann equation [24]. One can begin the iterative process by assuming that, for iteration '0', we can assume that $g_0(k) = 0$. The next iterate can then be found using the sequence

$$g_{r+1}(k) = \frac{S_{IN}(g_r') - v\frac{\partial f_0}{\partial z} - eEv\frac{\partial f_0}{\partial E}}{S_O(k) + 1/\tau_m}. \tag{6.87}$$

The energy derivative has also been introduced into the numerator to connect with results obtained in the relaxation time approximation. It is clear here that, if we do not have inelastic scattering, or if the inelastic scattering is isotropic, then only a single iteration is required. It is the anisotropic inelastic scattering that requires the need to continue to iterate. Once the iteration procedure converges, then the transport parameters can be found by integration over the g term. For example, the mobility is now found from

$$\mu = \frac{1}{3E}\frac{\int vk^2 g(k)dk}{\int k^2 f_0\, dk}. \tag{6.88}$$

The term in the denominator is merely the carrier density which is used to remove the latter from the result that normally arises, just as in the considerations in the relaxation-time approximation, and (6.20).

In figure 6.6, the mobilities for electrons in several semiconductors are plotted as a function of temperature for a doping level of 10^{16} cm^{-3}. These were calculated with

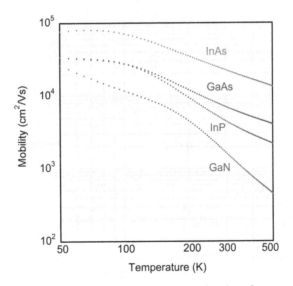

Figure 6.6. Mobilities of some representative semiconductors as a function of temperature. These mobilities are determined using Rode's method via a computational tool at NanoHUB.org [26], assuming an impurity concentration of 10^{16} cm^{-3}.

the Rode method tool [26] that exists at NanoHUB.org, a national center for computational nanotechnology. In general, all the mobilities decrease with temperature due to the increase in the phonon occupation and general increase in the average energy of the distribution of electrons. In the case of InAs, there is a drop-off at the lowest temperatures due to the stronger presence of impurity scattering in this narrow gap material.

6.2.2 Adding a magnetic field

The inclusion of a small magnetic field in the left-hand side of the Boltzmann equation is relatively straight-forward, just as in the relaxation-time approximation of section 6.1.4. However, we assume that there are two terms in the linear response formalism [27], by which the distribution function can be written as

$$f(\mathbf{k}) = f_0(E) + g(\mathbf{k})\cos\vartheta + h(\mathbf{k})\cos\varphi, \qquad (6.89)$$

where the angle φ is the angle between the momentum vector and the vector defined by $\mathbf{B} \times \mathbf{E}$, where the latter is the transverse force component of the Lorentz force. In this approach, we assume that the magnetic field is transverse to the electric field, as this is the standard configuration for studying the Hall effect, as discussed in section 6.1.4. If there were any longitudinal magnetic field, such as a field parallel to the electric field, it would have no effect on the transport along the electric field as is evident from (6.64).

We can now proceed just as for the case in the absence of a magnetic field by inserting (6.89) into the Boltzmann equation and finding the linear response terms. In fact, we can use our regular notation for the out-scattering and in-scattering of the inelastic scattering processes. This leads to

$$
\begin{aligned}
g(k) &= \frac{S_{IN}(g') - v\frac{\partial f_0}{\partial z} - \frac{eE}{\hbar}\frac{\partial f_0}{\partial k} + \beta S_O(k)h(k)}{S_O(k) + 1/\tau_m} \\
h(k) &= \frac{S_{IN}(h') - v\frac{\partial f_0}{\partial z} - \frac{eE}{\hbar}\frac{\partial f_0}{\partial k} - \beta S_O(k)g(k)}{S_O(k) + 1/\tau_m}
\end{aligned}
\qquad (6.90)
$$

where $\beta = \dfrac{\omega_c}{\left(S_O(k) + \frac{1}{\tau_m}\right)}$, with ω_c the cyclotron frequency defined just below (6.42). If we are dealing with non-parabolic bands, then the mass in the cyclotron frequency is defined by $\hbar k/v$, according to our definitions of effective mass in section 2.6. Now, it turns out that equation (6.90) do not converge for the iterative procedure, especially when $\mu B \geqslant 1$. The last term in the numerator of (6.90) increases as B^2 and, even for small values, leads to the problems of convergence. Rode [27] shows how this can be converted to a contraction mapping which will provide the needed convergence. This is achieved by removing the g and h terms in the numerator on the right hand sides of (6.90), leading to

$$g(k) = \frac{S_{IN}(g') - v\frac{\partial f_0}{\partial z} - \frac{eE}{\hbar}\frac{\partial f_0}{\partial k} + \beta S_{IN}(h')}{[S_O(k) + 1/\tau_m](1 + \beta^2)}$$

$$h(k) = \frac{S_{IN}(h') + \beta\left(v\frac{\partial f_0}{\partial z} + \frac{eE}{\hbar}\frac{\partial f_0}{\partial k}\right) - \beta S_{IN}(g')}{[S_O(k) + 1/\tau_m](1 + \beta^2)} \qquad (6.91)$$

It is shown by Rode [27] that these last two equations are a proper single *complex* function in which the two terms are the real and imaginary parts of the function. Hence, the two added terms in (6.89) are the real and imaginary parts of a single linear response function that is actually complex due to the magnetic field.

The results of the above treatment lead us to be able to define the Hall mobility from the two parts of the complex function as

$$\mu_H = R_H\sigma = \frac{\langle h/B \rangle}{\langle g \rangle}, \qquad (6.92)$$

where R_H is the Hall constant (6.47) and σ is the conductivity (6.18). Here, the brackets denote the average used, for example, as in (6.88).

In figure 6.7, we plot the Hall factor r, the ratio of the Hall mobility to the normal mobility, for two different carrier densities for the alloy $In_{0.65}Al_{0.35}As$ over a range of temperatures. This Hall factor was calculated using Rode's method as described above. An important point is that the higher doped material is completely degenerate at all temperatures and so the Hall factor remains close to unity as it would be in metals. The full degeneracy arises as the doping density is larger than the effective density of states, thus requiring the Fermi energy to sit within the conduction band even at 300 K. The lower density actually freezes out the carriers

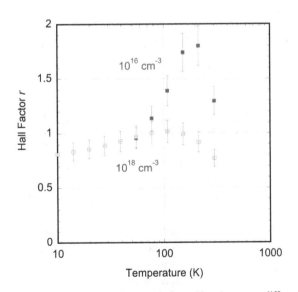

Figure 6.7. The Hall scattering factor r for electrons in $In_{0.65}Al_{0.35}As$ at two different doping densities.

at low temperature, but the peak that is observed is quite common in polar materials [24]. However, it is somewhat larger than usually found, and this may be due to the use of a single average polar optical phonon in the calculation when it is known to be a two mode system [28], as described in section 3.5. Whether or not this two-mode behavior will affect the Hall factor is currently unknown.

6.3 The effect of spin on transport

We introduced the appearance of spin-splitting in the measurements of transport at high magnetic fields already in section 6.1.5. Such spin-splitting is a result of the Zeeman effect [29], in which the energy of a free carrier is modified by the magnetic field interacting with the spin. Normally, the effect is most familiar with optical spectroscopy of impurity levels in a solid, where the complex spin structures lead to a splitting into a family of curves. For the free carrier in a semiconductor, however, the Zeeman effect leads to just two levels, given as the additional energy

$$E_z = g\mu_B \mathbf{S} \cdot \mathbf{B} = \pm \frac{1}{2} g\mu_B B_z, \tag{6.93}$$

where the magnetic field is oriented in the z-direction in the last form. Here, $\mu_B = e\hbar/2m_0$ is the Bohr magneton, 57.94 μV T^{-1}. The factor g is referred to as the Landé g-factor, which is mainly a 'fudge' factor that has a value of 2 for a truly free electron. It differs greatly from this value in semiconductors, and can even be negative ($\sim$−0.43 in GaAs at low temperatures [30]), which has the effect of reversing the ordering of the two spin-split energy levels. As can be seen in figures 6.4 and 6.5, the Zeeman effect leads to splitting of the Shubnikov-de Haas oscillations at high magnetic fields, which removes the spin degeneracy of the Landau levels. Interestingly enough, at very high magnetic fields, the g-factor changes with the magnetic field, especially in nano-structures [31–33].

We discussed the spin–orbit interaction in chapter 2 in connection with the band structure, where we encountered the Pauli spin matrices and the vector function

$$\boldsymbol{\sigma} = \sigma_x \boldsymbol{a}_x + \sigma_y \boldsymbol{a}_y + \sigma_z \boldsymbol{a}_z, \tag{6.94}$$

where

$$\sigma_x = \begin{bmatrix} 0 & 1 \\ 1 & 0 \end{bmatrix}, \ \sigma_y = \begin{bmatrix} 0 & -i \\ i & 0 \end{bmatrix}, \ \sigma_z = \begin{bmatrix} 1 & 0 \\ 0 & -1 \end{bmatrix} \tag{6.95}$$

are the Pauli spinors. It is clear from these matrices that the wave function is now more complicated when we add a spin component. In any quantum confinement problem such as the atom or a quantum well, the total wave function is a sum over a set of eigen-functions $\psi_i(\mathbf{r})$. For each of these functions, which are defined by a set of quantized values due to the confinement, one has not considered the spin. If we now want to also include the spin, then we need an additional part of each eigen-state wave function. Typically, this is a multiplicative term describing the spin state. Traditionally, this is a two component wave function called a spinor. From the Zeeman effect, we know that one typically denotes the extra energy for the spin up state by the value 1/2, and the value of the spin down state by the value $-1/2$. We use this to denote the two possible states and their spinors as

$$\varphi\left(\frac{1}{2}\right) = \begin{pmatrix} 1 \\ 0 \end{pmatrix}$$
$$\varphi\left(-\frac{1}{2}\right) = \begin{pmatrix} 0 \\ 1 \end{pmatrix}, \tag{6.96}$$

where the first row refers to the *up* state and the second row refers to the *down* state. Thus, the eigen-values correspond to those adopted in the Zeeman effect, as mentioned above.

Because the spin angular momentum has been taken to be oriented along the z-axis, we expect the spin matrix for the z-component of angular momentum must be diagonal for two reasons. First, it must commute with the total spin and with the Hamiltonian, and, second, it must produce the eigen-values found from the Zeeman effect. Thus, we simply state that

$$S_z = \frac{\hbar}{2}\sigma_z = \frac{\hbar}{2}\begin{bmatrix} 1 & 0 \\ 0 & -1 \end{bmatrix}, \tag{6.97}$$

and this gives

$$S_z\varphi\left(\frac{1}{2}\right) = \frac{\hbar}{2}\begin{bmatrix} 1 & 0 \\ 0 & -1 \end{bmatrix}\begin{pmatrix} 1 \\ 0 \end{pmatrix} = \frac{\hbar}{2}\varphi\left(\frac{1}{2}\right)$$
$$S_z\varphi\left(-\frac{1}{2}\right) = \frac{\hbar}{2}\begin{bmatrix} 1 & 0 \\ 0 & -1 \end{bmatrix}\begin{pmatrix} 0 \\ 1 \end{pmatrix} = -\frac{\hbar}{2}\varphi\left(-\frac{1}{2}\right) \tag{6.98}$$

as expected. We know from the study of normal angular momentum that the value of L^2 is given as $l(l + 1)\hbar^2$. Thus, we expect a similar result for the spin angular momentum, and using $s = 1/2$, we get $S^2 = s(s + 1)\hbar^2$. More properly, this value is for the square of the magnitude of the total spin angular momentum. Again, the matrix representation of this total spin angular momentum must be diagonal, and this is given by

$$[S]^2 = \frac{3\hbar^2}{4}\begin{bmatrix} 1 & 0 \\ 0 & 1 \end{bmatrix}. \tag{6.99}$$

To find the other components of the spin angular momentum, and the other spin matrices, we introduce rotating coordinates, as

$$S_+ = S_x + iS_y$$
$$S_- = S_x - iS_y, \tag{6.100}$$

Then, we can write the square magnitude of the total spin angular momentum as

$$S^2 = S_x^2 + S_y^2 + S_z^2 = S_+S_- + S_z^2 - i[S_x, S_y]. \tag{6.101}$$

The commutator is given by the normal commutators encountered in angular momentum. That is, we have

$$[S_x, S_y] = i\hbar S_z$$
$$[S_y, S_z] = i\hbar S_x. \qquad (6.102)$$
$$[S_z, S_x] = i\hbar S_y$$

Then, (6.101) becomes

$$S^2 = S_x^2 + S_y^2 + S_z^2 = S_+S_- + S_z^2 + \hbar S_z. \qquad (6.103)$$

Similarly, if we reverse the two rotating terms, we get

$$S^2 = S_x^2 + S_y^2 + S_z^2 = S_-S_+ + S_z^2 - \hbar S_z. \qquad (6.104)$$

We can combine these last two equations, and use the results of (6.97) and (6.99) to yield

$$S^2 - S_z^2 = \frac{1}{2}(S_+S_- + S_-S_+) = \frac{\hbar^2}{2}. \qquad (6.105)$$

The operators S_+ and S_- act as creation and annihilation operators for the spin angular momentum. That is, operating on a spinor with the first of these operators will raise the angular momentum, which can only occur if it acts on the spin down state and produces the spin up state, or

$$S_+\varphi\left(-\frac{1}{2}\right) = \hbar\varphi\left(\frac{1}{2}\right) \Longrightarrow S_+ = \hbar\begin{bmatrix} 0 & 1 \\ 0 & 0 \end{bmatrix}. \qquad (6.106)$$

Similarly, acting with the operator S_- removes a quantum of angular momentum and lowers the spin angular momentum. This can occur only if the operator acts upon the spin up state and produces the spin down state, or

$$S_-\varphi\left(\frac{1}{2}\right) = \hbar\varphi\left(-\frac{1}{2}\right) \Longrightarrow S_- = \hbar\begin{bmatrix} 0 & 0 \\ 1 & 0 \end{bmatrix}. \qquad (6.107)$$

We can easily invert equation (6.100) to give

$$S_x = \frac{1}{2}(S_+ + S_-) = \frac{\hbar}{2}\begin{bmatrix} 0 & 1 \\ 1 & 0 \end{bmatrix} = \frac{\hbar}{2}\sigma_x$$
$$S_y = \frac{1}{2}(S_+ - S_-) = \frac{\hbar}{2}\begin{bmatrix} 0 & -i \\ i & 0 \end{bmatrix} = \frac{\hbar}{2}\sigma_y \qquad (6.108)$$

Thus, we see that the Pauli spinors given in (6.95) are the matrix operator parts of the spin angular momentum. We will have reason to use these operators in our following developments of spin-based operations in the band structure and transport.

While the Zeeman effect is the best known of the spin effects on transport, there are other effects which have become better known since the intense interest in spin-based semiconductor devices arose a few decades ago. This interest was spawned by the idea of a spin based transistor [34], but has grown over the possibility of a plethora of spin-based logic gates which will not be subject to the capacitance limitations of charge-based switching circuits. Many of these new concepts are dependent upon the propagation of spin channels, and the use of the spin orientation as a logic variable, and this has fostered the term *spintronics* [35]. In depth coverage of this area is, of course, beyond what can be done in this book, but the basic concepts are described in the next few sub-sections.

The spin of an electron can be manipulated in a large number of manners, but in order to take advantage of current semiconductor processing technology, it would be preferred to find a purely electrical means of achieving this. For this reason, a great deal of attention has centered on the spin Hall effect in semiconductors, where in the presence of spin–orbit coupling a transverse spin current arises in response to a longitudinal charge current, without the need for magnetic materials or externally applied magnetic fields [36]. We have already encountered the spin–orbit interaction in section 2.4.3, but we also used it when we dealt with the $\mathbf{k} \cdot \mathbf{p}$ interaction in section 2.5. But, there are other forms of the spin–orbit interaction that are of interest in situations in which symmetries are broken in the semiconductor device. In the spin Hall effect, we achieve edge states, as in the quantum Hall effect, but which are spin polarized.

The spin Hall effect most commonly originates from the Rashba form of spin–orbit coupling [37], which is present in a two-dimensional electron gas (2DEG) formed in an asymmetric semiconductor quantum well. This is known as *structural inversion asymmetry*. Early studies showed that in the infinite two-dimensional sample limit, arbitrarily small disorder introduces a vertex correction that exactly cancels out the transverse spin current [38]. However, in finite systems such as quantum wires, the spin Hall effect survives in the presence of disorder and manifests itself as an accumulation of oppositely polarized spins on opposite sides of the wire [39]. This has led to the proposal for a variety of devices that utilize branched, quasi-1D structures to generate and detect spin-polarized currents through purely electrical measurements [40], and experiments have been performed to try to measure these effects [41]. In addition to Rashba spin–orbit coupling, a term due to the *bulk inversion asymmetry* of the host semiconductor crystal, known as Dresselhaus spin–orbit coupling [42], can also yield a spin Hall current. We will deal with these two asymmetries in the reverse order, treating the older one first.

6.3.1 Bulk inversion asymmetry

Bulk inversion asymmetry arises in crystals which lack an inversion symmetry, such as the zinc-blende materials. In these crystals, the basis at each lattice site is composed of two dissimilar atoms, such as Ga and As. Because of this, the crystal has lower symmetry than e.g. the diamond lattice, due to this lack of inversion symmetry. Without this inversion symmetry, one still can have symmetry of the energy bands $E(\mathbf{k}) = E(-\mathbf{k})$, but the periodic part of the Bloch functions no longer satisfy $u_k(\mathbf{r}) = u_k(-\mathbf{r})$. Thus, the normal two-fold spin degeneracy is no longer required throughout the Brillouin zone [42]. In fact, this interaction, when treated within perturbation theory, gives rise to the warped surface of the valence bands, as given in section 2.5.3. For the conduction band, the perturbing Hamiltonian can be written as

$$H^{BIA} = \eta\left(\{k_x, k_y^2 - k_z^2\}\sigma_x + \{k_y, k_z^2 - k_x^2\}\sigma_y + \{k_z, k_x^2 - k_y^2\}\right), \quad (6.109)$$

where k_x, k_y, and k_z are aligned along the [100], [010], and [001] axes, respectively, and the σ_i are the Pauli spin matrices, discussed above. The terms in curly brackets are modified anti-commutation relations given by

$$\{A, B\} = \frac{1}{2}(AB + BA), \quad (6.110)$$

while the parameter η is given by [43]

$$\eta = -\frac{4i}{3}PP'Q\left[\frac{1}{(E_G + \Delta)(\Gamma_0 - \Delta_c)} - \frac{1}{E_{G\Gamma_0}}\right], \quad (6.111)$$

where E_G and Δ have their meanings from section 2.5.1 and the quantities P, P', and Q are couplings along the line of (2.100). That is, E_G and Δ are the principal band gap at the zone center and the spin–orbit splitting in the valence band, and the others are various momentum matrix elements. Here, Γ_0 is the splitting of the two lowest conduction bands at the zone center and Δ_c is the spin–orbit splitting of the conduction band at the zone center. This interaction is stronger in materials with small band gaps, as may be inferred from the last equation. Note that (6.109) is cubic in the magnitude of the wave vector and is often referred to as the k^3 term.

While the above expressions apply to bulk semiconductors, much of the work of the past two decades has been applied to quasi-two-dimensional systems in which the carriers are confined in a quantum well such as exists at the interface between AlGaAs and GaAs. Often, this structure is then patterned to create a quantum wire. For example, a common configuration is with growth along the [001] axis, so that there is no net momentum in the z-direction, and $\langle k_z \rangle = 0$, while $\langle k_z^2 \rangle \neq 0$. Then, (6.109) can be written as

$$H^{BIA} \rightarrow \eta\left[\langle k_z^2 \rangle(k_y\sigma_y - k_x\sigma_x) + k_xk_y(k_y\sigma_x - k_x\sigma_y)\right]. \quad (6.112)$$

The prefactor of the first term in the square brackets is constant, and depends upon the material and the details of the quantum well. The average over the z-momentum corresponds to the appearance of sub-bands in the quantum well. But, this structure has now split (6.109) into a k-linear term and a k^3 term.

To explore (6.112) a little closer, let us chose a set of spinors to represent the spin up and down states as defined in (6.96). Then, the linear first term in (6.112) gives rise to an energy splitting according to

$$\Delta E_1 \sim -\eta\langle k_z^2 \rangle(k_x \pm ik_y). \quad (6.113)$$

In the rotating coordinates where $k_\pm = k_x \pm ik_y$, we see that the spin up state rotates around the z-axis in a right-hand sense, with the spin tangential to the constant energy circle. On the other hand, the spin down state rotates in the opposite direction, but with the spin still tangential to the energy circle (in two dimensions).

Let us ignore the cubic terms for the moment and solve for the energy eigenvalues for the two spin states. We assume that the normal energy bands are parabolic for convenience, so that the Hamiltonian can be written as

$$
H = \begin{bmatrix} \dfrac{\hbar^2 k^2}{2m_c} & -\eta\langle k_z^2\rangle(k_x + ik_y) \\ -\eta\langle k_z^2\rangle(k_x - ik_y) & \dfrac{\hbar^2 k^2}{2m_c} \end{bmatrix}. \tag{6.114}
$$

We may now find the energy as

$$
E = \frac{\hbar^2 k^2}{2m_c} \pm \eta\langle k_z^2\rangle k. \tag{6.115}
$$

Not only is the energy splitting linear in k, but it is also isotropic with respect to the direction of $\mathbf{k}$. Thus, the energy bands are composed of two inter-penetrating paraboloids, and a constant energy surface is composed of two concentric circles. The inner circle represents the positive sign in (6.115), while the outer circle corresponds to the negative sign. The two eigen-functions are given by

$$
\varphi_z = \frac{1}{\sqrt{2}} \begin{bmatrix} 0 \\ e^{\pm i\vartheta} \end{bmatrix}, \tag{6.116}
$$

where ϑ is the angle that $\mathbf{k}$ makes with the [100] axis of the underlying crystal, within the hetero-structure quantum well. This angle is the polar angle for the two coordinates. Hence, the root with the upper sign, which we take to be the net up spin, has the spin tangential to the inner circle and the lower sign is tangential to the outer circle. Now, let us add the cubic terms, and the energy levels now become

$$
E = \frac{\hbar^2 k^2}{2m_c} \pm \eta\langle k_z^2\rangle k \left[1 + \left(\frac{k^4}{\langle k_z^2\rangle^2} - 4\frac{k^2}{\langle k_z^2\rangle} \right) \sin^2\vartheta\,\cos^2\vartheta \right]^{1/2}. \tag{6.117}
$$

This has a much more complicated momentum and angle dependence, and it is no longer isotropic in the transport plane. Similarly, the phase on the down spin contribution to the eigen-function is no longer simply defined as a simple phase factor.

6.3.2 Structure inversion asymmetry

In the quantum well described above, the structure is asymmetric around the hetero-junction interface. In addition, there is a relatively strong electric field in the quantum well, and motion normal to this can induce an effective magnetic field. This is the structural inversion asymmetry. So both the previous version, as well as this one, can lead to spin splitting without any applied magnetic field. The spin–orbit interaction, discussed in section 2.4.3, can be rewritten, for this situation, as

$$H_{SO} = r\boldsymbol{\sigma} \cdot (\boldsymbol{k} \times \nabla V) \rightarrow H_R = \boldsymbol{\alpha} \cdot (\boldsymbol{\sigma} \times \boldsymbol{k})_z, \tag{6.118}$$

where

$$r = \frac{P^2}{3}\left[\frac{1}{E_G^2} - \frac{1}{(E_G + \Delta)^2}\right] + \frac{P'^2}{3}\left[\frac{1}{\Gamma_0^2} - \frac{1}{(\Gamma_0 + \Delta_c)^2}\right], \tag{6.119}$$

and $\alpha_z = r\langle E_z\rangle$. Taking just the z component of (6.118), we find that the Rashba Hamiltonian can be written as [37]

$$H_R = \alpha_z(k_y\sigma_x - k_x\sigma_y). \tag{6.120}$$

We note that the factor in parentheses is the same factor appearing in the cubic term above. If we use the same basis set for the spin wave functions, then the Rashba energy is given as

$$E_R = \mp i(k_x \pm ik_y). \tag{6.121}$$

While the spin states are split in energy, this does not simply add to the bulk inversion asymmetry. First, the two spin states are orthogonal to each other, and then they are phased shifted (with opposite phase shift) relative to the previous results. It is easier to understand the effect here of this Rashba term if we diagonalize the Hamiltonian for the two spin states. We can write the Hamiltonian, for parabolic bands considering only this structural inversion asymmetry as

$$H = \begin{bmatrix} \dfrac{\hbar^2 k^2}{2m_c} & \alpha_z(k_x + ik_y) \\ \alpha_z(k_x - ik_y) & \dfrac{\hbar^2 k^2}{2m_c} \end{bmatrix}. \tag{6.122}$$

The eigen-values can now be found for this Hamiltonian to be

$$E = \frac{\hbar^2 k^2}{2m_c} \pm \alpha_z k. \tag{6.123}$$

Not only is the energy splitting linear in k, but it is also isotropic (which the linear term in the previous section is not) with respect to the direction of $\boldsymbol{k}$. Thus, the energy bands are composed of two inter-penetrating paraboloids (figure 6.8), and a constant energy surface is composed of two concentric circles. The inner circle represents the positive sign in (6.123), while the outer circle corresponds to the negative sign. The two eigen-functions are given by

$$\varphi_\pm = \frac{1}{\sqrt{2}}\begin{bmatrix} 1 \\ \mp ie^{i\vartheta} \end{bmatrix} = \frac{1}{\sqrt{2}}\begin{bmatrix} 1 \\ \mp e^{i(\vartheta + \pi/2)} \end{bmatrix}, \tag{6.124}$$

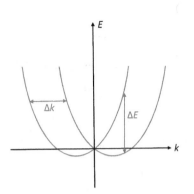

Figure 6.8. The two energy bands for the linear splitting of the Rashba effect. Here, $\Delta k = 2m_c\alpha_z/\hbar^2$ and $\Delta E = 2\alpha_z k$. The bands represent two interpenetrating paraboloids of revolution that are offset due to the Rashba contribution.

where ϑ is the angle that $\mathbf{k}$ makes with the [100] axis of the underlying crystal, within the hetero-structure quantum well described in the previous section. The second form of (6.124) clearly shows the phase shift relative to the bulk inversion asymmetry wave function. The spin direction is tangential to the two circles, but pointed in the negative angular momentum direction for the inner circle and in the positive angular momentum direction for the outer circle. When both spin processes are present, the spin behavior becomes quite anisotropic in the transport plane [44]. However, the Dresselhaus bulk inversion asymmetry is generally believed to be much weaker than the Rashba terms discussed here, especially as the strength of this latter effect can be modified by an electrostatic gate applied to the hetero-structure.

6.3.3 The spin Hall effect

One of the more remarkable features of the Rashba spin–orbit term (the structural inversion asymmetry terms) is that this effect gives rise to an intrinsic spin Hall effect in a nanowire, in which the longitudinal (charge) current along the nanowire gives rise to a transverse spin current. In this situation, one spin state will move to one side of the nanowire, while the other spin state moves to the opposite side. This spin Hall effect is considered to be intrinsic as it does not rely upon the presence of any impurities with their spin scattering. This spin effect can be illustrated with a simple approach, in which we take the spin orientation as the z-direction. For this we define a spin current via

$$J_s = \frac{\hbar}{2}\sigma_z\mathbf{v}, \tag{6.125}$$

where $\mathbf{v}$ is the velocity operator in the x-y plane. If we apply this to the Hamiltonian and wave functions of the previous section, we find that the spin current is given by

$$\langle J_s\rangle_\pm = \pm\frac{\alpha_z}{2}(\sin\vartheta\mathbf{a}_x - \cos\vartheta\mathbf{a}_y), \tag{6.126}$$

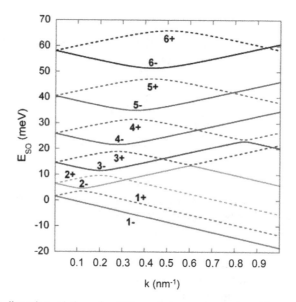

Figure 6.9. Electron dispersion relation of a 100 nm InAs quantum wire with a spin–orbit strength of $\alpha_z = 20$ meV · nm. This shows just the Rashba contribution to the energy subbands and the hybridization of these in the multi-subband case. Reprinted with permission from [44]. Copyright 2009 by the Institute of Physics Publishing.

where the upper and lower signs refer to the positive and negative branches of the energy in (6.123). The spin current lies in the plane of the two-dimensional electron gas and is always normal to the momentum direction down the nanowire (in fact, the spin current is always normal to the momentum whichever its direction). This can be utilized to try to create spin filters and other spintronics applications.

Let us consider the dispersion of the subbands in a 100 nm wide wire in InAs with a spin–orbit strength $\alpha_z = 20$ meV · nm. This wire will have several occupied subbands which are all spin split. However, the spin down level from subband 2 will cross and hybridize with the spin up level from subband 1 as we raise the momentum k. This produces mixed spin states in the various hybridized levels as shown in figure 6.9, where the Rashba contribution to the energy is plotted (the parabolic part of the energy has been subtracted for clarity) [44]. In each branch there remains some components of both spins, and this can be seen in figure 6.10 for the 1+ and 2− branches, which shows the content as a function of the momentum. One can clearly see that the spin polarization reverses as the momentum passes through the crossing and hybridization region in the 2− branch. On the other hand, the 1+ branch is predominantly spin up prior to the crossing, but after the crossing the two spin states are about equally occupied. Note that the two spin states prefer different sides of the wire in these two levels.

Because the spin up and down states are displaced away from the center of the waveguide, we can consider creating a spin filter [45]. If we have the wire lead into a Y junction where it splits into two other nanowires, one spin will preferentially go into a single arm of the Y junction. This allows us to achieve spin filtering, although

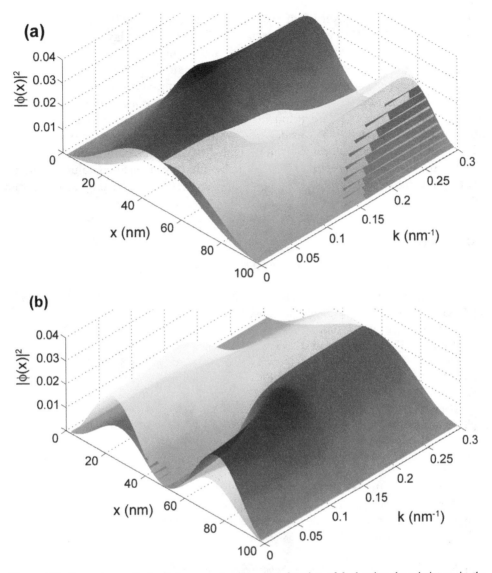

Figure 6.10. Squared magnitude of the wave function as a function of k showing the admixture in the hybridized, yet still spin-resolved, for (a) subband 1+ and (b) 2− levels. The yellow corresponds to spin up and the red to spin down. Reprinted with permission from [44]. Copyright 2009 by the Institute of Physics Publishing.

the presence of an induced magnetic field will cause precession of the spin in each of the two branches. There have been attempts to actually observe this spin selection in branched nanowires [41].

A somewhat different approach has been presented by Wunderlich *et al* [46]. In this experiment, a *p*-type layer is produced in an AlGaAs/GaAs heterostructure. This structure is then patterned into a ribbon so that *n*-type layers can be placed in close proximity. These layers then produce two light-emitting diodes at the edges of the ribbon. The spin Hall effect is especially strong in the hole layer, and this guides the spin polarized carriers in the wide nanowires to the diodes, so that each diode sees a different spin polarization. This is used to create optical circular polarization from the diodes due to the spin polarization of the injected holes. Because of the optical selection rules, the presence of a particular spin polarization will create a corresponding circular polarization in the light, and this can be measured. The expected spin polarized light was detected, with opposite polarization from each of the two LEDs. Finally, the group investigated whether the spin Hall effect has a universal value. The measured strength suggests that this measured value is roughly twice the value expected if it were universal. They term this a 'strong' intrinsic spin Hall effect.

References

[1] Ferry D K and Nedjalkov M 2018 *The Wigner Function in Science and Technology* (Bristol: IOP Publishing)

[2] Bogoliubov N N 1946 *J. Phys. Soviet Un.* **10** 256

[3] Born M and Green H S 1946 *Proc. Roy. Soc. (London)* **A188** 10

[4] Kirkwood J G 1946 *J. Chem. Phys.* **14** 180

[5] Yvon J 1937 *Act. Sci. Ind.* 542, 543 (Paris: Herman)

[6] Ferry D K 1991 *Semiconductors* (New York: Macmillan), pp 179–85

[7] Hess K and Vogl P 1972 *Phys. Rev.* B **6** 4517

[8] Ferry D K 2012 *Proceedings of IEEE Nanotechnology Materials Devices Conference* (New York: IEEE)

[9] Santos I, Sanchez S M and Quinones S 2014 *Mobility and Resistivity Tool*, available at NanoHub.org, DOI: 10.431/D3H98ZD97

[10] Price P 1965 *Fluctuation Phenomena in Solids* ed R E Burgess (New York: Academic), 355–80

[11] Nougier J and Rolland M 1973 *Phys. Rev.* B **8** 5728

[12] Ziman J M 1960 *Electrons and Phonons* (Oxford: Clarendon) ch 12

[13] Ferry D K 2001 *Quantum Mechanics* 2nd edn (Bristol: IOP Publishing)

[14] Landau L D and Lifshitz E M 1958 *Quantum Mechanics: Non-Relativistic Theory* (London: Pergamon) ch 16

[15] Kittel C 1963 *Quantum Theory of Solids* (New York: Wiley), 220

[16] von Klitzing K, Dorda G and Pepper M 1080 *Phys. Rev. Lett.* **45** 494

[17] Avron J E, Osadchy D and Seiler R 2003 *Phys. Today* **56** 38

[18] Laughlin R 1981 *Phys. Rev.* B **23** 5632

[19] Zak J 1964 *Phys. Rev.* **134** A1602

[20] Hofstadter D R 1976 *Phys. Rev.* B **14** 2239

[21] Tsui D, Störmer H L and Gossard A C 1982 *Phys. Rev. Lett.* **48** 1559

[22] Laughlin R B 1983 *Phys. Rev. Lett.* **50** 1395

[23] Ferry D K, Goodnick S M and Bird J P 2009 *Transport in Nanostructures* 2nd edn (Cambridge: Cambridge University Press)

[24] Rode D L 1975 *Semiconductors Semimetals* ed R K Willardson and A C Beer (New York: Academic), 10 1

[25] Paige E G S 1964 *The Electrical Conductivity of Germanium, Vol. 8 of Progress in Semiconductors* ed A F Gibson and R E Burgess (New York: Wiley)

[26] Mohamed M, Bharthuar A and Ravaioli U 2014 *Rode's Method Tool*, available at NanoHub.org, DOI: 10.4231/D3RV0D18V

[27] Rode D L 1973 *Phys. Stat. Sol. (b)* **55** 687

[28] Milekhin A G *et al* 2008 *J. Appl. Phys.* **104** 073516

[29] Zeeman P 1897 *Phil. Mag.* **43** 226

[30] Oestreich M and Rühle W W 1995 *Phys. Rev. Lett.* **74** 2315

[31] Narita S *et al* 1981 *Jpn. J. Appl. Phys.* **20** L447

[32] Nicholas R J *et al* 1983 *Sol. State Commun.* **45** 911

[33] Aoki N *et al* 2002 *Physica* B **314** 235

[34] Datta S and Das B 1990 *Appl. Phys. Lett.* **58** 665

[35] Žutic I, Fabian J and das Sarma S 2004 *Rev. Mod. Phys.* **76** 323

[36] Murakami S, Nagaosa N and Zhang S 2003 *Science* **301** 1348

[37] Bychov Y A and Rashba E I 1984 *J. Phys.* C **17** 6039

[38] Inoue J, Bauer G E W and Molenkamp L W 2004 *Phys. Rev.* B **70** 041303

[39] Nikolic B K, Souma S, Zarbo L B and Sinova J 2005 *Phys. Rev. Lett.* **95** 046601

[40] Cummings A W, Akis R and Ferry D K 2006 *Appl. Phys. Lett.* **89** 172115

[41] Jacob J *et al* 2009 *J. Appl. Phys.* **105** 093714

[42] Dresselhaus G 1955 *Phys. Rev.* **100** 580

[43] Sakurai J J 1967 *Advanced Quantum Mechanics* (Reading, MA: Addison-Wesley) pp 85–7

[44] Cummings A W, Akis R and Ferry D K 2011 *J. Phys. Condens. Matter.* **23** 465301

[45] Cummings A W, Akis R and Ferry D K 2009 *Appl. Phys. Lett.* **89** 172115

[46] Wunderlich J, Kaestner B, Sinova J and Jungwirth T 2005 *Phys. Rev. Lett.* **94** 047204

IOP Publishing

Semiconductors (Second Edition)
Bonds and bands
David K Ferry

Chapter 7

High field transport

In chapter 6, the one electron transport equation was obtained for low electric fields, that is essentially the very near to equilibrium situation. When the electric field is increased so that the system moves into the far from equilibrium situation, it becomes quite difficult to solve the Boltzmann equation, and the approximations for Rode's method are no longer applicable. As a result, the over-riding theoretical problem in such transport is the need to obtain the relevant distribution function in the presence of both the high electric field and the varying properties of the scatterers as the average energy of the carriers increases. For transport properties, the distribution function is not an end in itself, since integrals over this function must be performed in order to evaluate the transport coefficients, the velocity and the mobility, etc. This process is made more difficult by the need to specify whether we are in the transient regime or the steady-state high field regime, and at what electric field are we interested in the coefficients. While this may seem to be an easy question, it is important to note that in the nonlinear transport regime, the ergodic approximation fails. In particular, it fails because the distribution function itself is not a stationary process. It varies in time and its form arises from a delicate balance between the driving forces, the electric field usually, and the dissipative forces, the scattering.

We may recall that for a state at energy E, carriers can come into this state from a state at $E + \hbar\omega_0$ by the emission of a phonon of energy $\hbar\omega_0$. Or they can leave the state of energy E and transition to the state at $E + \hbar\omega_0$ by the absorption of a phonon of energy $\hbar\omega_0$. Similarly, the state at energy E is connected to the state of energy $E - \hbar\omega_0$ by emission or absorption of the appropriate phonon. The rate of absorption of phonons from the state at E (and moving upward to the higher energy level) may be expressed as

$$AN_q \exp\left(-\frac{E}{k_{\mathrm{B}}T_e}\right), \tag{7.1}$$

doi:10.1088/978-0-7503-2480-9ch7

provided that the distribution function remains a Maxwellian at an electron temper-ature of T_e. In (7.1), the quantity N_q is the Bose–Einstein distribution for the phonons and remains at the lattice temperature, which may not actually be the case and is discussed later. The rate of emission from the upper level to our level of interest may similarly be expressed as

$$A(N_q + 1)\exp\left(-\frac{E + \hbar\omega_0}{k_B T_e}\right). \tag{7.2}$$

It may be recalled that the Bose–Einstein distribution has the property of $(N_q + 1) = N_q \exp(-\hbar\omega_0/k_B T)$, where T is the lattice temperature. Detailed balance is achieved only if $T_e = T$. However, when streaming terms in the Boltzmann equation are driven by a high electric, the increase of the electron temperature comes from the high electric field and a steady-state equilibrium is reached only when

$$e\mathbf{v} \cdot \mathbf{E} = -\frac{dE}{dt}\bigg|_{\text{coll}}. \tag{7.3}$$

(The $\mathbf{E}$ on the left is the electric field, while that on the right is the energy. As usual, we must be aware of this confusion.) In order to satisfy this equation, there must be an increase in the emission terms relative to the absorption terms to thermalize the energy gain from the electric field, and combining (7.1) and (7.2) leads to

$$\exp\left[\frac{\hbar\omega_0}{k_B T}\left(1 - \frac{T}{T_e}\right)\right] > 0. \tag{7.4}$$

This argument is good only for the Maxwellian distribution, but it is often assumed in order to define an equivalent electron temperature, something that is often not possible. The statement itself is nothing more than a conservation of energy in the system, although this itself is not guaranteed, but this conservation of energy must be maintained to establish a non-equilibrium steady-state in the system. The simplest case in which steady-state is not obtained comes from electron transfer between non-equivalent valleys, such as transfer from the Γ valley to upper L and X valleys in the Gunn effect [1, 2]. In this case, the electrons transfer from a high mobility central valley to the low mobility satellite valleys, and this causes a negative differential mobility (conductance) to appear in the current-electric field relation. This negative differential mobility is a region of instability and leads to traveling high electric field domains [1, 3]. Hence, no steady-state exists in this situation. These coherent structures (the domains) can be maintained only through a sufficient flow of energy into the system, and have been termed dissipative structures [4]. Even so, the distribution function must be carefully determined if these effects are to be properly treated and understood.

7.1 Physical observables

Early studies of high-electric field transport in solids focused mainly on the break-down studies of dielectrics [5], although the electrical properties of silicon had been studied much earlier [6]. Studies of the variation of the velocity with high electric fields began with Ryder [7], who discovered that the velocity in Si and Ge saturated at high electric fields. The saturated (or nearly saturated) velocity is an important parameter for electron device considerations. For example, with a modern nanoscale MOSFET of say 20 nm gate length, an applied voltage of 1 V produces an average electric field of 0.5 MV cm^{-1} in the channel, which is an incredible electric field, especially when one considers that the saturation velocity sets in at about 20 kV cm^{-1} in Si at room temperature. Then, in many so-called figures of merit for high speed, high frequency, or high power devices, the saturated velocity is an important parameter as it affects the frequency response and power handling properties of the device. The saturated velocity is also important as a mirror into the electron–phonon interactions in the material. Consequently, it certainly affects the distribution function that is found at each electric field.

It turns out that the saturation velocity is thought to be a scalable parameter from material to material, precisely because it is a mirror into the carrier–lattice interactions. When we have electrons dominated by relative high energy phonons, then we can estimate the order of magnitude of the energy loss from (7.3) to be

$$ev_sE \sim \frac{3}{2}\hbar\omega_0(e^x - 1)N_q, \tag{7.5}$$

where the factor in parentheses arises from the difference between emission and absorption of a phonon and v_s is the saturated velocity. On the other hand, the balance of momentum which leads to the saturated velocity may be estimated by

$$eE \sim 2m_cv_s(e^x + 1)N_q, \tag{7.6}$$

since both emission and absorption of a phonon cause a loss of momentum. The numerical factors in these two equations are a property of non-polar optical phonon scattering [8]. These two equations may be solved together to yield an estimate of the saturated velocity as

$$v_s \sim \sqrt{\frac{3\hbar\omega_0}{4m_c}} \tanh^{1/2}\left(\frac{\hbar\omega_0}{2k_BT}\right). \tag{7.7}$$

As remarked, (7.7) is strictly correct only for non-polar optical phonons dominating the scattering, and even in this case (7.7) has appeared in slightly different forms in different derivations that appear in the literature. Nevertheless, the result (7.7) suggests that the velocity can be scaled among various semiconductors as it depends only upon material parameters and the lattice temperature. In figure 7.1, we show that the scaling doesn't work very well by plotting the observed (either experiment or calculation) versus the prediction of (7.7). The well documented experimental values are shown in red, while the calculated expectations are in blue. The green symbols are the peak velocities seen in material which shows negative differential conductance. These latter materials show electron transfer from the Γ to the L or X minima, and the transport in the Γ is typically dominated by the polar optical phonon.

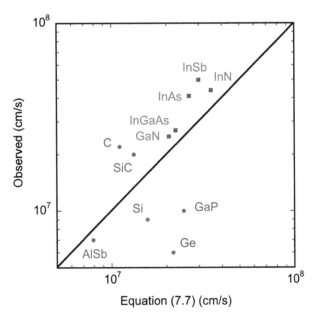

Figure 7.1. The observed saturated velocity from experiment (red) or calculation (blue) and the peak velocity (green) observed in many materials as a function of the scaling prediction (7.7).

Nevertheless, the peak velocity seems to reflect the expected result (7.7). We will return to these velocities later in this chapter.

7.1.1 Equivalent valley transfer

In chapter 2, we found that the minimum of the conduction band in Si lies approximately 85% of the way to X along the [100] direction. As there are six such directions in the cubic unit cell, there are going to be six equivalent minima, and a constant energy surface is an ellipsoid of revolution. These ellipsoids are characterized by a heavier longitudinal mass and a lighter transverse mass. The presence of these six ellipsoids of revolution pose an important point when we talk about transport away from thermal equilibrium. For example, if we orient the electric field in the [100] direction, the four ellipsoids lying in a plane that is normal to the electric field exhibit their transverse mass, while the two ellipsoids with their major axis parallel to the electric field exhibit their longitudinal mass. As a result, carriers in the 4 ellipsoids will be heated to a higher electron temperature than the other two as they exhibit the smaller mass. As a result, there will be a repopulation between the 'hot' and 'cold' valleys [8]. If we orient the electric field in a [110] direction, then there will be four valleys in the plane of the electric field that all make the same angle with the field. The two valleys normal to the electric field show the transverse mass, while the other four show a mass arising from the angle between the field and the major axes of the ellipsoids, thus a mixture of the longitudinal and transverse masses. Once again, it is possible for a repopulation between the 'hot' and

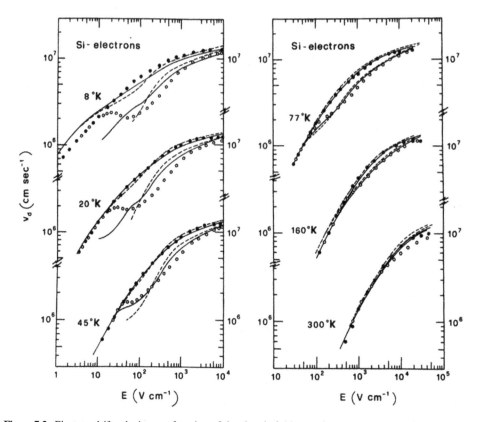

Figure 7.2. Electron drift velocity as a function of the electric field at various temperatures. Closed and open circles refer to experimental data with the field parallel to [111] and [100] directions, respectively. For the same directions, continuous and broken lines indicate the theoretical results from Monte Carlo simulations. The solid line includes the low energy phonons, while the broken lines omit these phonons. Reprinted with permission from Canali *et al* [10]. Copyright 1975 by the American Physical Society.

'cold' valleys to occur. Finally, if we orient the field in the [111] direction, it will make the same angle with all six of the ellipsoids and no repopulation will occur.

Experimental results using time of flight measurements [9] confirm these expectations, and are well observable at lattice temperatures below room temperature [10]. In figure 7.2, we show experimental and theoretical results for Si over a range of electric fields and temperatures. It is clear from the experimental data that different curves are obtained for the [111] and [100] directions, even at room temperature. At low temperature, at and below 20 K, the [100] curve even shows negative differential conductance. The authors used Monte Carlo simulations to study the repopulation and drift velocity to illuminate the physics involved in these measurements. The combination of experiment and theory and the resulting outstanding fit between the two is remarkable given that it occurred in what could be called the 'early days' of such studies. When the electric field gets very large and the current saturates, the difference in field direction tends to be washed out by the very high electron temperatures.

In section 5.4.3, the phonons involved for inter-valley scattering in Si were discussed for the *f* and *g* scattering between the six ellipsoids. For each of these two processes, it was found that a high energy optical mode and a low energy acoustical mode would satisfy the scattering process. Normally, the acoustic modes are considered to be forbidden for these scattering events. Yet, the results of Canali *et al* [10] found that the inclusion of these inter-valley acoustic modes was necessary to fit the drift-velocity data as a function of temperature, field strength, and field orientation at all temperatures. It was shown slightly later that, while forbidden, these inter-valley acoustic modes could be treated by a first-order interaction (section 5.4.2) [11, 12], and this was shown to yield a much better agreement with the temperature dependence of the conductivity in Si [11], as is shown in figure 5.5. The important point is that the energy dependence of the first-order interaction (5.41) is stronger than the normal zero-order interaction (5.38). As a result, the first-order interaction, while normally weaker, becomes dominant at energies above ~0.5 eV, and this can just be seen in figure 5.4. More recent studies, using a full empirical pseudo-potential band structure, coupled to the ensemble Monte Carlo technique, has confirmed that these inter-valley acoustic modes are important, and may even be dominant, in transport in Si [13].

If the electric field is not oriented along the principle axes, e.g., not one of the three axes discussed above, the current in each of the valleys will not be oriented parallel to the electric field, but will be at a different angle to the field in each pair of the six valleys. Hence, the heating can be different in each of the three pairs of valleys, and repopulation can occur as above. However, in this case, the current will not sum (over the six valleys) to yield a net current parallel to the field. Instead, a transverse current can be excited, just as in the Hall effect, which must be forced to zero by a transverse electric field. If the current is down the major axis of a Si wire, then the electric field will be rotated to produce not only the longitudinal excitation field, but also a transverse field, as in the Hall effect, but without the magnetic field. This is known as the Sasaki-Shibuya effect [14]. If we use a <110> plane, then it contains all three major axes, which may be oriented with [100] along the *y*-axis and [110] along the *x* axis. The [111] direction then lies in the plane at an angle of ~58° from the *y* axis. For *n*-type material, the total electric field will be rotated, from the current, towards the [110] direction if the current makes an angle (with the *y*-axis) larger than this 58°, and toward the [100] if the current direction is smaller than this angle. For *p*-type material, the reverse is found to be the case. The Sasaki-Shibuya effect tends to be observable at temperatures below room temperature and at electric fields near where saturation of the current sets in. Again, when the electric field gets very large and the current saturates, the difference in field direction tends to be washed out by the very high electron temperatures.

7.1.2 Non-equivalent valley transfer

The second major inter-valley transfer that occurs is when only a single conduction band minimum, with a small effective mass, is normally occupied, such as in the case of GaAs discussed in the introduction to this chapter (in connection with the Gunn effect [1]). With the application of a high electric field, carriers in this central Γ valley are heated to relatively high energies. Those carriers in the high energy tail of the distribution can even reach energies where they can scatter via inter-valley optical phonons to the satellite L and X valleys of the conduction band. Since these satellite valleys tend to have relatively high effective masses (similar to Si for X valleys and Ge for L valleys), the energetic carriers see a higher density of states in these satellite valleys and tend to prefer to scatter to them. For example, the band structure of GaAs was shown in figure 2.13. Empirical pseudo-potential calculations from Chelikowski and Cohen predict that the L valley lies some 0.31 eV above the Γ minimum and the X valley lies some 0.52 eV above the minimum, at 0 K [15]. Using electro-reflectance, experiments gave the L-Γ separation as 0.3 eV and the X-Γ separation as 0.39 eV, both at 77 K [16], although the latter was later corrected to 0.46 eV [17]. In a review, Blakemore suggested that the room temperature values should be 0.285 eV and 0.448 eV, respectively for these two separations [18]. More recently, high resolution photoemission studies of GaAs with a negative electron affinity at 120 K produced values of 0.3 eV and 0.46 eV for these two parameters [19]. Today, the most common values adopted are 0.29 eV and 0.45 eV at room temperature. Values for other semiconductor are not so well known.

The Gunn effect has been seen in InSb, InAs, PbTe, InGaAs, InP, GaN, perhaps in Ge, as well as in GaAs. It has also been seen in some transition metal di-chalcogenides, such as MoS_2 [20] and strained WS_2 [21]. In fact, it can generally be assumed to exist in all semiconductors in which the lowest conduction band lies at Γ or, as in the case of the transition metal di-chalcogenides, the lowest conduction band corresponds to a direct band gap. The general velocity, as a function of electric field, exhibits a peak at a modest electric field (about 2.5 kV cm^{-1} in GaAs), following which it becomes smaller as more and more electrons are transferred to the L valleys. And, of course, at high enough fields electrons can move to the X valleys as well. In figure 7.3, a calculated velocity field curve for InAs electrons at room temperature is shown. In InAs, the L valleys lie some 0.71 eV above the L minimum [22], but the very low value of effective mass leads to a higher mobility and more rapid heating of the carriers. Hence, the peak in the velocity occurs at only about 1.5 kV cm^{-1}. The fraction of the electrons that are transferred to these L valleys is also shown in the figure. Generally, the velocity peaks when only a few percent of the carriers have transferred to the satellite valleys. The mobility in these satellite valleys is so low compared to that of the Γ valley, that the transfer of just a few percent of the carriers is enough to lead to a drop in *average* current and velocity [23].

Typically, the velocity falls rapidly as more and more of the carriers are transferred to the L valleys (there is no significant population in the X valleys at these electric fields). The final velocity, at very high electric field, is governed by the satellite valleys and the scattering among the set of equivalent minima. The region between 1.5 kV cm^{-1} and ~15 kV cm^{-1} is where the strong negative differential conductance (NDC) is found, and experiments generally cannot probe this area. If domains are triggered by a single small region in the device, then current oscillations can be seen with a frequency that is usually given by the drift velocity divided by the length of the sample [3]. These are known as transit-time oscillations, and were

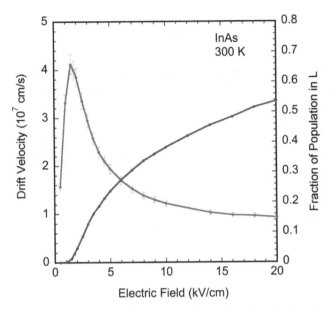

Figure 7.3. The velocity-field curve (red, and left label) and the fraction of carriers in the L valleys (blue, right label) determined from an ensemble Monte Carlo simulation for InAs at room temperature.

already found in Gunn's first studies [1]. If there are multiple sites where domains are triggered, then the device usually shows only noise in the current. The latter noise behavior is also the case in some semiconductors where the NDC region does not exist throughout the sample. For example, in InSb in a magnetic field, the electric field will be concentrated near the cathode end of the sample and will be inhomogeneous across the width of the sample due to the shorting of the Hall field at the metallic contact. Hence, the domains can arise in the high field region near the cathode contact, but they cannot be sustained in the major part of the sample, and the current shows only noise [24]. The near cathode domains have been measured using a capacitive probe and confirm this general behavior as the domains rapidly die out just a fraction of a micron from the cathode. As the point where they die is somewhat random, the sample exhibits strong microwave noise that has its origin in the Gunn effect [25].

The fact of electron transfer and the drop in velocity for a relatively low number of electrons transferring can be argued in a simple approach using the density of states discussed in section 4.1. The density of electronic states, per unit energy, was given by (4.10) for a parabolic band as

$$n(E) = \frac{1}{2\pi^2}\left(\frac{2m_c}{\hbar^2}\right)^{3/2}\sqrt{E}, \tag{7.8}$$

where m_c is the effective density-of-states mass for the valley in question. For the central valley, the total population is given by (4.37) (after expanding the reduced coordinates) as

$$n_\Gamma = \frac{1}{4}\left(\frac{2m_{c\Gamma}k_B T_{e\Gamma}}{\pi\hbar^2}\right)^{3/2} \exp\left(\frac{E_F}{k_B T_{e\Gamma}}\right), \tag{7.9}$$

with $E_F < 0$ for the non-degenerate statistics we are using. We have also assumed that the carriers are describable by a Maxwellian at an electron temperature greater than the lattice temperature. For the upper valleys, which we take to be the L valleys, we can modify (7.9) as

$$n_L = \frac{1}{4}\left(\frac{2m_{cL}k_B T_{eL}}{\pi\hbar^2}\right)^{3/2} \exp\left(\frac{E_F}{k_B T_{eL}}\right)\exp\left(-\frac{\Delta_{\Gamma L}}{k_B T_{eL}}\right), \tag{7.10}$$

where $\Delta_{\Gamma L}$ is the energy separation between the two valleys. It has also been assumed that the temperature is different in the two sets of valleys with $T_{e\Gamma} > T_{eL}$. Then the ratio of populations between the two sets of valleys is simply

$$\frac{n_L}{n_\Gamma} = 4\left(\frac{m_{cL} T_{eL}}{m_{c\Gamma} T_{e\Gamma}}\right)^{3/2} \exp\left(-\frac{\Delta_{\Gamma L}}{k_B T_{eL}}\right)\exp\left[\frac{E_F}{k_B T_{eL}}\left(1 - \frac{T_{eL}}{T_{e\Gamma}}\right)\right]. \tag{7.11}$$

For InAs, the appropriate values for the material parameters are thought to be $4^{2/3}m_{cL} = 0.83m_0$, $m_{c\Gamma} = 0.04m_0$, and $\Delta_{\Gamma L} = 0.71\,eV$. If the two temperatures are equal (that is, for the equilibrium case), the prefactor is approximately 95. Then the two populations would be equal if the two temperatures were equal and $T_e \sim 1800$ K, which is rather high, but not unreasonable. The question we are after, however, is what does the ratio of the two densities have to be for NDC to occur. The point is, that carriers that reach an energy of the valley separation are more likely to scatter to the L valleys. When this happens they give up a significant part of their kinetic energy to potential energy determined by $\Delta_{\Gamma L}$. Thus, they are unlikely to retain the same effective temperature that they had in the central valley. Hence, we have to carry forward a fuller analysis, but under the same approximations stated above.

The total conductivity of the sample, assuming a homogeneous electric field is given by the two contributions of density to be

$$\sigma = e(n_\Gamma \mu_\Gamma + n_L \mu_L). \tag{7.12}$$

The change in the conductivity with electric field is then given by

$$\frac{d\sigma}{dE} \approx e\mu_\Gamma \frac{dn_\Gamma}{dE} + e\mu_L \frac{dn_L}{dE} = e(\mu_\Gamma - \mu_L)\frac{dn_\Gamma}{dE}, \tag{7.13}$$

since the total number of carriers is constant. Here, we have ignored the variation of the mobility with the electric field, since it is usually small at the fields at which the peak in velocity is observed, as may be observed in figure 7.3 (the velocity is almost linear up to the peak). In addition, the mobility in the L valleys is quite small compared to that in the central valley and can be ignored in (7.13). In measurements, it is actually the current that is observed and not the velocity, so that we create the derivative of the current as

$$\frac{dJ}{dE} = \sigma + E\frac{d\sigma}{dE} = e(n_\Gamma\mu_\Gamma + n_L\mu_L) + eE\mu_\Gamma\frac{dn_\Gamma}{dE}. \tag{7.14}$$

If the differential conductivity is negative, we can rewrite (7.14) in terms of the dominant mobility to achieve

$$-\frac{dn_\Gamma}{dE} > \frac{n_\Gamma}{E}. \tag{7.15}$$

Now, we have to evaluate the left-hand side, which may be done with the help of (7.11) as

$$\frac{dn_\Gamma}{dE} = -\frac{dn_L}{dE} = -4\left(\frac{m_{cL}}{m_{c\Gamma}}\right)^{3/2}\exp\left(-\frac{\Delta_{\Gamma L}}{k_B T_e}\right)\left[\frac{dn_\Gamma}{dE} + \frac{n_\Gamma\Delta_{\Gamma L}}{k_B T_e^2}\frac{dT_e}{dE}\right], \tag{7.16}$$

where again we have assumed the two temperatures are equal as a worst case. We may now write the desired derivative term as

$$\frac{dn_\Gamma}{dE} = -\left(\frac{\Delta_{\Gamma L}}{k_B T_e}\right)\frac{n_L n_\Gamma}{n_L + n_\Gamma}\frac{1}{T_e}\frac{dT_e}{dE} \tag{7.17}$$

and (7.15) becomes

$$\left(\frac{\Delta_{\Gamma L}}{k_B T_e}\right)\frac{n_L}{n_L + n_\Gamma}\frac{E}{T_e}\frac{dT_e}{dE} > 1. \tag{7.18}$$

It is not a bad assumption at this point to assume that the electron temperature increases quadratically with the electric field, as this is often the case observed in simulations. Then, the last two factors yield a value of 2.

To simplify the discussion, let us introduce the normalized units $y = \Delta_{\Gamma L}/k_B T_e$, and we can now rewrite (7.18) as

$$y > \frac{1}{2}\left(1 + \frac{n_\Gamma}{n_L}\right) \sim \frac{1}{2}\left(1 + \frac{e^y}{95}\right). \tag{7.19}$$

One must be warned that there are at least two solutions, one for which almost all the carriers are in the L valleys, and then the other case (with larger y) where most of the carriers are in the central valley at larger values of y. Hence, we find the desired solutions when $y \sim 7.14$, or $T_e \sim 1100$ K. This would give us about 7% of the carriers in the L valleys, which remains larger than the observed values in figure 7.3, but not by much. This failure arises from assuming that the L valley temperature is the same as that in the central valley, and neglects the loss of potential energy in the transfer process. This neglect affects the transfer rate and gives us an estimate that is clearly too large. Nevertheless, it indicates that one does not have to have a massive transfer to provide the NDC behavior. We should remark that, in the case of GaAs, the prefactor is only about 70, and this leads to a temperature of ~ 500 K, for which one still requires about 7% of the carriers to transfer, but under the same conditions. This just serves to remind us that the carriers in InAs, with the smaller mass and higher mobility, gain energy much faster than in GaAs.

7.1.3 Transient velocity

In high electric fields, there are at least two different relaxation times corresponding to the relaxation of both momentum and energy. When we study the transient transport, we have to consider both of these times, and we show that both are required to see a transient velocity that is greater than the steady-state velocity. We may think about these two times as corresponding to the shift in the distribution that occurs upon applying the electric field (the momentum relaxation process) and the spread of the distribution to higher energies as the electrons gain energy (the energy relaxation process). If the energy relaxation process is slower than the energy relaxation process, as is often the case, the velocity can overshoot its steady-state value (the saturation velocity) in high fields. This is because the electrons respond with a momentum relaxation time corresponding to the lattice temperature. As the average increases, this leads to a reduction in the momentum relaxation time, which lowers the average velocity.

We can illustrate this overall process with the Langevin equation approach. Normally, it is known that the representative velocity for each particle of an ensemble of electrons may be written as [26]

$$\frac{dv}{dt} = \frac{eE}{m_c} - \frac{v}{\tau_m} + \frac{1}{m_c}R(t),\tag{7.20}$$

where the last term on the right-hand side is the *random force*. Scattering causes the relaxation time, which arises from the average effect of the scattering, while the random force arises from the variations in scattering from its average effect. The random force thus leads to noise from the scattering and this can be inter-valley noise as well as intra-valley noise [27]. When we now average (7.20) over the ensemble of electrons, the resulting drift velocity is given by

$$v_d = \langle v \rangle = \frac{e\tau_m}{m_c}E(1 - e^{-t/\tau_m}).\tag{7.21}$$

This is just the low-field Boltzmann equation approach in which the first set of terms defines the mobility. Hence, this cannot lead to any overshoot, because the equation has only two stable points (where $dv/dt = 0$): 0 (in the absence of E) and the steady-state drift velocity.

We know, for example from figure 5.6, that the scattering rate increases as the average energy of the carriers increases. Hence, the momentum relaxation time must get shorter, and this has to be added to the Langevin equation of (7.20). The new, retarded equation may be written as [28]

$$\frac{dv}{dt} = \frac{eE}{m_c} - \int_0^t \gamma(t - u)v(u)du + \frac{1}{m_c}R(t).\tag{7.22}$$

Here, the relaxation term has been replaced with a convolution term in which the velocity is convolved with the time varying relaxation function γ. As previously, we assume that the field is 'turned on' at $t = 0$. The relaxation function is a 'memory function' for the non-equilibrium system, and will be related to the correlation function for the velocity. The equation (7.22) is a retarded Langevin equation, as the rate of change of the velocity depends not only upon its current value, but all values in the past. To solve this, we Laplace transform it, so that the convolution integral becomes a simple product of the two transforms. From this transformed equation, we can find the characteristic function

$$X(s) = \frac{1}{s + \hat{\gamma}(s)}, \tag{7.23}$$

where $\hat{\gamma}(s)$ is the Laplace transform of $\gamma(t)$. The solution of (7.22) can then be written as

$$v(t) = v(0)x(t) + \frac{eE}{m_c} \int_0^t x(u)du + \frac{1}{m_c} \int_0^t R(t - u)x(u)du, \tag{7.24}$$

which is a general expression of the velocity of each carrier under the influence of the external field and the collisions. We now average this over the ensemble of electrons as previously. Here, we will still assume that the average of the random force is zero, and that it is uncorrelated to relaxation function x. This leads to

$$v_d = \frac{eE}{m_c} \int_0^t x(u)du, \tag{7.25}$$

which tells us that $x(t)$ is precisely the relaxation function discussed by Kubo [29] and (7.25) is a form of the Kubo equation. To understand this fact, consider the two-time correlation function

$$\phi_{\Delta v}(t', t) = \phi_{\Delta v}(t, t') = \langle v(t)v(t') \rangle - v_d(t)v_d(t'). \tag{7.26}$$

The need for a two-time function lies in the fact that the distribution function is non-stationary. If both sides of (7.24) are multiplied by $v(0)$ before the ensemble average is taken, with the assumption (already used) $\langle v(0) \rangle = 0$, and that the initial velocity is uncorrelated to the random force, we arrive at

$$\phi_{\Delta v}(0, t) = \langle v^2(0) \rangle x(t). \tag{7.27}$$

From this form, we recognize that the characteristic function is just the normalized (to the zero field rms velocity) velocity autocorrelation function. We can use this to explicitly write the Kubo form as

$$v_d = \frac{eE}{m_c \langle v^2(0) \rangle} \int_0^t \phi_{\Delta v}(0, t)du. \tag{7.28}$$

The drift velocity then describes the dissipation (via the driving electric field) while the correlation function describes the fluctuations. However, it is important to note that his so-called fluctuation-dissipation theorem holds only for the initial transient velocity, and fails when we do the full two-time behavior, since the expectation value in the denominator becomes fully time dependent and varies nonlinearly with the field. Moreover, one should note that if an overshoot velocity is observed, the correlation function *must go negative* over a portion of its time variation.

To go to the full two-time behavior, we introduce a time variation in the momentum relaxation process through the ansatz

$$\gamma(t) \equiv \frac{1}{\tau_m \tau_e} e^{-t/\tau_e}. \tag{7.29}$$

Here, we note that if we integrate (7.29) over time, we retrieve the normal momentum scattering rate $1/\tau_m$. This leads to the characteristic function (7.23) being described by

$$X(s) = \frac{1}{s + \dfrac{1}{\tau_m}\left(\dfrac{1}{s\tau_e + 1}\right)}. \tag{7.30}$$

This has the inverse transform

$$x(t) = \frac{1}{2}\left(1 - \frac{1}{\sqrt{1 - 4\tau_e/\tau_m}}\right)\exp\left[-\frac{t}{2\tau_e}(1 + \sqrt{1 - 4\tau_e/\tau_m})\right] \\ + \frac{1}{2}\left(1 + \frac{1}{\sqrt{1 - 4\tau_e/\tau_m}}\right)\exp\left[-\frac{t}{2\tau_e}(1 - \sqrt{1 - 4\tau_e/\tau_m})\right]. \tag{7.31}$$

This last equation exhibits the non-monotonic behavior required to achieve velocity overshoot provided $\tau_e > \tau_m/4$.

These results illustrate a number of important points for high field transport. First, the dynamics become retarded with a memory effect because of the extra time variation arising from the evolution of the distribution function. This, in turn, opens the door for velocity overshoot to occur. Moreover, this entire process is coupled to the onset of velocity saturation, which itself tells us about the far-from-equilibrium behavior of the carrier system in the presence of the high electric field. In this sense, the steady-state distribution usually has very little in common with the equilibrium distribution, but the form of the high field distribution function is determined by a balance between the driving force, the high field, and the relaxation forces from the collisions. There is a caveate to the above discussion, and that is the Langevin equations used here are for non-interacting particles. When the interaction between the electrons is included, for example in the Monte Carlo simulations, the Kubo formula (7.28) is no longer valid, as the nonlinear interaction between the particles voids a large number of the underlying assumptions. To my knowledge, no equivalent formula has ever been found.

In figure 7.4, we plot the population of the carriers in the Γ valley as a function of the energy. The population is a product of the distribution function and the density

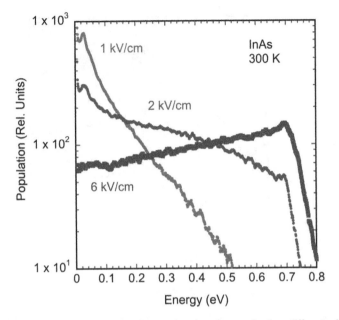

Figure 7.4. The population of carriers in the Γ valley as a function of energy for three different values of the electric field. The sharp break at 0.7 eV is the onset of inter-valley transfer. The population differs from the distribution function by the density of states at each energy. Nevertheless, the dramatic change in the distribution as the field is raised is quite obvious.

of available states (7.8), although here the non-parabolic form is used. The plot is for three different values for the electric field, one below the peak velocity in figure 7.3, and two above the peak velocity. In the latter two cases, one can clearly see the effect of inter-valley scattering at energies above 0.7 eV, as the population in the Γ valley drops dramatically. The average kinetic energy of the carriers is 100 meV at 1 kV cm^{-1}, 250 meV at 2 kV cm^{-1}, and 320 meV at 6 kV cm^{-1}. The much smaller rise in energy in the last value is because of the loss of kinetic energy to potential energy by those electrons which have transferred to the L valleys. The population in the L valleys at these three fields is 1%, 5%, and 28%, respectively. There is, of course, some uncertainty in these numbers due to the statistical nature of the Monte Carlo process by which they have been determined.

Of course, the discussion of figure 7.4, and the data of the last paragraph, refer to the steady state situation. In this section, we are discussing the transient and overshoot velocities for the carriers, to which we now return. In figure 7.5, we plot the transient velocities for the carriers in InAs at the same fields for which the population was plotted in figure 7.4. The curves for the two higher values of field clearly exhibit the property of overshoot, with some rather high values for this quantity. From curves such as these, one could fit (7.31) and try to obtain values for the two relaxation times. But, this is not such an important exercise. Now, what one can infer from figure 7.5 is interesting. For example, at 6 kV cm^{-1}, one can estimate that the average carrier will have traveled about 0.8 micron in the first picosecond. If

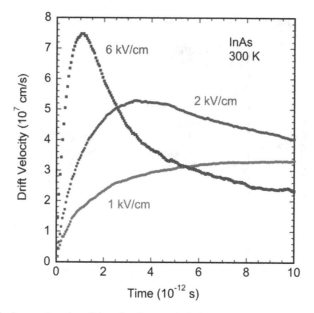

Figure 7.5. The velocity as a function of time for electrons in InAs at 300 K, and for three different values of the electric field. The two higher fields are for values of the field above the peak velocity, and exhibit the phenomenon of 'overshoot.'

this were a 100 nm gate length transistor, the carrier would have exited the device well before it reached this velocity. In a high-electron mobility transistor, there is usually a space between the actual source and the gate, and similarly between the gate and the drain, so that the source-drain distance is larger than the gate length. As a result, the carriers are accelerating prior to entering the gate region, and hence the higher overshoot velocities can be used to effect a higher cutoff frequency for the device, as f_T usually depends on the transit time through the gate region [30]. Typically, the field is highest at the drain end of the gate, and carriers will transfer to the satellite valleys already under the gate. In very high frequency devices, the carriers will usually remain in the satellite valleys until they enter the drain. What this all means is that proper engineering of the transistor design means that the overshoot velocities can be used to impact the design and improve the performance of very high frequency transistors.

An important aspect of this above discussion is that there are techniques by which the higher transient velocities and the distribution function can be measured experimentally. One such approach is based upon single-particle Raman scattering [31]. In the first studies, a sample of GaAs was cooled to 1.6 K, although the pulsed laser heated the sample to about 25 K, as determined by the anti-Stokes TO phonon scattering of a pulsed laser without any electric field applied. The material was illuminated by a pulsed laser, with the scattered light collected at a right angle to the incident pulse, and the pulse was timed with the application of the electric field. Significant changes in the free carrier scattering appeared for electric fields above

2 kV cm^{-1} at room temperature, so that most studies were done at the low temperature mentioned. The scattering from the free particles results in a Stokes shift from the distribution when the particle velocity is positive and an anti-Stokes shift when the velocity is negative. Hence, by measuring the frequency shift of these two scattering processes, the population at each velocity can be measured, and the distribution function determined. These early experiments clearly exhibited the electron heating from the applied electric field.

Later experiments used the back-scattering arrangement, in which a 3 ps dye laser pulse illuminated the GaAs sample, which was cooled to 80 K [32]. The back-scattered Raman signal for the single particle scattering was collected and passed through a double spectrometer to determine the frequency shifts from the hot carrier motion. Here, the incoming laser pulse was collinear with the electric field, and the sample was a *p-i-n* structure. At high electric fields, the measured distribution function began to exhibit the sharp breaks shown in figure 7.4 as the particles moving in the field direction began to reach the satellite valley energy. However the estimated average velocity did not exhibit the high peak velocity or the overshoot effects indicated above, although the results did compare well with estimates from Monte Carlo simulations. It appears that the 3 ps half-width of the laser pulse was too large to exhibit the full nature of the transient behavior. But, the photo-excited carrier density in the intrinsic region was 2.2×10^{18} cm^{-3} and this is likely too large a value to show these effects, as the density makes the sample heavily degenerate, especially at 80 K. Later experiments, using a 0.6 ps half-width laser, and by holding the carrier density to $\sim 10^{17}$ cm^{-3}, these experiments did show much higher velocities and significant numbers of carriers with $\sim 10^{8}$ cm s^{-1} and an estimated drift velocity for the entire ensemble of $\sim 5 \times 10^{7}$ cm s^{-1} at 20–25 kV cm^{-1} [33]. Hence, there was clearly a significant overshoot appearing at these fields. In addition, luminescence thought to derive from the upper L valleys in GaAs has also been seen in these fast measurements [34]. These experiments have been extended to many other semiconductors, all of which confirmed the presence of the overshoot effects and the high average velocities that could be attained.

In figure 7.6, we plot the simulated back-scattering measurement of the carrier distribution function for In$_{0.6}$Ga$_{0.4}$N at 300 K. The dramatic difference between the two curves arises from the fact that the peak velocity occurs just above 100 kV cm^{-1}, and that the back-scattering signal comes predominantly from the central valley electrons (due to the optical momentum being much smaller than any carrier momentum). The experimental data may be compared to that from [35]. The InGaN layer in this experiment was about 0.15 micron thick and grown upon a GaN substrate. The laser pulse was 0.6 ps half-width at 2.17 eV. The Γ valley electrons exhibit a saturated velocity of about 3.5×10^{7} cm s^{-1} above the overall peak velocity, so that the observed experimental peak is already near the saturated velocity in this valley. The single data point at 0 in both curves is the result of direct back scattering from the laser signal, rather than a signal from the carrier distribution function.

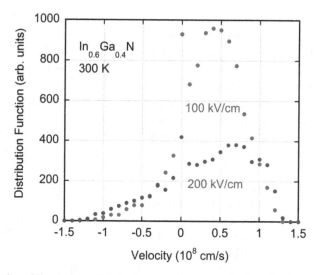

Figure 7.6. Simulation of the single particle back-scattered spectrum for $In_{0.6}Ga_{0.4}N$. The drop-off in total population arises from the plot containing only those electrons in the central conduction band valley.

7.1.4 Impact ionization

As we remarked at the beginning of this chapter, the early studies of high field effects in semiconductors were aimed at understanding the breakdown of the material. In the device configuration, this breakdown is usually called avalanche breakdown, which is an exponential growth process. For example, when an electron in the conduction band has enough energy, it can interact with a bound electron in the valence band, and excite it across the energy gap. This creates an additional electron in the conduction band and a hole in the valence band. This process is often called the *inverse Auger effect*. Now, there are three particles, all of which are further accelerated under the high electric field, and these three can create six new particles. This, of course, leads to the exponential growth in the number of particles contributing to current. The actual collisional event that starts this process is known as *impact ionization*. At low temperatures, collisions with an electron bound to the donor can lead to the same process with the exception that the 'hole' remains bound as an excited donor. Generally, the entire process of impact ionization and avalanche breakdown are described by two parameters: the ionization rate α (in units of cm^{-1}) and the generation rate g (in units of s^{-1}). The first of these is expressed as

$$\alpha = \frac{1}{n_0 v_d} \sum_{E > E_T} n(E) \Gamma_{ion} f(E), \qquad (7.32)$$

where Γ_{ion} is the impact ionization scattering rate (discussed below), n_0 is the low field carrier density, and the other parameters have their normal meaning. The generation rate is also described by this form with an added factor, the velocity, as

$$g = \frac{1}{n_0 v_d} \sum_{E > E_T} n(E) v(E) \Gamma_{\text{ion}} f(E), \tag{7.33}$$

where here E is the energy. In both of these equations, E_T is the threshold energy for the ionizing collision (also discussed below).

Exact calculations for the ionization coefficient and the generation rate must involve knowledge of the distribution function of the carriers. As discussed above, this is in itself a difficult task to achieve. Most of the so-called exact theories really boil down to finding the distribution function first. Our desire here is really describing the impact ionization process itself and to understand a reasonable view of how it depends upon energy and/or the electric field. In the early days, there were two competing ideas on how impact ionization occurred. In one of these, it was assumed that 'lucky' electrons occasionally escaped some scattering events and ballistically traveled to high energies where they underwent impact ionization. The second approach assumed that the distribution function was mainly spherically symmetric and the diffusion of particles to higher energy provided the source for impact ionization. We will first examine these views before undertaking a more general approach.

Wolff [36], in one of the earliest theories applicable to semiconductors, assumed that a strong scattering interaction was always present in the semiconductor, and that this led to a quasi-Maxwellian distribution function, dominated by the spherically symmetric part. Nearly all of the discussion in this chapter contradicts this approach, so it is fair to ask whether Wolff's assumptions could be relevant. At very high electric fields, and carrier energies, the diffusive effect (in energy space) dominates any anisotropy arising from the scattering processes, so that one may assume something like a Rode distribution, but with neither of the two terms being anywhere near equilibrium form. So, the symmetric part can become a dominant term and lead to the majority of impact ionization. As we will see, Wolff's approximations are an important part of the result. This diffusion approximation leads to the assumption of a quasi-Maxwellian at an electron temperature related to the mean free path L as

$$T_e \sim \frac{(eEL)^2}{3\hbar\omega_0}, \tag{7.34}$$

where L is the mean free path for optical phonon emission, and, since our two coefficients follow from an integration over the distribution function, we obtain the ionization coefficient to be

$$\alpha(E) \cong A \exp\left(-\frac{C}{E^2}\right), \tag{7.35}$$

where here, and in (7.34), E is the electric field.

Shockley [37], on the other hand, assumed a very weak interaction between the electron and the lattice, so that the distribution function retained a strongly peaked form arising from the electric field streaming term in the Boltzmann equation. He considered that impact ionization arose from a few 'lucky' electrons in the tail of the asymmetrical part of the distribution that streamed almost ballistically to high enough energy to initialize the impact ionization. So, in his case, the ionization coefficient would vary as

$$\alpha(E) \cong B \exp\left(-\frac{C'}{eEL}\right). \tag{7.36}$$

In general, the fields are neither so high as to validate Wolff's approach nor so low as to as to validate Shockley's approach.

A general approach. It is perhaps more rational to inquire as to whether the two approaches above can be combined in some way. This task was pursued by Baraff [38]. He assumed that the distribution function could be expanded into two terms, one of which is spherically symmetric and the second of which is almost a delta function in space oriented along the electric field. Hence, the distribution function is assumed to definable by

$$f(\boldsymbol{p}, \cos\vartheta) = A(\boldsymbol{p}) + B(\boldsymbol{p})\delta(1 - \cos\vartheta). \tag{7.37}$$

To obtain the two functions A and B, the distribution function can be expanded in a Legendre polynomial series. For example, we can write an arbitrary term in this series as

$$f_i(\boldsymbol{p}) = \frac{2i+1}{2} \int_0^\pi f(\boldsymbol{p}, \cos\vartheta) P_i(\cos\vartheta) d(\cos\vartheta), \tag{7.38}$$

where the P_i are the Legendre polynomials. If we now use (7.37) in this last equation, we find that the lowest order coefficients can be written as

$$
\begin{aligned}
f_0(\boldsymbol{p}) &= A(\boldsymbol{p}) \\
f_1(\boldsymbol{p}) &= \frac{3}{2} B(\boldsymbol{p}) \\
f_2(\boldsymbol{p}) &= \frac{5}{2} B(\boldsymbol{p}) = \frac{5}{3} f_1(\boldsymbol{p})
\end{aligned}
\tag{7.39}
$$

The result for the spherically symmetric part is very similar to a well-known distribution, the Druyvesteyn distribution, which may be written as

$$f_0(\boldsymbol{p}) = \frac{1}{(p^2)^a} \exp\left(-\frac{p^2/2m_c}{k_B T_e}\right), \tag{7.40}$$

with

$$
\begin{aligned}
\frac{1}{a} &= \frac{2}{3} + \frac{eEL}{3\hbar\omega_0} \\
T_e &= \frac{(eEL)^2}{3\hbar\omega_0} + \frac{2}{3} eEL
\end{aligned}
\tag{7.41}
$$

Thus, we see that both forms appear in the solution, and that moving from one to the other is merely due to the size of the electric field. If $eEL = \hbar\omega_0$, we get a solution quite similar to the Shockley form. On the other hand, if $eEL \gg \hbar\omega_0$, we get a form which is basically the Wolff form. So, the initial onset of impact ionization seems to arise from the quasi-ballistic carriers streaming to high energy, while for higher fields the diffusion from the main body of the distribution takes over. It is worth remarking that, especially in Shockley's regime, the band structure details are expected to be exceedingly important. This is especially true in Si, due to the six ellipsoids in a constant energy surface with the anisotropic masses. Hence, not only L, but also the threshold energy E_T, is orientation dependent. In fact, this direction dependence has been observed in GaAs [39] and Si [40]. In these studies, it was found that the highest impact ionization occurred in the [110] direction in GaAs and in the [100] direction in Si. In both cases, it was observed that the directional dependence was greatly reduced as the electric field was increased, in keeping with the above expectations.

The electric fields of interest, as found in the above mentioned experiments, are generally not so low as to validate the Shockley approach, nor so high as to fully support the Wolff approach. They generally show data that lies in the transition region between these two extremes. Baraff [38] actually determined the impact ionization rates from his distribution function and showed that it could be characterized by a set of reduced parameters, as may be inferred from (7.40) and (7.41). That is, the normalized ionization rate αL could be plotted as a function of the reduced threshold energy E_T/eEL, with a single remaining parameter, the normalized dominant phonon energy $\hbar\omega_0/E_T$. The curves are not the result of any analytic form, but of numerical integration over the distribution functions. However, a fairly simple analytic expression to describe the curves has been obtained [41], which is given as

$$\alpha(E) = \frac{eE}{E_T} \exp(b - \sqrt{b^2 + x^2}), \tag{7.42}$$

with

$$x = \frac{E_T}{eEL}, \quad b = 0.217\left(\frac{E_T}{\hbar\omega_0}\right)^{1.14}. \tag{7.43}$$

Here, E remains the electric field. A reasonable fit to impact ionization in most materials can be obtained using these generalized curves.

The threshold energy E_T. There are two major factors that will affect the shape exhibited by either the generation rate (7.33) or the ionization coefficient (7.32) and their functional dependence upon the electric field. These are the threshold energy for ionization E_T and the energy dependence of the ionizing collision. We will deal with the first of these in this section and the second in the next section.

We consider a typical ionizing collision in which an electron in the conduction band (a similar process can occur from a hole in the valence band for p-type material) interacts with an electron in the valence band (e.g., an electron which is not free to travel). Prior to the collision, the incident electron lies at a relatively high energy $E_c(\mathbf{p}_1)$ and the bound electron lies in the valence band at $E_v(\mathbf{p}_2)$. After the collision, the incident electron has created an electron-hole pair, giving us two electrons in the conduction band and a hole in the valence band. The two electrons in the conduction band now lie at $E_c(\mathbf{p}_3)$ and $E_c(\mathbf{p}_4)$. The hole lies at $E_v(\mathbf{p}_2)$, where the original electron was situated. There are several complications to consider. First, the initial electron and the two final electrons do not have to be in the same conduction band valley, and certainly do not satisfy this for Si, where a phonon has to be involved for the indirect process. Hence, phonons may be involved in the various transitions.

During the ionizing collision, both the total energy and total momentum have to conserved [42]. The conservation of energy may be expressed as

$$E_c(\mathbf{p}_1) = E_c(\mathbf{p}_3) + E_c(\mathbf{p}_4) - E_v(\mathbf{p}_2) + \sum_j c_j \hbar \omega_0, \tag{7.44}$$

where the last sum runs over the phonons involved, and the sign on the valence band energy arises from the fact that a hole is left in state $\mathbf{p}_2$. The equivalent form for momentum conservation is

$$\mathbf{p}_1 = \mathbf{p}_3 + \mathbf{p}_4 - \mathbf{p}_2 + \sum_j c_j \hbar \mathbf{q}_j. \tag{7.45}$$

In both of these equations, the coefficients c_j are $\pm n$, where n is an integer, since multiple phonons may be involved in the various transitions. The threshold for ionization is the minimum energy for the incident particle that can still satisfy these two equations. Thus, phonon absorption processes are more likely to be involved, as the emission process would lead to a larger requirement on the energy of the incident particle. To proceed, we minimize the equations by working with the differential forms

$$d\mathbf{p}_1 = 0 = d\mathbf{p}_3 + d\mathbf{p}_4 - d\mathbf{p}_2 + \sum_j c_j \hbar d\mathbf{q}_j \tag{7.46}$$

and

$$\begin{aligned} d\mathbf{p}_1 \cdot \frac{dE_c(\mathbf{p}_1)}{d\mathbf{p}_1} = 0 = d\mathbf{p}_3 \cdot \frac{dE_c(\mathbf{p}_3)}{d\mathbf{p}_3} + d\mathbf{p}_4 \frac{dE_c(\mathbf{p}_4)}{d\mathbf{p}_4} \\ - d\mathbf{p}_2 \cdot \frac{dE_v(\mathbf{p}_2)}{d\mathbf{p}_2} + \sum_j c_j \hbar d\mathbf{q}_j \cdot \frac{d\omega_0}{d\mathbf{q}_j}. \end{aligned} \tag{7.47}$$

We recognize in each term of the last equation the velocity at each energy, and so the last equation can be written as

$$d\mathbf{p}_2 \cdot \mathbf{v}_2 = d\mathbf{p}_3 \cdot \mathbf{v}_3 + d\mathbf{p}_4 \cdot \mathbf{v}_4 + \sum_j c_j \hbar d\mathbf{q}_j \cdot \mathbf{v}_j. \tag{7.48}$$

This last equation can now be combined with (7.46) if the latter is multiplied (vector dot product) by v_2. This leads to the stability condition for the threshold energy as

$$dp_3 \cdot (v_3 - v_2) + dp_4 \cdot (v_4 - v_2) + \sum_j c_j \hbar dq_j \cdot (v_j - v_2) = 0. \tag{7.49}$$

A sufficient condition for stability is thus having all the velocities equal to one another. This means that all the final particles must have the same velocity if the threshold energy is to be minimum. If phonons are involved this is a very restrictive condition as it requires the charge particles to have the same velocity as the phonons. Since phonon velocities are usually quite small when compared to electron velocities, this restricts the charge carriers to lie very near the band extrema for phonon involved processes. This also restricts the phonon processes to be for cases in which the incident electron makes a band transition.

In a direct-gap semiconductor with spherical, parabolic bands, the effective masses can be introduced by m_e and m_h. The threshold criterion requires that v_e and v_h are equal in the absence of the phonon processes. This leads to

$$k_v = \gamma k_c, \ \gamma = m_h / m_e. \tag{7.50}$$

Thus, the energy equation (7.44) becomes

$$\begin{aligned} E_T = E_c(p_1) &= 2\left(\frac{\hbar^2 k_c^2}{2m_e}\right) + \frac{\hbar^2 k_v^2}{2m_h} + E_G \\ &= (2 + \gamma)\frac{\hbar^2 k_c^2}{2m_e} + E_G \end{aligned} \tag{7.51}$$

and (7.45) becomes

$$p_1 = 2p_c - p_v = \hbar k_c(2 + \gamma). \tag{7.52}$$

We have assumed here that the two final electron states are in the same band. From the last equation, we may also determine the threshold energy to be

$$E_T = E_c(p_1) = (2 + \gamma)^2 \frac{\hbar^2 k_c^2}{2m_e}. \tag{7.53}$$

This can now be used to eliminate the momentum from (7.51), and finally find the threshold energy to be

$$E_T = E_G\left(\frac{2 + \gamma}{1 + \gamma}\right). \tag{7.54}$$

This is for the electron initiated process. When $\gamma = 1$, we note that both threshold energies are given by $1.5E_G$. On the other hand, if γ is very large, the ionization threshold for electrons is just the gap itself, while if this parameter is very small, it becomes twice the gap. If we are using holes, we merely replace γ by $1/\gamma$, and the two situations are reversed.

Table 7.1. Some ionization threshold energies (eV).

Material	(100)	(110)	(111)
Electron initiated			
Si	1.18 (U)	2.1 (U)	3.1 (U)
Ge	1.01 (U)	1.11 (D)	0.76 (U)
GaAs	2.1 (D)	1.7 (D)	3.2 (D)
GaP	2.6 (U)	2.8 (U)	3.0 (D)
InSb	0.2 (D)	0.2 (D)	0.2 (D)
Hole Initiated			
Si	1.73 (D)	1.71 (D)	2.73 (D)
Ge	0.97 (D)	0.91 (D)	0.88 (D)
GaAs	1.7 (D)	1.4 (D)	1.6 (D)
GaP	2.4 (D)	2.3 (D)	2.9 (D)
InSb	0.2 (D)	0.2 (D)	0.4 (D)

With the foregoing discussion, it is clear that one can determine the threshold energy quite easily with the equal velocity criterion *if sufficiently accurate band structures are available*. In table 7.1, we give the lowest threshold energies in the various directions for some common semiconductors [42, 43]. In the table, the phonon assisted transitions are denoted as Umklapp (U), while the direct transitions are denoted by (D).

The ionizing collision. The collision rate for the ionizing collision can be calculated in a similar manner to that used for normal momentum relaxation times. The interaction potential is the screened Coulomb interaction between the incident electron, which we take to be in the conduction band, and the bound electron in the valence band. Further, we will assume that the bands are spherical and parabolic, although the approach is easily extended to a non-parabolic band [44]. It will also be assumed that the dielectric function is just a simple value and the frequency and momentum dependence of this quantity will be ignored. We may then write the probability of scattering as

$$\Gamma_{\text{ion}} = \frac{1}{\tau_{\text{ion}}} = \frac{2\pi}{\hbar} \sum_{k_2, k_4} |M(k_1, k_2, k_3, k_4)|^2 \delta(E_1 - E_2 - E_3 - E_4). \tag{7.55}$$

Here, we have used the previous notation for consistency and clarity. For the perturbing potential, it is assumed that the electrons interact via the screened Coulomb potential, as mentioned. This potential may be written in momentum space as

$$\Phi(r) = \sum_q \frac{e^2}{\varepsilon(q^2 + q_{\text{D}}^2)} e^{iq \cdot r}, \tag{7.56}$$

where q_D is the Debye screening wave vector. The wave functions are taken to be Bloch functions in the appropriate band as

$$
\begin{aligned}
\psi(k_1) &= u_c(k_1)\exp(ik_1 \cdot r_1) \\
\psi(k_2) &= u_v(k_2)\exp(ik_2 \cdot r_2) \\
\psi(k_3) &= u_c(k_3)\exp(ik_3 \cdot r_1) \\
\psi(k_4) &= u_c(k_4)\exp(ik_4 \cdot r_2)
\end{aligned}
\tag{7.57}
$$

Thus, the two electrons make the transitions discussed in the previous section. The matrix element for the scattering event is then

$$
M = \sum_q \frac{e^2}{\varepsilon(q^2 + q_D^2)} I_1 I_2,
\tag{7.58}
$$

where the two overlap integrals are

$$
\begin{aligned}
I_1 &= \int u_c^\dagger(k_3)u_c(k_1)\exp[i(q + k_1 - k_3) \cdot r_1]dr \\
I_2 &= \int u_c^\dagger(k_4)u_v(k_2)\exp[i(-q + k_2 - k_4) \cdot r_2]dr
\end{aligned}
\tag{7.59}
$$

Using the closure of a complete set in Fourier transforms, the integration may be split into a summation over the lattice vectors and an integration over the unit cell. The summation leads to the conservation of momentum condition so that the summation over **q** leads to

$$
q = -k_1 + k_3 = k_2 - k_4.
\tag{7.60}
$$

With this limitation, the overlap integrals are more easily evaluated. The first integral in (7.59) is just the overlap integral that appears in phonon scattering and is unity in parabolic bands. The second integral is one that we will encounter again in chapter 8, and is common for transitions across the band gap. In this case, the squared magnitude of the overlap integral is given by the f-sum rule [45] as

$$
|I_2|^2 = 1 + \frac{m_h}{m_e}.
\tag{7.61}
$$

The relaxation time for an ionizing collision can now be calculated by summing over the three energies of the hole and the final two particles in the conduction band. To facilitate this, a number of simplifying assumptions will be made. It will explicitly be assumed that the hole comes from the heavy-hole band, so that $E_2 = -E_G$, due to the large mass of this band. Second, it will be assumed that $k_4 \ll k_2$, which follows from the large size of the hole mass, especially when compared to the electron mass. Then, $q^2 = k_2^2 = 2m_h E_G/\hbar^2$. The delta function can then be used to set E_4 in terms of the other energies, and a first integration can be performed over E_2 using the joint density of states functions as

$$\Gamma_{\text{ion}} = \frac{m_e^3 e^4}{\hbar^3 \pi^3 \varepsilon^2 m_h^2}\left(1 + \frac{m_h}{m_e}\right)\int_0^{E-E_G} \frac{\sqrt{E'(E - E' - E_G)}}{(E_G + E_D)^2}dE', \tag{7.62}$$

where $E_D = \hbar^2 q_D^2/2m_h$ is an energy corresponding to a hole with the Debye screening wave vector. Equation (7.62) can now be readily integrated to give the resulting total scattering rate for the ionizing collision as

$$\Gamma_{\text{ion}}(E) = \frac{m_e^3 e^4}{8\hbar^3 \pi^2 \varepsilon^2 m_h^2}\left(1 + \frac{m_h}{m_e}\right)\left(\frac{E - E_G}{E_G + E_D}\right)^2, \tag{7.63}$$

and, of course, the energy must be larger than the energy gap, which here is the threshold energy due to the assumption of the very large hole mass. If a more reasonable hole mass is used, relative to the electron mass, the energy gap term should be replaced by the actual threshold energy (7.54).

The use of (7.63) in an ensemble Monte Carlo program is straight forward. In figure 7.7, we compare such an ensemble Monte Carlo calculation of the ionization coefficient [46] with several sets of data [47–49] for electrons in Si. The comparison between the calculated values and the experimental data varies significantly, which is more due to the difficulty of measuring the data than that of the calculation. Significantly, the calculated data here was from a non-parabolic, analytic band approach which agreed well with a more extensive full band Monte Carlo simulation [50]. The use of impact ionization through an ionizing collision corresponds to what could be called a *soft* threshold for the process, as it introduces only a probability the

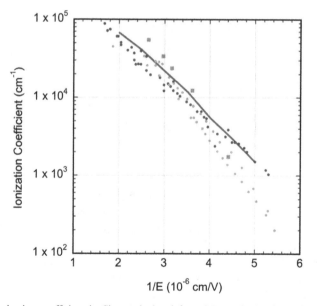

Figure 7.7. The ionization coefficient in Si as calculated from Monte Carlo simulations (red curve) and measured in some early devices. The dark blue dots are from [49], the light blue dots are from [47], and the grey symbols are from [48].

collision will occur. A *hard* threshold would assume that the collision occurred immediately upon the particle reaching the threshold condition. In some sense the Shockley approach has a hard threshold.

In low dimensional systems, the problem can become more complicated. In graphene, for example, there is no real band gap and the nature of the ionizing collision depends more on the effective gap between possible electron and hole states near the Fermi energy. Nevertheless, the impact ionizing collision has been calculated for graphene [51–55], with good results. It is somewhat more problematic in the transition metal di-chalcogenides, where the conduction and valence bands are primarily derived from the metal *d* states. Here, it is found that these principle bands are rather narrow in energy and certainty not as wide as the band gap. As a result, impact ionization is not really expected, yet it occurs [21]. On the other hand, these materials suffer from significant numbers of chalcogenide vacancies [56], that introduce defect states near mid-gap. Thus, impact ionization could excite electrons trapped in these states to the conduction band. Also, there could arise a multi-step process where the valence band electron is excited to the defect level by one hot electron and then is ionized to the conduction band by a second hot electron. This latter two-step process would be the inverse of Shockley–Read–Hall recombination. Simulations with an appropriately modified ionizing collision model, using the Monte Carlo technique, have yielded an ionization rate roughly corresponding to that found in the experiments. This rate is shown in figure 7.8 for both MoS_2 and WS_2 [57].

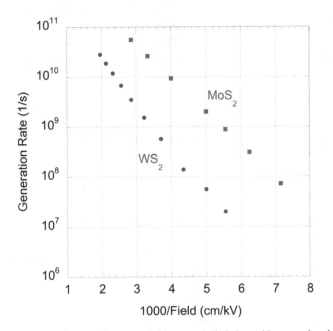

Figure 7.8. The carrier generation rate for two transition metal di-chalcogenides assuming that the secondary electrons come from mid-gap defect electron traps. The generation rate is computed using an ensemble Monte Carlo simulation for the two materials. Figure is reprinted with permission from [57]. Copyright 2017 by IOPP.

7.2 The ensemble Monte Carlo technique

The use of the Boltzmann transport equation above, and techniques found in use by various groups to solve this equation, are quite difficult to evaluate carefully in the real situation of a semiconductor with non-parabolic energy bands and complicated scattering processes. An alternative approach is to use the computer to completely solve the transport problem with a stochastic methodology. The ensemble Monte Carlo (EMC) technique has been used now for more than five decades as a numerical method to simulate far-from-equilibrium transport in semiconductor materials and devices. It has been the subject of many reviews [58–60]. The approach taken here is to introduce the methodology, and illustrate how it is implemented. Many people believe that the EMC approach actually solves the Boltzmann equation, but this is true only in the long-time limit. For short times, the EMC is actually a more exact approach to the problem.

The EMC is built around the general Monte Carlo technique, in which a random walk is generated to simulate the stochastic motion of particles subject to collision processes. These collisions provide both the momentum relaxation process and the random force that appears in e.g. (7.24). Random walks, and stochastic techniques, may be used to evaluate complicated multiple-dimensional integrals [61, 62]. In the Monte Carlo transport approach, we simulate the basic free flight of a carrier, and randomly interrupt this flight with instantaneous scattering events, which shift the momentum (and energy) of the carrier. Here, the length of each free flight, and the selection of the appropriate scattering process, are selected by weighted probabilities, with the weights adjusted according to the physics of the transport process. In this way, very complicated physics can be introduced without any additional complexity of the formulation (albeit at much more extensive computer time in most cases). At appropriate times through the simulation, averages are computed to determine quantities of interest, such as the drift velocity, average energy, and so forth. By simulating an ensemble of carriers, rather than the single carrier normally used in a Monte Carlo procedure, the non-stationary time-dependent evolution of the carrier distribution, and the appropriate ensemble averages, can be determined quite easily without resorting to any need for time averages.

7.2.1 The path integral

To begin, the Boltzmann equation will be written in terms of a path integral as a method to illustrate the steps in the EMC process. In this, the streaming terms on the left-hand side will be written as partial derivatives of a general derivative of the time motion along a 'path' in a 6-dimensional phase space; this is then used to develop a closed-form integral equation for the distribution function. This integral has itself been used to develop an iterative technique [63], but provides one basis of the connection between the Monte Carlo procedure and the Boltzmann equation. To begin, the Boltzmann equation is written as

$$\left(\frac{\partial}{\partial t} + \boldsymbol{v} \cdot \nabla + e\mathbf{E} \cdot \frac{\partial}{\partial \boldsymbol{p}}\right) f(\boldsymbol{p}, \mathbf{r}, t) = -\Gamma_0 f(\boldsymbol{p}, \mathbf{r}, t)$$
$$+ \int d^3\boldsymbol{p}' P(\boldsymbol{p}', \boldsymbol{p}) f(\boldsymbol{p}', \mathbf{r}, t) \tag{7.64}$$

where

$$\Gamma_0 = \int d^3\boldsymbol{p} P(\boldsymbol{p}, \boldsymbol{p}') \tag{7.65}$$

is the total *out-scattering rate*. That is, (7.65) provides the entire rate of decrease of population from the state described by $f(\mathbf{p},\mathbf{r},t)$ due to scattering of particles out of this state. The last term on the right in (7.64) provides the complementary scattering of particles into the state.

At this point it is convenient to transform to a variable that describes the motion of the distribution function along a trajectory in phase space. It is usually difficult to think of the motion of the distribution function, but perhaps easier to think of the motion of a typical particle that characterizes the distribution function. For this, the motion is described in a six-dimensional phase space, which is sufficient for the one-particle distribution function being considered here [64]. The coordinate along this trajectory is taken to be s, and the trajectory is rigorously defined by the semi-classical trajectory, which can be found by any of the techniques of classical mechanics (i.e., it corresponds to that path which is an extremum of the action). It is as easy to remember, however, that it follows Newton's laws, where the forces arise from all possible potentials—including any self-consistent ones that arise in device simulations. Each normal coordinate can be parameterized as a function of this variable as

$$\boldsymbol{r} \to \boldsymbol{x}^*(s), \quad \boldsymbol{p} = \hbar\boldsymbol{k} \to \boldsymbol{p}^*(s), \quad t \to s, \tag{7.66}$$

and the partial derivatives are constrained by the relationships

$$\frac{d\boldsymbol{x}^*}{ds} = \boldsymbol{v}, \quad \boldsymbol{x}^*(t) = \boldsymbol{r}, \quad \frac{d\boldsymbol{p}^*}{ds} = e\mathbf{E}, \quad \boldsymbol{p}^*(t) = \boldsymbol{p}. \tag{7.67}$$

With these changes, the Boltzmann equation becomes simply

$$\frac{\partial f}{\partial s} + \Gamma_0 f = \int d^3\boldsymbol{p}^{*\prime} P(\boldsymbol{p}^{*\prime}, \boldsymbol{p}^*) f(\boldsymbol{p}^{*\prime}, \boldsymbol{x}^*, s). \tag{7.68}$$

This is now a relatively simple equation to solve. It should be recalled at this point that $P(\boldsymbol{p}^{*\prime}, \boldsymbol{p}^*)$ is the probability per unit time that a collision scatters a carrier from state $\boldsymbol{p}^{*\prime}$ to $\boldsymbol{p}^*$, and these variables will be retarded due to the phase-space variations described above. The form (7.68) immediately suggests the use of an integrating factor $\exp(\Gamma_0 s)$, so that this equation becomes

$$\frac{d}{ds}(f(\boldsymbol{p}^*)e^{\Gamma_0 s}) = \int d^3\boldsymbol{p}^{*\prime} P(\boldsymbol{p}^{*\prime}, \boldsymbol{p}^*) f(\boldsymbol{p}^{*\prime}, \boldsymbol{x}^*, s)e^{\Gamma_0 s}, \tag{7.69}$$

where the individual momenta evolve in time as the energy increases in time along the path s due to the acceleration of the external fields. But, this depends upon the gauge we employ. Normally, in the frame of the particle, it moves ballistically as the potential falls away due to the field. But as the 'laboratory' coordinates are restored, this energy increase will appear (again, this is just a choice of gauge for the field and momentum). Indeed, the major time variation lies in the momenta themselves. The Boltzmann equation can now be rewritten as

$$f(\boldsymbol{p}^*, t) = f(\boldsymbol{p}^*, 0)e^{-\Gamma_0 t} + \int_0^t ds \int d^3\boldsymbol{p}^{*\prime} P(\boldsymbol{p}^{*\prime}, \boldsymbol{p}^*) f(\boldsymbol{p}^{*\prime}, \boldsymbol{x}^*, s)e^{-\Gamma_0(t-s)}, \quad (7.70)$$

and, if we restore the time variables appropriate to the laboratory space, we arrive at

$$f(\boldsymbol{p}, t) = f(\boldsymbol{p}, 0)e^{-\Gamma_0 t} + \int_0^t ds \int d^3\boldsymbol{p}' P(\boldsymbol{p}' - e\mathbf{E}, \boldsymbol{p}) f(\boldsymbol{p}' - e\mathbf{E}, \boldsymbol{r}, s)e^{-\Gamma_0(t-s)}. \quad (7.71)$$

This last form is often referred to as the Chambers–Rees path integral [65], and is the form from which an iterative solution or the Monte Carlo solution can be developed.

The integral (7.71) has two major components. The first is the scattering process by which the carriers described by $f(\boldsymbol{p}')$ are scattered (by the processes within P). The second is the ballistic drift under the influence of the field, with a probability of the drift time given by $\exp[-\Gamma_0(t - t')]$. These are the two parts of the Monte Carlo algorithm, and it is from such an integral that we recognize that the Monte Carlo method is merely evaluating the integral stochastically. The problem with it is that there is no retardation in the scattering process, so that the scattering rate and energy are supposed to respond instantaneously to changes in the momentum along the path $\boldsymbol{p}' - eFt'$. That is, the number of particles represented by the distribution function within the integral instantaneously responds during the previous drift. In essence, this is the Markovian assumption, and is true only in the long-time limit. The EMC process can be used in the short-time, transient regime without modification, but then it is a solution of a non-Markovian version of the Boltzmann equation, which is called the Prigogine–Resibois equation [66].

7.2.2 Free flight generation

The second part of the integral (7.71) mentioned above addresses the dynamics of the particle motion that is assumed to consist of free flights interrupted by instantaneous scattering events. The latter change the momentum and energy of the particle according to the physics of the particular scattering process. Of course, we cannot know precisely how long a carrier will drift before scattering, as it continuously interacts with the lattice and we only approximate this process with a scattering rate determined by first-order time-dependent perturbation theory, as discussed in chapter 5. Within our approximations, we may simulate the actual transport by introducing a probability density $P(t)$, where $P(t)dt$ is the joint probability that a carrier will both arrive at time t without scattering (after its last scattering event at $t = 0$), and then will actually suffer a scattering event at this time (i.e., within a time interval dt centered at t). The probability of actually scattering within this small time interval at time t may be written as $\Gamma[\mathbf{k}(t)]dt$, where $\Gamma[\mathbf{k}(t)]$ is the total scattering rate of a carrier of wave vector $\mathbf{k}(t)$ (we use almost exclusively the wave vector, rather than the velocity or momentum, in this section). This scattering rate represents the sum of the contributions of each scattering process that can occur for a carrier of this wave vector (and energy). The explicit time dependence indicated is a result of the evolution of the wave vector (and energy) under any accelerating electric (and magnetic) fields. In terms of this total scattering rate, the probability that a carrier has not suffered a collision after time t is given by

$$\exp\left[-\int_0^t \Gamma_0(k(t'))dt'\right]. \tag{7.72}$$

Thus, the probability of scattering within the time interval dt after a free flight time t, measured since the last scattering event, may be written as the joint probability

$$P(t) = \Gamma_0(k(t))\exp\left[-\int_0^t \Gamma_0(k(t'))dt'\right], \tag{7.73}$$

where the pre-factor provides the normalization so that the integral over all time of (7.73) gives unity. However, when we want to generate random flight times, we have to use the random number generator which gives us a uniform distribution of real numbers between zero and one (including these two values). Hence, we do not need the normalization factor, and the random flight time is sampled from $P(t)$ according to the random number r as

$$r = \int_0^t P(t')dt' = 1 - \exp\left[-\int_0^t \Gamma_0(k(t'))dt'\right]. \tag{7.74}$$

The exponential term varies from 0 to 1, and since $1 - r$ is statistically the same as r, this latter expression may be simplified as

$$\ln(r) = -\int_0^t \Gamma_0(k(t'))dt'. \tag{7.75}$$

Since $0 \leqslant r \leqslant 1$, the logarithm term is negative, and this will cancel the negative sign on the right-hand side of (7.75).

Equation (7.75) is the fundamental equation used to generate the random free flight for each carrier in the ensemble. If there is no accelerating field, the time dependence of the wave vector vanishes, and the integral is trivially evaluated. In the general case, however, this simplification is not possible, and it is expedient to resort to a mathematical *trick*. Here, we will introduce a fictitious scattering process that has no effect on the carrier. This process is called *self-scattering*, and the energy and momentum of the carrier are unchanged under this process [67]. We will assign an energy dependence to this process in just such a manner that the total scattering rate is a constant, as

$$\Gamma_T - \Gamma_{\text{self}}(E) = \sum_i \Gamma_{0,i}(E), \tag{7.76}$$

where the sum runs over all the *real* scattering processes as a function of the energy E (which we use instead of the wave number). The energy dependence of the self-scattering term is determined so that the quantity Γ_T is independent of energy and therefore independent of the acceleration time. Because the self-scattering process has no effect on the energy, momentum, or position of the particle, it will have no effect upon the transport properties being evaluated by the Monte Carlo process. But, the introduction of this imaginary scattering process simplifies the temporal integration in (7.75) so that we can determine the time of flight to be

$$t = -\frac{1}{\Gamma_T}\ln(r). \tag{7.77}$$

In an ensemble Monte Carlo process, each individual particle in the ensemble carries its own flight time, and this time is selected at the start of the flight. This creates another problem in that the ensemble averages have to be computed at particular time steps, and the manner in which the particles are synchronized is discussed below.

7.2.3 Final state after scattering

The other half of (7.71) is the scattering process, which occurs at the end of each free flight. A typical electron arrives at time t (arbitrarily selected by the methods of the previous section) in a state characterized by momentum $\mathbf{p}_a$, position $\mathbf{x}_a$, and energy E. At this time, the duration of the accelerated flight has been determined from the probability of not being scattered, given above with a random number r, that lies in the interval [0, 1]. At this time, the energy, momentum, and position are updated according to the energy gained from the field during the accelerative period to the values mentioned above—that is, each particle gains additional momentum and energy according to their acceleration in the applied field during the time t. Once these new dynamical variables are known, the various scattering rates can now be evaluated for this particle's energy (in practice, these rates are usually stored as a table to enhance computational speed). A particular rate is selected as the germane scattering process according to a second random number r_2, which is used in the following approach: All scattering processes are ordered in a sequence with process 1, process 2, ..., process $n - 1$, and finally the self-scattering process. The ordering of these processes does not change during the entire simulation. Hence, at time t, we can use this new random number r_2 to select the process according to

$$\frac{1}{\Gamma_T}\sum_{i=1}^{S}\Gamma_{0,i}(E) < r_2 \leqslant \frac{1}{\Gamma_T}\sum_{i=1}^{S+1}\Gamma_{0,i}(E), \tag{7.78}$$

which selects the Sth scattering process. Once this process is selected, the energy and momentum conservation relations for this particular scattering process are used to determine the post-scattering momentum and energy $\mathbf{p}_2$ and E_2 (that is, $E_2 = E \pm \hbar\omega_0$, depending upon whether the process is absorption or emission, respectively, and the momentum is suitably adjusted to account for the phonon momentum).

Additional random numbers are used to evaluate any individual parts of the momentum that are not well defined by the scattering process, such as the angles θ, ϕ associated with the process. In isotropic scattering processes such as non-polar optical phonon and acoustic phonon scattering, all states on the final energy shell are equally probable. Hence, we can select the azimuthal angle with a third random number, so that $\phi = 2\pi r_3$. A fourth random number is now used to select the polar angle with $\theta = \pi r_4$. However, in Coulomb scattering events, such as impurity scattering and polar-optical phonon scattering, the scattering is anisotropic in the polar angle. The probability of scattering through a polar angle θ is provided by the square of the matrix element weighted delta function, which gives the angular probability to be proportional to $1/q^2 = 1/|\mathbf{k} - \mathbf{k'}|^2$. This is just the un-normalized function

$$P(\theta) = \frac{\sin\theta}{2E \pm \hbar\omega_0 - 2\sqrt{E(E \pm \hbar\omega_0)}\cos\theta} \tag{7.79}$$

for phonons. For impurity scattering, the energy is conserved, and one has to work in momentum space to obtain the effective probability. In this elastic scattering, the magnitude of $q = 2k\cos\theta$, and one includes the screening wave vector to avoid this singularity. Returning to polar optical phonon scattering, the distribution function (7.79) is then used to select the scattering angle θ with the random number r_4 through the equation

$$r_4 = \frac{\int_0^\theta P(\theta')d\theta'}{\int_0^\pi P(\theta')d\theta'} = \frac{\ln[(1 - \xi\cos\theta)/(1 - \xi)]}{\ln[(1 + \xi)/(1 - \xi)]}, \tag{7.80}$$

where

$$\xi = \frac{2\sqrt{E(E \pm \hbar\omega_0)}}{(\sqrt{E} - \sqrt{E \pm \hbar\omega_0})^2}. \tag{7.81}$$

Now, (7.80) can be inverted to give

$$\theta = \cos^{-1}\left[\frac{(1 + \xi) - (1 + \xi)^{r_4}}{\xi}\right]. \tag{7.82}$$

A word of caution is required here. The polar angle θ in this last equation is the angle between $\mathbf{k}$ and $\mathbf{k'}$, not between $\mathbf{k}$ and the electric field $\mathbf{E}$. One has to determine the last angle from the three components of $\mathbf{k}$, and then use the polar angle between the momentum and the electric field to do a coordinate transformation on $\mathbf{k'}$ because the anisotropy of the distribution function is around the field, not the momentum, whereas the anisotropy of the scattering is around $\mathbf{k}$. Once this is corrected properly, the azimuthal angle ϕ change is not specified by the matrix element, so that ϕ is randomly selected by the third random number as $2\pi r_3$.

The final set of dynamical variables obtained after completing the scattering process are now used as the initial set for the next iteration, and the process is continued for several hundred thousand cycles. This particular algorithm is one that is amenable to full vectorization (and/or parallelization), and this includes the hyper-threading and multi-cores of modern microchips in most PCs. In one general variant, the program begins by creating the large scattering matrix in which all of the various scattering processes are stored as a function of the energy; that is, this scattering table may be set up with e.g. 1 meV increments in the energy. This includes the self-scattering process. The energy is discretized, and the size of each elemental step in energy is set by the dictates of the physical situation that is being investigated. The initial distribution function is then established—the N electrons actually being simulated are given initial values of energy and momentum corresponding to the equilibrium ensemble, and they are given initial values of position and other possible variables corresponding to the physical structure being simulated. At this point, $t = 0$. If the inter-carrier forces are being computed in real space by a molecular dynamics interaction (discussed in chapter 9), the initial values of these forces, corresponding to the initial distributions in space are also computed. Part of the initialization process is also to assign to each of the N electrons a free flight time according to (7.77), which is its individual time at which it ends its free flight and undergoes scattering. Then each electron undergoes its free flight and a scattering process, which may be self-scattering. New times are selected for each particle and the process is repeated as long as desired.

7.2.4 Time synchronization of the ensemble

The key problem in treating an ensemble of particles is that each particle has its unique time scale. However, we want to compute ensemble averages for such quantities as the drift velocity and average energy, with the former defined as

$$v_d(t) = \frac{1}{N}\sum_i v_i(t). \tag{7.83}$$

For the best accuracy, all the particles need to be aligned at the same time t, which here runs from the beginning of the simulation. Thus, we need to overlay the system with a global time scale, with which each local particle time scale can be synchronized. In practice, this is achieved by introducing the global time variable T, which is discretized into steps as $n\Delta T$. Then at integer multiples of this time step, all particles are stopped in their free flights, and ensemble averages are computed. As described by (7.71), each particle has a flight which is composed of accelerations and scattering processes. Usually, the arrival at a synchronization time T' is during one of the free flights. Thus, the free flight is stopped at this time point, and the parameters computed, and then included in the averages over the particle. In this sense, the imposition of the second time scale synchronizes the distribution and gives the global, or laboratory, time scale of interest in experiments. Hence, this time scale can be called the *laboratory* time scale. We illustrate this process in figure 7.9. Once this is done, the particle is sent on its way to continue its free flight until it reaches its scattering time. This process is quite efficient, but this comes at the expense of more book-keeping in the algorithm. In figure 7.9, the horizontal lines represent the

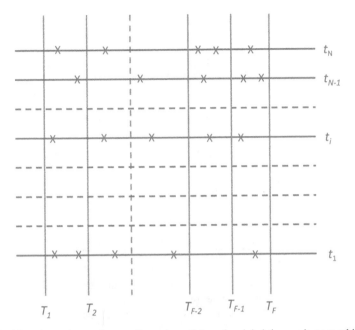

Figure 7.9. The idea of two time scales, one for each particle and a global time scale upon which all particles are synchronized.

individual particle time scales, while the vertical lines are the global time steps at which the motion is stopped and the ensemble averages computed. The crosses on the particle time lines are scattering events. While, these global time steps are shown as being longer than the individual scattering times, this does not have to be the case.

If one is incorporating non-linear effects, such as the electron–electron interaction in real space via molecular dynamics, or non-equilibrium phonons or degeneracy induced filling of the final states after scattering, then these processes are updated on the pauses of the T time scale as well. Moreover, in device simulations, the global time steps are the points where the particle density is projected onto the simulation grid (for Poisson's equation and the resulting potential). In many, if not most cases, the time steps required for Poisson's equation are the shortest time steps in the entire simulation [68]. Similarly, if we introduce the carrier-carrier scattering via molecular dynamics (chapter 9), the time step for up-dating the inter-particle forces is also one of the shortest time scales in the problem, and this also is done on the global time scale. Hence, it is not at all uncommon for the particle to flow through a great many global time steps before actually encountering a scattering process.

7.2.5 The rejection technique for nonlinear processes

In the case of polar optical phonon scattering, it was possible to actually integrate the angular probability function (7.79). This is not always the case for some unusual scattering processes, and one has therefore to resort to other statistical methods. One of these is the so-called rejection technique. Suppose the probability density function

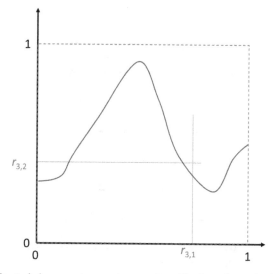

Figure 7.10. The rejection technique uses two random numbers. The first selects a horizontal value at which to evaluate the nonlinear function. The second determines whether or not to use the selected value, depending upon whether or not this value is larger or smaller than the second random number.

for the process, such as (7.79), is quite nonlinear and not easily integrated to get the total probability. Then one can use a pair of random numbers $(r_{3,1}, r_{3,2})$ to evaluate the angle. Consider figure 7.10, in which we plot a complicated probability density function. Here, the abscissa could, for example be the polar angle θ, which has been normalized by π. In this manner, the maximum of coordinate x is unity, so that the range of the function's argument is from zero to one. The maximum value of the function is also normalized to unity, and one can always renormalize this to the span of the random numbers. Now, the first random number is taken to correspond to the span of the function (the x axis). This determines the argument of the function that is to be evaluated. For example, let us assume that this is $r_{3,1} = 0.75$ (this is indicated by the vertical dashed line in the figure). We now use the second random number to determine whether $f(r_{3,1}) > r_{3,2}$. If this relationship holds, then the value $r_{3,1}$ is accepted for the argument of the function, and the scattering process proceeds with this value. If, on the other hand, the relationship is not valid, then two new random numbers are chosen and the process repeated. Certainly, values of $r_{3,1}$ for which the function is large are more heavily weighted in this rejection process. We consider this in more detail for two important processes: (1) state filling due to the degeneracy of the electron gas in this section, and (2) non-equilibrium phonons in the next section.

Let us now turn to the case of state filling. We do this by considering an example of the EMC process to illustrate its efficacy. In figure 7.11, we show the velocity and the average energy of electrons in InSb at 77 K. The dominant scattering process is the polar optical phonon, and no real saturation of the drift velocity occurs, even though the average energy rapidly increases above about 400 V cm^{-1}. It has been observed that impact ionization begins in this material at about 200 V cm^{-1} at 77 K [69, 70]. As another example, in figure 7.12, the transient velocity and average

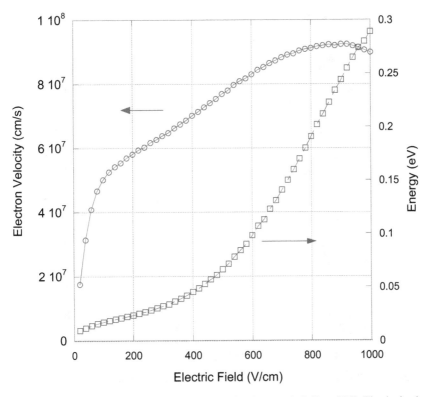

Figure 7.11. The drift velocity and average kinetic energy for electrons in InSb at 77 K. The doping here is in the low 10^{13} cm^{-3} range.

energy are shown for electrons in InGaN. The composition is such that the energy gap is 1.9 eV. The electrons show a small velocity overshoot at about 50 fs after the start of the electric field pulse. Here, the average electric field is 200 kV cm^{-1}.

In both of these simulations, degeneracy and Fermi–Dirac statistics have been introduced through the concept of a secondary self-scattering process [71, 72] based upon the rejection technique. We call it secondary self-scattering, because if the condition $f(r_{3,1}) > r_{3,2}$ is not satisfied, we treat the rejection exactly as a self-scattering process, which was introduced earlier. In this situation, failure of this inequality indicates that the state was already full, and the carrier cannot scattering into it. We understand this in the following way. Each of the scattering processes must include a factor of $[1 - f(E)]$, where $f(E)$ is the dynamic distribution function, and represents the probability that the final state after scattering is empty. Rather than recompute the scattering rates as the distribution function evolves in order to incorporate the degeneracy, all scattering rates are computed as if the final states were always empty. A grid in momentum space is maintained and the number of particles in each state is tracked (each cell of this grid has its population divided by the total number of states in the cell, which depends on the cell size, to provide the value of the distribution function in that cell). The scattering processes themselves

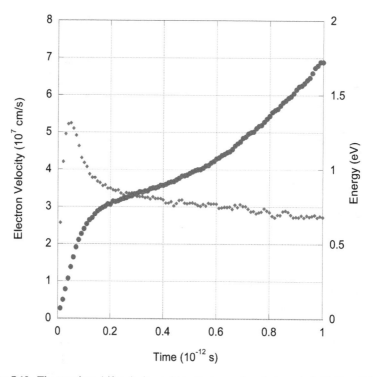

Figure 7.12. The transient drift velocity and kinetic energy for electrons in InGaN at 300 K.

are evaluated, but the acceptance of the process depends on a rejection technique. That is, an additional random number is used to accept the process if

$$r < 1 - f(\mathbf{p}_{\text{final}}, t). \tag{7.84}$$

Thus, as the state fills, most scattering events into that state are rejected and treated as a self-scattering process.

The most delicate point of the degeneracy method involves the normalization of the distribution function $f(\mathbf{p})$. The extension of the secondary self-scattering method to the ensemble Monte Carlo algorithm involves the fact that there are N electrons in the simulation ensemble, which represent an electron density of n. The effective volume V of 'real space' being simulated is N/n. The density of allowed wave vectors of a single spin in $\mathbf{k}$-space is just $V/(2\pi)^3$. In setting up the grid in the three-dimensional wave vector space, the elementary cell volume is given by $\Omega_k = \Delta k_x \Delta k_y \Delta k_z$. Every cell can accommodate at most N_c electrons, with $N_c = 2\Omega_k V/(2\pi)^3$, where the factor of 2 accounts for the electron spin. For example, if the density is taken to be 10^{17} cm^{-3}, $N = 10^4$, and $\Delta k_x \Delta k_y \Delta k_z = (2 \times 10^5$ cm$^{-1})^3$ ($k_F = 2.4 \times 10^6$ cm^{-1} at 77 K), then $V = 10^{-13}$ cm^3 and $N_c = 6.45$. N_c constitutes the maximum occupancy of a cell in the momentum space grid. (Obviously, a more careful choice of parameters would have N_c come out to be an integer, for convenience.) A distribution function is defined over the grid in momentum space by counting the number of electrons in each cell. The distribution function is

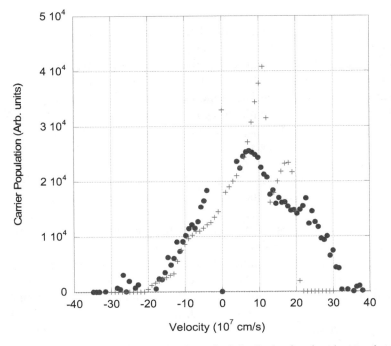

Figure 7.13. Single-particle Raman back-scattering determined distribution function along an electric field of 75 kV cm^{-1} in InN (red dots). A Monte Carlo determination of the distribution function along the field is shown by the light-blue dots.

normalized to unity by dividing the number in each cell by N_c for use in the rejection technique. It should be noted that N_c should be sufficiently large that round-off to an integer (if the numbers do not work out properly, as in the case above) does not create a significant statistical error.

7.2.6 Nonequilibrium phonons

Another application of the EMC technique using the presence of rejection techniques lies in the study of picosecond (or shorter) excitation of electron-hole carrier densities in semiconductors. These can be compared with the measured properties of the carriers by a variety of methods [73]. In figure 7.13, we show such a comparison. Here, single-particle Raman scattering is used to measure the distribution function along the propagation direction, as described earlier in this chapter. This is done by using the Raman shift introduced by the carriers, instead of by phonons. The back-scattered signal at a particular shift is then proportional to the number of carriers with that velocity. The results of figure 7.13 are for such a measurement in InN [74]. The scattering signal is predominantly from the electrons, due to their lighter mass. The effective mass of the electrons used in the EMC calculation is $0.045m_0$. This value is the generally accepted value, although there is some debate about this. However, under a number of circumstances, such as the excitation of the semiconductor by this intense laser pulse, the carriers are created high in the energy band, and then decay by a cascade of phonon emission processes. As a result of this cascade, the phonon distribution is driven out of equilibrium, and this affects both the emission and absorption processes by which the carriers interact with the phonons. In the case discussed, the phonon population can be measured by direct experimental determination of the phonon Stoke's and anti-Stoke's Raman lines. The former is governed by phonon absorption, hence is proportional to N_q, while the latter is governed by phonon emission, hence is proportional to $N_q + 1$. By taking the ratio of the two signals, we produce a Maxwellian at the phonon temperature. Hence, both the phonon and electron distributions can be measured simultaneously.

We can actually consider the non-equilibrium phonon distributions in our Monte Carlo process [75], through the use of a rejection technique. In the derivations presented earlier in this chapter, the assumption was made that the phonons are in equilibrium and characterized by N_q. As before, we use $\mathbf{q}$, rather than $\mathbf{k}$, for the momentum of the phonons. Once again, the momentum space is discretized for the phonon distribution, so that an individual cell in this discretized space has volume $\Delta q_x \Delta q_y \Delta q_z$. This small volume has available a number of states given by $2V/(2\pi)^3$, where V is determined by the effective simulation volume N/n, as previously, and the factor of 2 is for spin degeneracy. The difference between state filling for carrier degeneracy and phonon state filling is that there is no limit to the number of phonons that can exist within a given state, as they obey the Bose–Einstein distribution. The basic approach assumes that the phonons are out of equilibrium, and the carrier scattering processes are evaluated with an assumed $N_{max}(\mathbf{q})$. Then, within the simulation, the number of phonons emitted, or absorbed, with wave vector $\mathbf{q}$ is carefully monitored. At the synchronization times of the global time scale, the phonon population in each cell of momentum space is updated from the emission/absorption statistics that have been gathered during that time step. One must also include phonon decay, which is through a 3-phonon process to other modes of the lattice vibrations, so that the update algorithm is simply (see section 3.6.2)

$$N(\boldsymbol{q}, t + \Delta t) = N(\boldsymbol{q}, t) + G_{\mathrm{net},\Delta t}(\boldsymbol{q}) - \left[\frac{N(\boldsymbol{q}, t) - N_0}{\tau_{\mathrm{phonon}}}\right]\Delta t, \qquad (7.85)$$

where N_0 is the equilibrium phonon distribution, $G_{\mathrm{net},\Delta t}(\boldsymbol{q})$ is the net (emission minus absorption) generation of phonons in the particular cell *during the time step*, and τ_{phonon} is the phonon lifetime. During the simulation, each phonon scattering process is evaluated as if the maximum assumed phonon population were present. Then, a rejection technique is used, by which the phonon scattering process is rejected (and assumed to be a secondary self-scattering process) if

$$r_{\mathrm{test}} > \frac{N(\boldsymbol{q}, t)}{N_{\mathrm{max}}}. \qquad (7.86)$$

Here, N_{max} is the peak value that was assumed in setting up the scattering matrices. While, this is assumed here to be a constant for all phonon wave vectors, this is not required. A more sophisticated approach would use a momentum-dependent peak

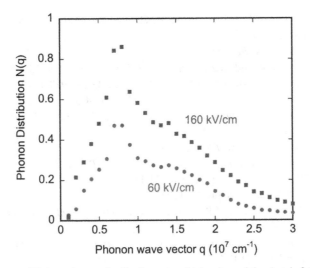

Figure 7.14. The non-equilibrium phonon distribution at two high values of the electric field as determined by Monte Carlo simulations for a GaN/AlGaN two-dimensional layer at the heterojunction interface. Adapted from [76].

occupation, but this is just one of a number of approaches to speeding up the calculation.

In figure 7.14, we plot the phonon distribution function at two values of the electric field for a GaN/AlGaN heterojunction device [76] at room temperature. Electrons in these structure show an experimental peak velocity near 3.2×10^7 cm s^{-1} at a field of 150 kV cm^{-1}, which is supported by detailed ensemble Monte Carlo simulations [76, 77]. The results in figure 7.14 indicate that the electrons interact only with a set of relatively low wave number phonons, but these can be driven quite far out of equilibrium.

References

[1] Gunn J B 1963 *Sol. State Commun.* **1** 88
[2] Ridley B K and Watkins T B 1961 *Proc. Phys. Soc.* **78** 293
[3] Shaw M P, Grubin H L and Solomon P R 1979 *The Gunn-Hilsum Effect* (New York: Academic)
[4] Nicolis G and Prigogine I 1977 *Self-Organization in Nonequilibrium Systems* (New York: Wiley-Interscience)
[5] Fröhlich H and Seitz F 1950 *Phys. Rev.* **79** 526
[6] Wick F G 2008 *Phys. Rev. Ser. I* **27** 11
[7] Ryder E J 1953 *Phys. Rev.* **90** 766
[8] Conwell E M 1967 *High Field Transport in Semiconductors* (New York: Academic)
[9] Norris C B and Gibbons J F 1967 *IEEE Trans. Electron. Dev.* **14** 38
[10] Canali C, Jacoboni C, Nava F, Ottaviani G and Alberigi-Quaranta A 1975 *Phys. Rev.* B **12** 2265
[11] Ferry D K 1976 *Phys. Rev.* **B14** 1605
[12] Siegel W, Heinrich A and Ziegler E 1976 *Phys. Stat. Sol.* a **35** 269

[13] Fischetti M V, Yoder P D, Khatami M M, Gaddemane G and Van de Put M L 2019 *Appl. Phys. Lett.* **114** 222104

[14] Shibuya M 1955 *Phys. Rev.* **99** 1189
Sasaki W, Shibuya M and Mizuguchi K 1958 *J. Phys. Soc. Jpn.* **13** 456
Sasaki W, Shibuya M, Mizuguchi K and Hatoyama G M 1959 *J. Phys. Chem. Sol.* **8** 250

[15] Chelikowsky J R and Cohen M L 1976 *Phys. Rev.* B **14** 556

[16] Aspnes D E, Olson C G and Lynch D W 1976 *Phys. Rev. Lett.* **37** 766

[17] Aspnes D E 1976 *Phys. Rev.* B **14** 5331

[18] Blakemore J S 1982 *J. Appl. Phys.* **53** R123

[19] Drouhin H-J, Hermann C and Lampel G 1985 *Phys. Rev.* B **31** 3859

[20] Ferry D K 2017 *Semicond. Sci. Technol.* **31** 11LT01

[21] He G *et al* 2017 *Scientific Repts.* **7** 11256

[22] Vurgaftman I, Meyer J R and Ram-Mohan L R 2001 *J. Appl. Phys.* **89** 5815

[23] Ridley B K 1963 *Proc. Phys. Soc.* **82** 954

[24] Ferry D K, Young R A and Dougal A A 1965 *J. Appl. Phys.* **36** 3684

[25] Porter W A and Ferry D K 1972 *Physica* **60** 155

[26] Kubo R 1974 *Lecture Notes in Physics: Transport Phenomena* (Berlin: Springer), 31 75

[27] Price P 1960 *J. Appl. Phys.* **31** 949

[28] Zwanzig R W 1961 *Lectures in Theoretical Physics* ed J De Boer and G E Uhlenbeck (New York: Interscience), 3 106

[29] Kubo R 1957 *J. Phys. Soc. Jpn.* **12** 570

[30] Akis R *et al* 2008 *IEEE Electron Dev. Lett.* **29** 306

[31] Mooradian A and McWhorter A L 1970 *Proceedings of the 10th International Conference on Physical Semiconductors* (Springfield, VA: National Bureau of Standards) 380

[32] Grann E D *et al* 1994 *Appl. Phys. Lett.* **64** 1230

[33] Grann E D *et al* 1995 *Appl. Phys. Lett.* **67** 1760

[34] Grann E D *et al* 1995 *Phys. Rev.* B **51** 1631

[35] Liang W *et al* 2003 *Appl. Phys. Lett.* **82** 1413

[36] Wolff P A 1954 *Phys. Rev.* **95** 1415

[37] Shockley W 1961 *Sol.-State Electron.* **2** 36

[38] Baraff G A 1962 *Phys. Rev.* **128** 2507

[39] Pearsall T P *et al* 1978 *Sol.-State Electron.* **21** 297

[40] Tang J Y, Shichijo H, Hess K and Iafrate G J 1981 *J. Phys. Colloques.* **42** C7–C63

[41] Okuto Y and Crowell C R 1972 *Phys. Rev.* B **6** 3076

[42] Anderson C L and Crowell C R 1972 *Phys. Rev.* B **5** 2267

[43] Czajkowski I K *et al* 1990 *Proc. IEE: J. Optoelectron.* **137** 79

[44] Curby R C and Ferry D K 1973 *Phys. Stat. Sol. (a)* **15** 319

[45] Harrison W A 1970 *Sol. State Theory* (New York: McGraw-Hill)

[46] Yamada T, Zhou J-R, Miyata H and Ferry D K 1994 *IEEE Trans. Electron. Dev.* **41** 1513

[47] Lee C A *et al* 1964 *Phys. Rev.* A **134** 761

[48] Crowell C R and Sze S M 1966 *Appl. Phys. Lett.* **9** 242

[49] van Overstraeten R and DeMan H 1970 *Sol.-State Electron.* **13** 583

[50] Fischetti M V and Laux S E 1988 *Phys. Rev.* B **38** 9721

[51] Rana F 2007 *Phys. Rev.* B **76** 155431

[52] Winzer T, Knorr A and Malic E 2010 *Nano Lett.* **10** 4839

[53] Girdhar A and Leburton J P 2011 *Appl. Phys. Lett.* **99** 043107

[54] Pirro L, Girdhar A, Leblebici Y and Leburton J P 2012 *J. Appl. Phys.* **112** 093707

[55] Tani S, Blanchard F and Tanaka K 2012 *Phys. Rev. Lett.* **109** 166603

[56] Zhu W *et al* 2013 *Nature Commun.* **10** 1038

[57] Ferry D K 2017 *Semicond. Sci. Technol.* **32** 085003

[58] Jacoboni C and Reggiani L 1983 *Rev. Mod. Phys.* **65** 645

[59] Jacoboni C and Lugli P 1089 *The Monte Carlo Method for Semiconductor Device Simulation* (Vienna: Springer)

[60] Hess K 1991 *Monte Carlo Device Simulation: Full Band and Beyond* (Boston: Kluwer)

[61] Binder K (ed) 1979 *Monte Carlo Methods in Statistical Physics* (Berlin: Springer)

[62] Kalos M H and Whitlock P A 1986 *Monte Carlo Methods* (New York: Wiley)

[63] Budd H F 1967 *Phys. Rev.* **158** 798
     Rees H D 1968 *Phys. Lett.* A **26** 416

[64] Budd H 1966 *J. Phys. Soc. Jpn. Suppl.* **21** 424

[65] Rees H D 1972 *J. Phys.* C **5** 64

[66] Kreuzer H J 1981 *Nonequilibrium Thermodynamics and Its Statistical Foundations* (London: Oxford University Press)

[67] Rees H D 1969 *J. Phys. Chem. Sol.* **30** 643

[68] Vasileska D, Goodnick S M and Klimeck G 2010 *Computational Electronics: Semiclassical and Quantum Device Modeling and Simulation* (Boca Raton, FL: CRC Press)

[69] McGroddy J C and Nathan M I 1966 *J. Phys. Soc. Jpn. Suppl.* **21** 437

[70] Ferry D K and Heinrich H 1968 *Phys. Rev.* **169** 670

[71] Bosi S and Jacoboni C 1976 *J. Phys.* C **9** 315

[72] Lugli P and Ferry D K 1985 *IEEE Trans. Electron. Dev.* **32** 2431

[73] Alfano R R (ed) 1984 *Semiconductors Probed by Ultrafast Laser Spectroscopy* (Orlando, FL: Academic)

[74] Liang L W *et al* 2005 *Phys. Stat. Sol.* c **2** 2297

[75] Lugli P, Jacoboni C, Reggiani L and Kocevar P 1987 *Appl. Phys. Lett.* **50** 1251

[76] Barker J M *et al* 2004 *J. Vac. Sci. Technol.* B **22** 2045

[77] Yu T-H and Brennan K F 2002 *J. Appl. Phys.* **91** 3730

IOP Publishing

Semiconductors (Second Edition)
Bonds and bands
David K Ferry

Chapter 8

Optical properties

The properties of excess carriers in semiconductors are some of the more interesting aspects of nano-electronics, and indeed of nearly all optically motivated devices today, from the vast and growing field of solar cells to the advanced lasers used in most scanning devices. The role of the excess carriers is crucial to the operation of such devices. In this chapter, the role of the optically created, non-equilibrium carriers is discussed. While the approach depends to a large extent upon the Boltzmann equation, and the continuity equation, it is quite possible to use the Monte Carlo approach even for these optically generated carriers, and full-band Monte Carlo techniques have been used to evaluate solar cell performance [1].

In our previous treatment of transport in chapters 6 and 7, it was assumed that the distribution function was generally not an explicit function of time, although we did discuss the transient transport situation. If, however, the force to which the carriers are responding is a periodic function of time, we must modify the treatment to directly include this periodic behavior. Such behavior arises from all sinusoidal sources, from relatively low frequency, up through the microwave frequencies and on to the optical frequencies. Certainly, excess carriers can be generated without this periodic behavior, such as by the direct injection processes in p-n junctions. Our aim is to study the methods by which carriers in the semiconductor can absorb photons, and the transport response that follows such absorption. We can break this response into two very different regimes: (1) $\hbar\omega < E_G$, for which the number of carriers is conserved, and we can use a simple Boltzmann equation approach (or a more suitable Monte Carlo technique), and (2) $\hbar\omega > E_G$, for which one must account for the new electron and hole pairs that are produced, the electron in the conduction band and the hole in the valence band. The quantity E_G, is the primary energy gap between the conduction band and the valence band. As we discussed in chapter 2, this gap may be direct or indirect, for which the latter case requires the transition to be accompanied by a phonon absorption or emission. However, at low temperatures, the absorption could be from a neutral donor (acceptor) to the conduction (valence) band.

For $\hbar\omega < E_G$, the absorption of high frequency energy is carried out by the free carriers already present in the bands, and their motion can be handled by a straight-forward extension of the transport equations. We treat this process first in section 8.1, and the process is called *free carrier absorption*. In some cases, the absorption process can be made resonant by the addition of a magnetic field, which allows a direct measurement of the effective mass of the carriers.

If, however, $\hbar\omega > E_G$, one must take account the inter-band transitions that will occur and this means consideration of the quantum mechanical processes that are involved. We turn to this topic in section 8.2, where both direct and indirect processes are discussed. In the direct process, the carrier momentum is conserved in the transition. In indirect processes, the transition must be accompanied by the emission or absorption of a phonon. While we will treat the primary cross-gap transitions, the approach can be superficially extended to the excitation of carriers from impurities, as mentioned above.

8.1 Free-carrier absorption

For incident high-frequency waves of energy insufficient to produce an electron-hole pair, such as microwave radiation, the absorption of the photon can occur either due to the phonons or to free carriers. We discussed the case of absorption by the polar optical phonons already in section 3.4 in regard to the contribution to the dielectric function that occurred from the lattice polarization. And, we will return to this polarization in chapter 9, when we discuss the full dielectric function and electron–electron interactions. In this section, however, we want to discuss the absorption of the microwave photon by the carriers. This absorption can only occur in a partially filled band, where there are free electrons or holes capable of gaining energy from the microwave field, although we will see that this absorption also requires scattering processes as well in order to conserve both energy and momentum. This process will lead to the microwave conductivity, although this 'microwave' frequency may well extend into the infrared region.

8.1.1 Microwave absorption

In order to discuss the free-carrier absorption, we need to retain explicitly the time variation of the distribution function in the Boltzmann equation. We do this in the relaxation time approximation, although it is quite easy to extend this to detailed consideration of any and all scattering processes. Our Boltzmann equation then becomes

$$\frac{\partial f}{\partial t} + \frac{1}{\hbar} \boldsymbol{F} \cdot \frac{\partial f}{\partial \boldsymbol{k}} = -\frac{f - f_0}{\tau_m}, \tag{8.1}$$

where $\boldsymbol{F}$ is the general force, rather than merely the electric field. This will allow us to include a magnetic field as well in the next section. Here, however, we will replace this with just the electric field, which varies as $\mathbf{E} = \mathbf{E}_0\exp(-i\omega t)$. We assume that, even with this sinusoidal excitation, a steady-state is reached in which both the velocity and the distribution exhibit this same temporal behavior. Then, if we write $f_1 = f - f_0$, (8.1) becomes

$$(1 - i\omega\tau_m)f_1 = e\tau_m(v \cdot \mathbf{E_0})\frac{\partial f_0}{\partial E}, \tag{8.2}$$

where the exponential temporal variation has been omitted from the equation as it is common in all terms. (Hopefully, the presence of both the field $\mathbf{E}$ and the energy E will not cause confusion.) It is important to note that the frequency ω and the momentum relaxation time τ_m appear as a product here. This implies that the absorption process *requires* a scattering event in order to occur. In fact, this requirement arises from the need to conserve both energy and momentum in the absorption process. As we have discussed in previous chapters, the microwave photon has almost zero momentum on the scale of the carrier. For example, a 10 GHz photon has a wave number $k \sim (2\pi/3)$ cm^{-1}, while the carrier has a wave number $>10^6$ cm^{-1}. Because the energy band is dispersive, the absorption cannot occur unless a scattering event provides the momentum shift that is needed to reach the higher energy $E + \hbar\omega$. As a result, free-carrier absorption requires both scattering and the photon in order to occur.

We note that the right-hand side of (8.2) is of the same form as (6.11), from which we produced the normal conductivity in the presence of an electric field. Using the results from (6.11), we can now write the microwave current as (we assume the electric field is along the x axis)

$$J_x = e^2 E_0 \int_0^\infty \frac{v_x^2 \tau_m}{1 - i\omega\tau_m} n(E)\frac{\partial f_0}{\partial E}dE. \tag{8.3}$$

By straight-forward integration, following the approach of chapter 6, we find that the high-frequency conductivity is given as

$$\sigma(\omega) = \frac{ne^2}{m_c}\left(\left\langle\frac{\tau_m}{1 + \omega^2\tau_m^2}\right\rangle + i\omega\left\langle\frac{\tau_m^2}{1 + \omega^2\tau_m^2}\right\rangle\right). \tag{8.4}$$

We have only considered the electron contribution here, but this result is readily extended to holes as well. The first important point from (8.4) is that the effectiveness of an electric field in producing a current is reduced at high frequencies, that is at frequencies that satisfy $\omega\tau_m > 1$. In chapter 5, it was found that the scattering time is of the order of 5×10^{-13} s, so that the high frequency regime is typically at frequencies of the order of 10^{12} Hz, or a THz.

The actual reduction of the conductivity is contained in the first term of (8.4), or what is known as the real part of the equation. We can write this reduction effect in the form

$$\text{Re}\left\{\frac{\sigma(\omega)}{\sigma(0)}\right\} = \frac{1}{\langle\tau_m\rangle}\left\langle\frac{\tau_m}{1 + \omega^2\tau_m^2}\right\rangle. \tag{8.5}$$

The second term in (8.4) is imaginary, and in fact contributes to the dielectric function. We can see this by writing the *total* current, the conduction current plus the displacement current, as

$$\boldsymbol{J} = \sigma\boldsymbol{E} - i\omega\varepsilon(\omega)\boldsymbol{E}, \tag{8.6}$$

and we recognize that we can write the dielectric function, using the imaginary part of (8.4) as

$$\varepsilon(\omega) = \varepsilon_s - \text{Im}\left\{\frac{\sigma}{\omega}\right\} = \varepsilon_s - \frac{\sigma(0)}{\langle\tau_m\rangle}\left\langle\frac{\tau_m^2}{1 + \omega^2\tau_m^2}\right\rangle. \tag{8.7}$$

A remarkable effect is that the dielectric 'constant' can, in fact, go negative at low frequencies due to the effect of the free carriers. This is a well-known effect in microwaves and leads to what is known as the 'skin depth,' that is, the depth to which a microwave will penetrate into a conductor.

Now, let us examine the region $\omega\tau \gg 1$, the very high frequency limit of (8.7). The large bracket term can be written out using the definition of the conductivity to obtain

$$\varepsilon(\omega) \sim \varepsilon_s\left(1 - \frac{\omega_p^2}{\omega^2}\right), \tag{8.8}$$

where

$$\omega_p^2 = \frac{ne^2}{m_c\varepsilon_s} \tag{8.9}$$

is the square of the free-carrier plasma frequency. In polar material, one has to decide which region's dielectric 'constant' to use; the high frequency dielectric constant if the frequency is above the Restrahlen range, or the low frequency dielectric constant if the frequency is below this range. Hence, (8.8) tells us that the semiconductor is reflecting below the plasma frequency.

The absorption coefficient for microwave fields may now be calculated using the results obtained above. First, we extend the propagator to be $\boldsymbol{E} = \boldsymbol{E}_0\exp(i\boldsymbol{k}_0 \cdot \boldsymbol{r} - i\omega t)$, so as to be able to address the spatial propagation. Here,

$$k_0 = \frac{\omega}{c}\sqrt{\frac{\varepsilon_s}{\varepsilon_0}} \tag{8.10}$$

Is the wave propagation vector. If the frequency is below the plasma frequency, the wave is heavily attenuated. Our interest here is in the weak free-carrier absorption at frequencies well above the plasma edge (a relatively small free carrier population), so that the imaginary part of the conductivity may be ignored. Still, the dielectric function is complex, and

$$\varepsilon(\omega) = \varepsilon_s \left(1 + i \frac{\text{Re}\{\sigma\}}{\omega \varepsilon_s} \right). \tag{8.11}$$

The ratio $\text{Re}\{\sigma\}/\varepsilon_s$ is termed the dielectric relaxation frequency ω_D. The absorption is determined from the imaginary part of the wave vector (8.10), and

$$k_i = \text{Im}\{k_0\} \sim \frac{\omega_p^2}{2c\omega^2} \sqrt{\frac{\varepsilon_s}{\varepsilon_0}} \left\langle \frac{1}{\tau_m} \right\rangle. \tag{8.12}$$

The absorption coefficient is just $\alpha = 2k_i$. This result is for the high frequency situation ($\omega\tau \gg 1$). At low frequencies, it is somewhat harder to determine the absorption coefficient due to the plasma reflection edge. We will return to more in depth properties of the dielectric function in chapter 9.

8.1.2 Cyclotron resonance

We have found so far that, at high frequencies, there is a reduction in the conductivity and also a modification of the permittivity found due to the interaction of the free carriers, primarily from the inductive lag of these free carriers in response to the microwave electric field. There are other effects which arise from this interaction, and we want to begin here to study the propagation of an electromagnetic wave along the applied magnetic field. In many cases, these effects are known in plasma physics and are collective effects of the electron gas as a whole, but these effects are useful in determining certain properties of the semiconductor as well.

We recall in chapter 6, that the determination of the conductivity in the presence of a magnetic field is more complicated. In this approach, (6.11) was replaced by (6.37). With the modification of (8.2), this becomes

$$f_1 = \frac{e\tau_m(\boldsymbol{v} \cdot \mathbf{A})}{1 + i\omega\tau_m} \frac{\partial f_0}{\partial E}, \tag{8.13}$$

where A was given by (6.40). If we include the denominator of (8.13) with the factor $\eta = 1 + i\omega\tau_m$, then we can rewrite the vector force (6.40) as

$$\mathbf{A} = \frac{\mathbf{E} + (\omega_c\tau_m/\eta)\boldsymbol{a_B} \times \mathbf{E} + (\omega_c\tau_m/\eta)^2 \boldsymbol{a_B}(\boldsymbol{a_B} \cdot \mathbf{E})}{1 + (\omega_c\tau_m/\eta)^2}, \tag{8.14}$$

where $\boldsymbol{a_B}$ is a unit vector along the applied magnetic field and the cyclotron frequency $\omega_c = eB/m_c$ has been introduced. With this latter form of A, we need only now deal with (6.37), but to do this we will first expand this quantity into its various components to make the approach somewhat clearer. For this, we take the magnetic field in the z direction for convenience, and this leads to

$$\begin{bmatrix} A_x \\ A_y \\ A_z \end{bmatrix} = \frac{\eta^2}{\eta^2 + \omega_c^2 \tau_m^2} \begin{bmatrix} 1 & -\dfrac{\omega_c \tau_m}{\eta} & 0 \\ \dfrac{\omega_c \tau_m}{\eta} & 1 & 0 \\ 0 & 0 & \dfrac{\eta^2 + \omega_c^2 \tau_m^2}{\eta^2} \end{bmatrix} \begin{bmatrix} E_x \\ E_y \\ E_z \end{bmatrix}. \tag{8.15}$$

This result can now be used in (6.37) to find the shifted distribution function, and in turn to find the components of the current. The latter are given by

$$J_x = \frac{ne^2}{m_c} \left\{ \left\langle \frac{\eta \tau_m}{\eta^2 + \omega_c^2 \tau_m^2} \right\rangle E_x - \left\langle \frac{\omega_c \tau_m^2}{\eta^2 + \omega_c^2 \tau_m^2} \right\rangle E_y \right\}$$

$$J_y = \frac{ne^2}{m_c} \left\{ \left\langle \frac{\omega_c \tau_m^2}{\eta^2 + \omega_c^2 \tau_m^2} \right\rangle E_x + \left\langle \frac{\eta \tau_m}{\eta^2 + \omega_c^2 \tau_m^2} \right\rangle E_y \right\}. \tag{8.16}$$

$$J_z = \frac{ne^2}{m_c} \left\langle \frac{\tau_m}{\eta} \right\rangle E_z$$

For an electromagnetic plane wave, the wave vector $\mathbf{k}$ is perpendicular to the components $\mathbf{E}$ and $\mathbf{B}$, which are in turn perpendicular to each other and related in magnitude through the wave impedance. In the formulation leading to (8.16), these components reduced to just the electric field terms themselves. We can now use the current densities in Maxwell's equations to relate the two types of current, conduction and displacement, to an effective dielectric permittivity.

Consider first the case in which there is only a linear polarized wave, with the electric field vector along the z-axis parallel to the applied magnetic field. This must obviously be a wave propagating across the applied magnetic field ($\mathbf{k} \perp \mathbf{B}$). Then only the third line of (8.16) is needed, and

$$\varepsilon(\omega) = \varepsilon_s + \frac{1}{\omega} \text{Im}\{\sigma_z\} = \varepsilon_s \left(1 + \frac{ne^2}{m_c} \left\langle \frac{\tau_m^2}{1 + \omega_c^2 \tau_m^2} \right\rangle \right)$$

$$\Longrightarrow \varepsilon_s \left(1 - \frac{\omega_p^2}{\omega^2} \right), \quad (\omega\tau \gg 1). \tag{8.17}$$

In the last line, the free-carrier plasma frequency has been introduced, and this result agrees with that of the last section.

In a similar manner, we can consider the case for a plane wave propagating across the magnetic field with $\mathbf{k} \perp \mathbf{B}$ and $\mathbf{E} \perp \mathbf{B}$. Again, this is for a linearly polarized wave, with both the propagation direction and the polarization normal to the applied magnetic field. This now involves several lines from (8.16), and, after some algebra, one obtains for the same high frequency situation $\omega\tau \gg 1$, the result

$$\varepsilon(\omega) = \varepsilon_s \left[1 - \frac{\omega_p^2(\omega^2 - \omega_p^2)}{\omega^2(\omega^2 - \omega_p^2 - \omega_c^2)} \right].$$

(8.18)

This leads to a coupled mode propagation, in which the electromagnetic wave interacts with both the plasma oscillations and the cyclotron motion.

Our main interest in this section, however, is for waves which are propagating *along* the magnetic field, $k \| \mathbf{B}$. While we discussed linearly polarized waves above, here we want to turn to the expression of a linearly polarized wave as two counter-rotating circularly polarized waves, in which we may describe the two electric field components as

$$E_\pm = E_x \pm iE_y,$$

(8.19)

where the upper sign if for *right*-circularly polarized waves and the lower sign is for *left*-circularly polarized waves. The current may now be written in terms of this wave electric field, where we will assume the relaxation time is energy independent so that we can simplify the equations by dropping the ensemble averages. Then

$$J_\pm = \frac{ne^2\tau_m}{m_c} \frac{\eta \pm i\omega_c\tau_m}{\eta^2 + \omega_c^2\tau_m^2} E_\pm$$

(8.20)

and

$$\sigma_\pm = \frac{\sigma_0}{1 + i(\omega \mp \omega_c)\tau_m}.$$

(8.21)

Examination of this equation shows that the right-circularly polarized wave will show a large absorption (the conductivity will rise to the low-frequency value) at the resonance condition $\omega = \omega_c$. This resonance occurs when the field is rotating about the magnetic field at the exact same period as the electrons themselves. Thus, the electrons see a constant electric field that continually accelerates them, and this is the basis for the resonant absorption. This is known as *cyclotron resonance*.

In figure 8.1, we show the results of cyclotron resonance studies of Si-doped GaAsN films [2]. The films were epitaxially grown on GaAs and were formed into GaAsN quantum wells with GaAs barriers. The nitrogen contents were 1.2% and 2%. The N incorporation reduces the band gap of GaAs, which makes it attractive for optical applications [3, 4]. The microwave source used was a free-electron laser with photons at 41.4 ± 0.5 meV. Panel (a) shows the cyclotron resonance and impurity-shifted cyclotron resonance at two low temperatures for a reference sample that doesn't contain N. Panels (b)–(d) then show the cyclotron resonance for three samples containing 0%, 0.1%, and 0.2% N. The red lines are fits to the classical theory. It is clear that incorporating N really leads to much broader resonances. From these fits, one can ascertain the effective masses for each sample, as will be discussed below.

In figure 8.2, the cyclotron resonance in ultra-pure diamond is shown [5]. Diamond has a band structure much like Si, with six constant energy ellipsoids at

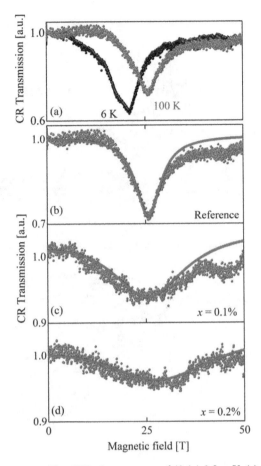

Figure 8.1. Transmission spectra with a FEL photon energy of 41.4 ± 0.5 meV. (a) Green CR and black ICR signal from reference sample at 100 K and 6 K, respectively. (b)–(d) CR signals from undoped (reference) and N-doped GaAs. The red lines are fits to the classical theory. The addition dip in (c) at 45 T is an artefact from noise in the system. Reprinted with permission from Esser *et al* [2], copyright 2015 by the American Institute of Physics.

the Δ points along (100). A microwave signal at 9.64 GHz was used to detect the resonance, while carriers were excited using a short pulse laser from an optical parametric amplifier at 5.494 eV (as the sample has significant B impurities, no free carriers were present at low temperature without the optical excitation). The sample was mounted in a dielectric cavity which could be accessed by both the microwave and optical signals. As may be observed, two electron resonances were observed due to the fact that the sample was not aligned in a manner for which all six valleys made the same angle with the wave propagation vector. Also apparent are resonances due to the light- and heavy-hole bands. Both the real and imaginary parts of the absorption spectra are shown and the signals may be well described by the classical theory.

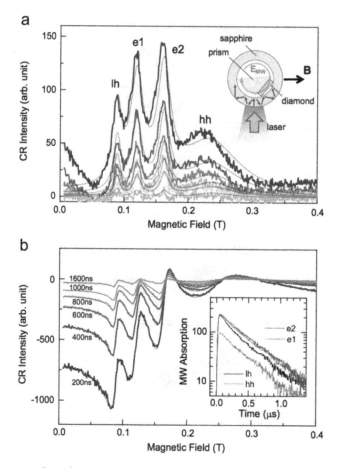

Figure 8.2. Microwave absorption at 9.64 GHz due to cyclotron resonance in diamond at 10 K. The magnetic field was applied in a direction 40° from the (001) axis. The incident laser pulse was 3.2 μJ. (a) Real part and (b) imaginary part of the signal for different delay times. The inset in (a) shows the sample configuration. The inset in (b) shows the temporal decay of the various signals. Reprinted with permission from Naka *et al* [5], copyright 2014 by Elsevier B. V.

As mentioned above, cyclotron resonance is used to determine the effective mass of the carriers in the semiconductor. The rotating electric field stays in phase with the rotation of the carriers around the magnetic field, and can thus accelerate the carriers. In this sense, cyclotron resonance measures the area of the carrier orbit by measuring the period (inverse frequency) required for the carrier to move around the magnetic field. At low temperatures, the carrier velocity is mainly the Fermi velocity of the carriers at the Fermi energy. Consider two orbits that lie in the same plane perpendicular to the magnetic field, but which have slightly different energies, differing by δE. These two orbits are separated in momentum space by the amount

$$\delta k = \frac{\delta E}{|\nabla_k E|} = \frac{\delta E}{\hbar v_k}, \qquad (8.22)$$

where the velocity has been introduced as the gradient of the energy (the normal group velocity). If an electron is moving on one of these orbits, the rate at which it is sweeping out the area between the orbits due to the Lorentz force is just

$$\frac{dk}{dt}\delta k = \left| \frac{e}{\hbar} v \times \mathbf{B} \right| \delta k = \frac{eB}{\hbar^2}\delta E. \tag{8.23}$$

The period of the orbit may be taken to be T, and the product of the period and the quantity in (8.23) is just the area between the two orbits. If the *enclosed* area of the orbit at energy E is termed $S(E)$, then

$$T\frac{eB}{\hbar^2}\delta E \equiv \frac{dS}{dE}\delta E \tag{8.24}$$

or

$$\omega_c = \frac{2\pi}{T} = \frac{2\pi eB}{\hbar^2}\left(\frac{dS}{dE}\right)^2. \tag{8.25}$$

In two dimensions, the area of the orbit is just πk^2, so that it is possible to write directly

$$\omega_c = \frac{eB}{\hbar^2 k}\frac{dE}{dk} = \frac{eBv_k}{\hbar k}, \tag{8.26}$$

from which it is now possible to write the cyclotron effective mass as

$$\frac{1}{m_c} = \frac{1}{\hbar^2 k}\nabla_k E = \frac{v_k}{\hbar k}. \tag{8.27}$$

This result defines the effective mass exactly as was done in section 2.6, where we discussed the effective mass theorem. The proper effective mass is the dynamic mass, and not the density of states mass, and it arises from the group velocity of the wave packet representing the carrier.

In figure 8.3, we plot the effective mass measured from cyclotron resonance for graphene. The data is taken from the review [6], and the solid curve is (8.27), which may be rewritten in terms of the density for graphene carriers as

$$m_c = \frac{\hbar k_F}{v_F} = \frac{\hbar}{v_F}\sqrt{\pi n}. \tag{8.28}$$

The fit shown in the figure uses the Fermi velocity as 8.3×10^7 cm s^{-1}. It is clear that the dynamic mass exhibited in transport is the wave packet group velocity as given in section 2.6. While graphene may have no rest mass (due to the zero band gap), it certainly has a dynamic mass that conforms to standard semiconductor theory.

8.1.3 Faraday rotation

From the conductivity (8.21), we can also find the permittivity of the carriers and the semiconductor under the conditions of cyclotron resonance to be

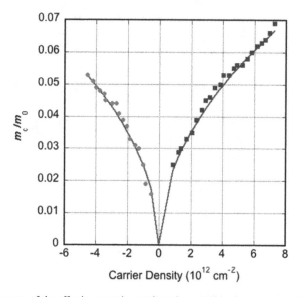

Figure 8.3. Measurements of the effective mass in graphene by cyclotron resonance (red dots). The line is the effective mass determined from (8.28) using $v_F = 8.3 \times 10^7$ cm s^{-1}.

$$\varepsilon(\omega) = \varepsilon_s \left[1 - \frac{\omega_p^2}{\omega(\omega \mp \omega_c)} \right]. \tag{8.29}$$

Since the permittivities of the two circularly polarized waves are different, these two waves will propagate with different velocities. This means that the polarization of the combination linearly polarized wave will rotate as it propagates along the magnetic field. This is called *Faraday rotation*. The right-hand circularly polarized wave has the smaller permittivity, as it has the negative sign in the denominator of (8.29), and will therefore propagate faster and undergo a smaller rotation over a given distance. This 'pulls' the linear polarization in the direction of the left-hand circularly polarized component. If propagation is over a distance d, the angles through which the two circularly polarized wave propagate are given by

$$\theta_\pm = k_\pm d = \frac{2\pi d}{\lambda_0} \kappa_\pm, \tag{8.30}$$

where $\kappa_\pm$ are the indices of refraction for the two waves and λ_0 is the free space wavelength of the electromagnetic wave. Thus, the effective rotation of the linearly polarized wave is related to the difference in the two components, and we can use (8.29) to find this angle as

$$\theta = \frac{2\pi d}{\lambda_0}(\kappa_+ - \kappa_-) \sim -\frac{dne^3}{\varepsilon_s c m_c^2 \omega^2} B, \tag{8.31}$$

where (8.29) has been expanded for the condition $\omega \gg \omega_p, \omega_c$. The rotation is a linear function of the magnetic field and can therefore be used as another tool to

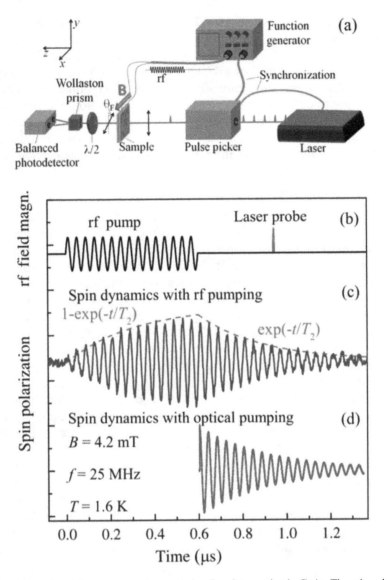

Figure 8.4. (a) Experimental arrangement for measuring Faraday rotation in GaAs. The spin polarization is probed by Faraday rotation of the linear polarization of the optical pulses. (b) Temporal profile of the rf field pulse applied to the sample. (c) dynamics of spin rotation. (d) Measured dynamics of spin polarization in the absence of the rf signal. Reprinted from Belykh *et al* [7], copyright 2019 by the American Physical Society.

determine the effective mass of the carriers, if the density is known, or of the density, if the effective mass is known. Because of the frequency in the denominator, Faraday rotation measurements are more often made in the microwave range rather than the optical range of frequencies, but the limits of the expansion have to be kept in mind.

In figure 8.4, we illustrate the use of Faraday rotation in experiments, in this case to determine the electron spin relaxation time [7]. Here, a Si-doped GaAs sample with an electron concentration of 1.4×10^{16} cm^{-3}, and a thickness of 350 μ, is used for the study. The experimental setup is shown in panel (a), and the sample is excited with a pulsed Ti:sapphire laser and a pulsed rf source producing 600 ns pulses of 25 MHz (curve b). Curve (c) shows the spin dynamics with rising amplitude during the rf pulse and decaying amplitude after the rf pulse. The time constants of these rising and falling parts of the signal give us the spin precession times and relaxation times. Curve (d) illustrates the Faraday rotation signal in an all-optical pump-probe experiment on the spin dynamics in the absence of the rf signal. From here, we can clearly see the use of Faraday rotation in modern studies of semiconductors.

8.2 Direct transitions

At sufficiently high frequencies, such as in the optical region, we achieve the situation where $\hbar\omega > E_G$, and the electromagnetic field can induce electrons from the valence band into the conduction band. This creates a new electron-hole pair that can contribute to the conduction process. This is the usual optical absorption that provides a method to study the various transitions possible from valence to conduction band, or even between different valence or different conduction bands. The requirement is that the lower energy state must be occupied and the upper energy state empty, to satisfy the Fermi statistics. Of interest in this section are the direct transitions, such as between the top of the valence band and the bottom of the conduction band in GaAs at Γ, which is known as the E_0 transition (the direct band gap). But these direct transitions can also occur at the X and L points. These transitions are strong primarily because of the high density of states near them.

The quantity that we want to finally obtain is the absorption coefficient, that describes how much of the optical electromagnetic wave is turned into electron-hole pairs. To do this, we must first determine the transition probability for the inter-band transition. Here, we will show how these matrix elements, needed for the transition probability, arise in a natural manner. Once the transition probability is determined, we can then calculate the number of electrons that make the transition for a given electromagnetic field power, and this will lead us to the absorption coefficient. In figure 8.5, the basic transition for the direct absorption is shown. This transition is vertical, as discussed earlier, because the momentum of the photon is

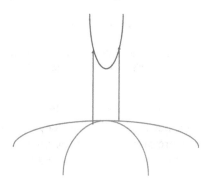

Figure 8.5. Direct optical absorption indicated for electrons coming from the heavy- and light-hole bands.

orders of magnitude smaller than the momentum of the electron and/or hole. Note that the transition is made away from $k = 0$. This is because the densities of states for both the valence band and the conduction are zero at this point (see chapter 4), so there are neither full states in the valence band nor empty states in the conduction band exactly at $k = 0$. Note also that a given excitation can make a transition from either the heavy hole band or the light hole band. If the photon has a sufficiently high energy, it can also transition from the spin-split-off valence band. For $\hbar\omega > E_G$, the energy in excess of the gap energy will split as kinetic energy between the valence and conduction band, with the factors for each determined by the effective masses of the two bands.

The first step of creation of an electron-hole pair is the absorption and subsequent annihilation of a photon resulting in the creation of the pair. This step is no different in a semiconductor from what occurs in an atom. Contrary to many beliefs in quantum mechanics, the electron does not just 'jump' into the conduction band via a quantum jump. It transitions through a standard probability function from quantum mechanics in a well-defined process and takes a few femtoseconds to occur. Perhaps the earliest study was by Geltman [8], who wished to find out just how long it took to ionize an atom. In ionizing an atom with an incident electromagnetic field (the photon field), the field first creates a correlation between the initial state (ground state) wave function and the final state (excited state) wave function. In fact, this correlation can lead to Rabi oscillations between these two states [9]. In other words, the unoccupied quantum state in the conduction band becomes entangled with the ground state in the valence band. This correlation is referred to as the polarization within the atom or the semiconductor. The first step of the electromagnetic wave of the photon is to create this polarization. The conjugate field then destroys this polarization (the field and its conjugate appear in the magnitude squared of the matrix element). Once the polarization is destroyed, the process is complete. But, the electron may remain in the valence band with the photon proceeding on its merry way, or the electron may wind up in the conduction band with the photon being destroyed. The fraction of these processes that lead to actual absorption of the photon is known as the quantum efficiency. Although it is the electromagnetic field (photon) that creates the interaction, conservation of the energy leads to the photon energy being divided between the energy gap and the kinetic energy of the electron and hole. The entire process takes only a few femtoseconds and is governed by the semiconductor optical Bloch equations [10–12]. We will not go into these equations here, but they are discussed in some detail in the previous references.

The process by which the above transition is determined is just our standard time-dependent perturbation theory (in which, however, we omit the details of the polarization as we only involve the initial and final states). Here, we take the transition of electrons from the essentially full valence band to the essentially empty conduction band, and write the transition coefficient as

$$a_{12} = -\frac{H_{12}}{\hbar}\left[\frac{e^{i(\omega_{21}+\omega)t} - 1}{\omega_{21} + \omega} - \frac{e^{i(\omega_{21}-\omega)t} - 1}{\omega_{21} - \omega}\right], \tag{8.32}$$

where $\hbar\omega_{21} = E_c(k_2) - E_v(k_1)$ and

$$H_{12} = \langle 2|H_F|1 \rangle = \int \psi_v^\dagger(k_1) H_F \psi_c(k_2) d^3 r \qquad (8.33)$$

is the matrix element for the transition. As is standard in the Fermi golden rule, if the long time limit is taken, the square brackets reduce to the energy-conserving delta function $2\pi\delta(\hbar\omega_{21} \pm \hbar\omega)$. Note that the two signs give us either absorption or emission. Because we are interested only in the absorption, we will use the negative sign in the following. Again, in the standard Fermi golden rule approach, the two momenta are equal by conservation of this quantity and we may write the transition rate as

$$c_{21} = \frac{|a_{12}|^2}{t} = \frac{4|H_{12}|^2}{\hbar^2} \frac{\sin^2[(\omega_{21} - \omega)t/2]}{(\omega_{21} - \omega)^2 t^2}. \qquad (8.34)$$

Now, we turn to the matrix element H_{12}. In the presence of an electromagnetic wave, the Hamiltonian for the electrons is given by

$$H = \frac{1}{2m_0}(-i\hbar\nabla - e\mathbf{A})^2 + V(r) + \phi(r), \qquad (8.35)$$

where A and ϕ are the vector and scalar potential, respectively, of the wave, and the other two terms have their normal meaning. While we have shown a mass for this equation, one should be careful about the meaning of this term, as we are discussing at least two bands which each have their own effective mass; one cannot define a single mass for this inter-band situation. Nevertheless, the mass shown comes from the original Hamiltonian and should be the free electron mass. This will be discussed further below. The two potentials arise from the components of the electromagnetic wave and these are all related (in the Lorentz gauge) via

$$\mathbf{E} = -\nabla\phi - \frac{\partial \mathbf{A}}{\partial t}, \quad \mathbf{B} = \nabla \times \mathbf{A}. \qquad (8.36)$$

Further gauge choices are certainly possible and have important usage for special applications in which these latter two equations are slightly modified. In general, the scalar potential is relatively slowly varying over a unit cell of the lattice (it is generally found to lead to certain electrostatic effects which are not important here), and the leading term in (8.35), the vector potential term, is the more important one in optical absorption situations. Hence, if we keep only the leading terms of importance, then (8.35) can be rewritten as

$$H = -\frac{\hbar^2}{2m_0}\nabla^2 + V(r) + \frac{i\hbar e}{m_c}\mathbf{A} \cdot \nabla. \qquad (8.37)$$

The last term in this equation is the perturbing potential that will be used in the matrix element. For an electromagnetic wave, this term oscillates at frequency ω.

For the wave functions, we use the standard normalized Bloch functions for each of the two bands

$$\psi_{c,v}(\mathbf{r}) = \frac{1}{\sqrt{N}} u_{c,v}(\mathbf{r}) e^{i\mathbf{k} \cdot \mathbf{r}}. \tag{8.38}$$

The transition of interest is that of an electron from near the top of the valence band to near the bottom of the conduction band. Hence, the appropriate subscript in (8.38) will be used in the following. We take the vector potential to have the form

$$\mathbf{A} = \frac{1}{2} \mathbf{A}_0 e^{i(\omega t - \mathbf{k}_0 \cdot \mathbf{r})} + c.c. \tag{8.39}$$

We may now write the matrix element as

$$\begin{aligned}
H_{12} &= \frac{i\hbar e A_0}{2m_0 N} \int d^3r \, u_c^{\dagger}(\mathbf{r}) e^{-i\mathbf{k} \cdot \mathbf{r}} e^{-i\mathbf{k}_0 \cdot \mathbf{r}} \mathbf{a} \cdot \nabla u_v(\mathbf{r}) e^{i\mathbf{k}' \cdot \mathbf{r}} \\
&= \frac{i\hbar e A_0}{2m_0 N} \int d^3r \, u_c^{\dagger}(\mathbf{r}) e^{-i(\mathbf{k}+\mathbf{k}_0-\mathbf{k}') \cdot \mathbf{r}} [\mathbf{a} \cdot \nabla u_v(\mathbf{r}) + i\mathbf{a} \cdot \mathbf{k}' u_v(\mathbf{r})]
\end{aligned} \tag{8.40}$$

where $\mathbf{a}$ is a unit vector in the direction of the vector potential. Since the central terms in the Bloch functions are periodic over a unit cell, we can write the integration as an integral over a unit cell plus a summation over all the unit cells. This provides the closure condition for the Fourier transforms, and we find that, within a reciprocal lattice vector $\mathbf{G}$, we have

$$\mathbf{k} + \mathbf{k}_0 - \mathbf{k}' = 0. \tag{8.41}$$

Since the optical wave vector is orders of magnitude smaller than either the valence band or conduction band wave vectors, it may be ignored, which leads to the two particle wave vectors being equal. This gives us the vertical transition which we have discussed previously above. Using this selection rule and summing over the unit cells in the lattice, we arrive at

$$H_{12} = \frac{i\hbar e A_0}{2m_0} \int d^3r \, u_c^{\dagger}(\mathbf{r}) [\mathbf{a} \cdot \nabla u_v(\mathbf{r}) + i\mathbf{a} \cdot \mathbf{k} u_v(\mathbf{r})]. \tag{8.42}$$

Normally, the first term in the brackets dominates the matrix element, and is called the 'allowed' transition, since it requires certain properties of the wave functions within the crystal. The second term is termed the 'forbidden' transition as it can occur whatever the relationship between the wave functions, because it basically breaks the required symmetry although it is a weaker term. We will treat these two terms separately in the following.

8.2.1 Allowed transitions

Allowed transitions arise from the first term in the bracket in (8.42). The operator involving the gradient of the Bloch function core is an operator basically on the atomic nature of the atomic wave functions in the unit cell. Because of the gradient operator, this produces essentially a dipole interaction, and this requires a change of angular momentum between the valence wave function and the conduction wave function. We may recall that, in the direct gap III-V semiconductors (see chapter 2), the top of the valence band is mainly composed of anion p-functions, while the bottom of the conduction is composed of cation s-functions. These two types of functions differ by a single unit of angular momentum, so that the transition in these materials is allowed and therefore is very strong. This leads to a high absorption coefficient. When this dipole transition is allowed, we may write the matrix element as

$$H_{12} = \frac{i\hbar e A_0}{2m_0} \int d^3 r u_c^\dagger(r) a \cdot \nabla u_v(r) = -\frac{e A_0}{2m_c} a \cdot P_{12}, \tag{8.43}$$

where we have introduced the 'dipole contribution' which represents the integral over the unit cell of the central parts of the Bloch functions. For a given semiconductor, the wave functions can be found exactly and the integral done easily to give this dipole contribution.

We now introduce the matrix element into the transition rate (8.34) which leads us to the result

$$c_{12} = \frac{e^2 A_0^2 t}{m_0^2 \hbar^2} |a \cdot P_{12}|^2 \frac{\sin^2[(\omega_{21} - \omega)t/2]}{(\omega_{21} - \omega)^2 t^2}, \tag{8.44}$$

where

$$\hbar\omega_{21} = E_c(k) - E_v(k) = E_G + \frac{\hbar^2 k^2}{2}\left(\frac{1}{m_{c,e}} - \frac{1}{m_{c,h}}\right) = E_G + \frac{\hbar^2 k^2}{2m_R}. \tag{8.45}$$

and the 'relative' mass has been introduced in the last term. For monochromatic radiation, a sum must be done over the available states k that are allowed in the process. (Even for broad-band radiation, this sum must still be carried out, but is much more complicated in form as it involves the joint density of states [13].) In this integration, we find the angle-averaged dipole moment as

$$\int d\Omega \, |a \cdot P_{12}|^2 = 4\pi \langle P_{12}^2 \rangle. \tag{8.46}$$

Now, we may carry out the integration over the allowed states in the transition as

$$\begin{aligned} P(\omega) &= \frac{e^2 A_0^2 t}{m_0^2 \hbar^2} \int d^3 k \frac{|a \cdot P_{12}|^2}{4\pi^3} \frac{\sin^2[(\omega_{21} - \omega)t/2]}{(\omega_{21} - \omega)^2 t^2} \\ &= \frac{e^2 A_0^2}{m_0^2 \hbar^3} \frac{2k\langle P_{12}^2 \rangle m_R}{\pi^2} \int \frac{\sin^2 x}{x^2} dx \\ &= \frac{2e^2 A_0^2 k \langle P_{12}^2 \rangle m_R}{\pi m_0^2 \hbar^3}. \end{aligned} \tag{8.47}$$

At this point, we make connection with the oscillator strength that has been used earlier in this chapter and elsewhere. This is defined to be the ratio of the kinetic energy in the transition to the photon energy, as

$$f_{osc} = \frac{2\langle P_{12}^2 \rangle}{m_0 \hbar \omega}. \tag{8.48}$$

Now, we reintroduce the energy from the wave vector so that the absorption strength (8.47) can be rewritten as

$$P(\omega) = \frac{e^2 A_0^2 (2m_R)^{3/2} \omega f_{osc}}{2\pi m_0 \hbar^3} (\hbar \omega - E_G)^{1/2}. \tag{8.49}$$

The oscillator strength is normally of order unity for the allowed transitions and varies with the mass as $(1 + m_0/m_h)$.

Finally, it is necessary to normalize (8.49) to the number of incident photons arriving per second in order to calculate the absorption coefficient. The number of photons absorbed within the semiconductor sample, for a thickness of d, is just $P(\omega)d$, which is related to the absorption coefficient by $\alpha F_0 d$, where F_0 is the flux of incoming photons. The latter quantity is readily calculated from the power incident on the sample, given by the Poynting vector S of the plane wave, as

$$F_0 = \frac{S \cdot a_k}{\hbar \omega} = \frac{\omega A_0^2}{2\hbar} \sqrt{\frac{\varepsilon_\infty}{\mu_0}} = \frac{\omega A_0^2}{2\hbar} G_\omega, \tag{8.50}$$

where G_ω is the wave admittance. Finally, the absorption coefficient can be found to be

$$\alpha = \frac{e^2 (2m_R)^{3/2} f_{osc}}{\pi m_0 \hbar^2 G_\omega} (\hbar \omega - E_G)^{1/2}. \tag{8.51}$$

The square-root variation in (8.51) is just that expected from the density of final states to be found in the conduction band. The numerator mass involved is the reduced mass accounting for both the valence and conduction states. The denominator mass, however, arose from the Hamiltonian, and is prior to the effective mass approximation, as it still contains the full potential of the crystal, and thus should be the free electron mass. Excellent tabulations of the properties of materials, including the optical absorption coefficient (as a function of energy), can be found at [14, 15]. In figure 8.6, we illustrate the allowed absorption coefficient for InAs doped to 1×10^{16} cm^{-3} [16], semi-insulating GaAs [17], and wurtzite GaN doped to 1×10^{17} cm^{-3} [18]. For InAs and GaAs, the data is at room temperature and the GaN data is at 77 K.

8.2.2 Forbidden transitions

If the dipole transition above is forbidden by symmetry of the wave functions in the Bloch functions of the conduction and valence bands, the absorption can still occur via the last term in the bracket of (8.42). Consider, for example, the transition metal di-chalcogenides, which were discussed in section 2.3.2. Both the conduction and valence band extrema, located at the K points in the two-dimensional Brillouin zone, are formed primarily from the metal atom d levels. The d levels all have

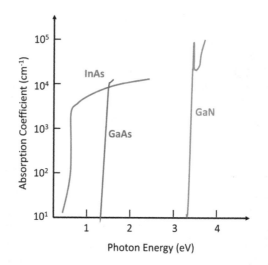

Figure 8.6. The allowed direct transition absorption coefficient for some typical semiconductors at room temperature.

angular momentum $l = 2$. Hence, one would expect the dipole term to vanish. However, in both cases there is a small admixture of p states, from which one can develop the dipole term (between the p and d levels). But, this is still going to be weak due to the weak contribution from the p states. Hence, one might expect that the forbidden absorption term might well be the significant driver in these materials.

When the dipole interaction vanishes, then the optical absorption matrix element is composed of the second term in (8.42), and the matrix element can be written as

$$H_{12} = -\frac{e\hbar A_0}{2m_0} \boldsymbol{a} \cdot \boldsymbol{k} \int u_c^\dagger u_v d^3\boldsymbol{r}. \tag{8.52}$$

We can now define an oscillator strength

$$f' = \left| \int u_c^\dagger u_v d^3\boldsymbol{r} \right|^2, \tag{8.53}$$

and averaging over the angle in the dot product, the power absorbed is found to be

$$\begin{aligned}
P(\omega) &= \frac{4\pi e^2 A_0^2 f'}{3m_0^2} \int \frac{k^4 t}{4\pi^3} \frac{\sin^2[(\omega_{21} - \omega)t/2]}{(\omega_{21} - \omega)^2 t^2} dk \\
&= \frac{e^2 A_0^2 k^3 m_R f'}{6\pi^2 m_0^2 \hbar},
\end{aligned} \tag{8.54}$$

under the same approximations as those used in obtaining (8.47). Reinserting the energy dependence of the momentum terms, one arrives at

$$P(\omega) = \frac{e^2 A_0^2 (2m_R)^{5/2} f'}{12\pi^2 m_0^2 \hbar^4} (\hbar\omega - E_G)^{3/2} \tag{8.55}$$

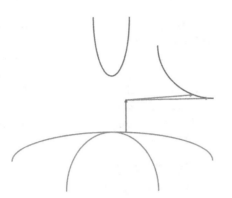

Figure 8.7. Indirect optical transition through an evanescent state and phonon assisted.

and

$$\alpha = \frac{e^2(2m_R)^{5/2}f'}{6\pi^2 m_0^2 \hbar^2 G_\omega} \frac{(\hbar\omega - E_G)^{3/2}}{\hbar\omega} \tag{8.56}$$

Is the absorption coefficient for the forbidden transition in the direct optical gap. The net dependence on the frequency still varies as the square root of the photon energy, but this variation is somewhat more complicated. In general, the absorption for the forbidden transition is weaker than that of the allowed transition.

8.3 Indirect transitions

In the above discussion, it was clear that the optical transition must be vertical, as the momentum of the photon is much smaller than the electron or hole momentum. In indirect band gap semiconductors, however, this would lead to very large energy photons being required, unless a phonon becomes involved in the momentum conservation. Hence, we can write the momentum conservation (8.41) for the indirect band gap as

$$\boldsymbol{k} + \boldsymbol{k}_0 - \boldsymbol{k}' \pm \boldsymbol{q} = 0. \tag{8.57}$$

Here, **q** is the phonon momentum vector. The energy conservation condition then becomes

$$E_c(k') - E_v(k) = \hbar\omega \pm \hbar\omega_q. \tag{8.58}$$

We illustrate this transition in figure 8.7 for an indirect band gap. We note that we can either absorb or emit the phonon; both processes are allowed, although it is the absorption of a phonon that requires the lower energy photon of the two processes.

8.3.1 Complex band structure

Now, an astute observer will immediately comment that there are no band structure states at the point where the photon connects with the phonon. That is, when we discussed the band structure in chapter 2, there were no allowed states at this energy and momentum site. But, we have to remember that the band structure was computed for those wave functions which had propagating behavior; e.g., allowed waves could exist with real momentum. Now, we understand that impurities and defects can introduce states in the gap regions, but there are other states that exist in this region as well. These states actually come from the band structure, but are not propagating states. Let us illustrate this with a simple two band model. For this, we adopt the model of (2.104) for mirror image bands within the $\mathbf{k} \cdot \mathbf{p}$ model, which we may write as (ignoring the free carrier mass term and the split-off valence band)

$$E(E - E_G) - k^2P^2 = 0. \tag{8.59}$$

In our treatment of chapter 2, we always considered $k^2 > 0$, that is the case of real k. This leads to the two mirror image bands

$$E = \frac{E_G}{2}\left[1 \pm \sqrt{1 + \frac{4k^2P^2}{E_G^2}}\right], \tag{8.60}$$

where the zero of energy lies at the top of the valence band. These bands are shown in figure 8.8 as the blue curves.

It should be understood that there is no physical reason to use $k^2 > 0$. Of course, this gives propagating waves. But, we could equally choose to use $k^2 < 0$. If we make this selection, then we can rewrite (8.59) as

$$k = \pm i(1/P)\sqrt{E(E_G - E)}, \tag{8.61}$$

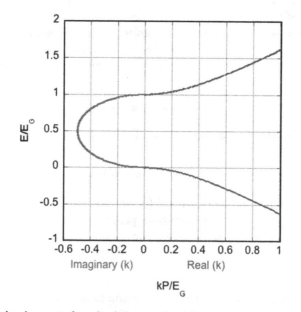

Figure 8.8. Complex band structure for a simple $\mathbf{k} \cdot \mathbf{p}$ two band model. The curves are for real k, while the red curve (on negative axis) is for imaginary k.

which gives us the so-called *complex* band structure. Now, k is imaginary, so that the state decays away exponentially from its site in real space. This imaginary k is plotted in figure 8.8 as a negative k (red curve). The complex band structure is much more complicated than the real band structure, because every pair of bands, such as the simple bands of figure 8.8, is connected by the imaginary k curves, and this leads to enormous complexity of crossing bands. Yet, they are important as they give us localized states throughout the gap.

These are so-called evanescent states, which are highly localized and are certainly not propagating waves. But, a photon can pump an electron into one of these states, where the electron can subsequently make a transition to the conduction band through phonon scattering. And this is the process of the indirect transition—it is a two-step process.

8.3.2 Absorption coefficient

The process shown in figure 8.7 involves the absorption of a photon and the emission or absorption of a phonon. The final state of the electron is near the minimum of the conduction band that is the lowest of these conduction bands. This minimum is along Δ, near X, for silicon and at L for germanium. As discussed above, the absorption of the photon ends at a state that lies in the forbidden bap, but is an evanescent state in which the electron can reside for a short period of time. Some call this a virtual state, but it arises from the complex band structure. The electron resides in this state sufficiently long to eventually either absorb a phonon or emit a phonon in order to complete the indirect transition.

The perturbation treatment must be expanded to account for the two quite distinctly different processes that compose the indirect transition. Now, the perturbing potential not only includes the extra term in (8.37), but also a term arising from the electron–phonon interaction (see chapter 5). We can proceed as we did earlier if we write the total vector potential as

$$\mathbf{A} = \frac{1}{2}\mathbf{A}_0 e^{i(\omega t - k_0 \cdot r)} + \frac{1}{2}\mathbf{A}_s e^{i(\omega t - q \cdot r)} + c.c., \tag{8.62}$$

where the first term is the optical field and the second term is the lattice field. The theory of the second order time-dependent perturbation theory is found in some advanced quantum theory books, but it still boils down to calculating matrix elements. We will not go through the details, but merely outline the process. The second order process here involves both an optical matrix element and an electron–phonon matrix element. Then, the transition probability involves the magnitude squared of each of these two matrix elements. This must now be integrated over the initial and final states of the hole and the electron and have the phonon properties inserted. The latter, of course, involve the Bose–Einstein distribution. In general, all the unknown sins are incorporated in the magic oscillator strength. Here, only the final results are presented, which are expressed in terms of the energy dependence of the absorption coefficient [19], which gives

$$\alpha \sim \frac{x}{e^x - 1}(\hbar\omega - E_G + \hbar\omega_q)^2, \quad x = \frac{\hbar\omega_q}{k_B T} \tag{8.63}$$

for the absorption of a phonon, and

$$\alpha \sim x(\hbar\omega - E_G - \hbar\omega_q)^2 \tag{8.64}$$

for the emission of a phonon. By comparison with (8.56), the indirect transition has a very significant temperature dependence which arises from the phonon contribution. In addition, there is a much stronger dependence upon photon energy, although the pre-factor is considerably smaller. In general, the second-order process is usually at least an order of magnitude weaker than the first-order process. The absorption coefficient for Si is shown in figure 8.9. It may be seen that near the threshold, the absorption is more than an order of magnitude smaller than for the direct gap III-Vs. There is also an increase in the absorption near 4 eV that arises from direct transitions into a secondary conduction band at Γ (the second conduction band at Γ in figures 2.14 and 2.16, which is the allowed transition).

8.4 Recombination

If the carriers excited by the inter-band optical absorption are left to themselves, they eventually will recombine, returning the system to an equilibrium state. In this process, the excited electron drops across the energy gap to the empty valence band state (the hole that was created by the absorption process). This cross gap process can be a direct process in which a photon or group of phonons are emitted, or an indirect process that is the inverse of indirect absorption, or through a series of trap/defect states lying in the energy gap. Just as in impact ionization (section 7.1.4), the ionization coefficient defined above has a corresponding generation rate G_0 (gives the number per second that are created as opposed to the number per centimeter that

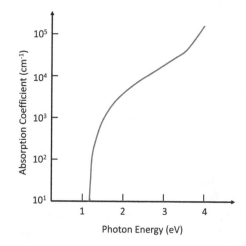

Figure 8.9. Absorption coefficient for Si at 300 k. The rise at ~4 eV is due to direct transitions to the second conduction band at Γ.

are created). This allows us to write a continuity equation for the electron density in the form

$$\frac{dn}{dt} = \frac{d(\Delta n)}{dt} = G_0 - \frac{n}{\tau_n}, \tag{8.65}$$

where τ_n is a quantity that describes the rate of recombination and may be a function of the electron (or hole) density. In equilibrium, with no excess generation, the time derivative term vanishes, and $G_0 = n_0/\tau_n$. Then, we can write the excess density as

$$\frac{d(\Delta n)}{dt} = -\frac{\Delta n}{\tau_n}. \tag{8.66}$$

An equivalent equation can be written for the holes that are produced by the absorption process. In fact, the recombination is traditionally discussed in terms of the *minority* carrier concentration, as the excess carriers may be the dominant part of the minority carrier population but a negligible part of the majority carrier population.

8.4.1 Radiative recombination

In creating the recombination rate, we have to begin with the absorption coefficient discussed above and the relation just above (8.66). If α is the absorption coefficient at a particular frequency, then the mean free path of the photon in the semiconductor is just $1/\alpha$, and the mean photon lifetime is just $(\alpha c/\sqrt{\varepsilon_\infty})^{-1}$ in the semiconductor. This lifetime can be used to determine the rate of carrier generation in the semiconductor, and this fact can then be used to balance the generation and recombination of excess carriers through the continuity equation above. To begin, the photon lifetime above is written as

$$\frac{1}{\tau_{ph}} = \frac{\alpha c}{\sqrt{\varepsilon_\infty}}. \tag{8.67}$$

In order to calculate the generation rate, this must be multiplied by the number of photons actually present at a given frequency and small increment of frequency, which arises from the blackbody radiation formula, as

$$N_{ph}(\nu)d\nu = \frac{8\pi\nu^2\varepsilon_\infty^{3/2}}{c^3}\frac{d\nu}{e^{h\nu/k_BT} - 1}. \tag{8.68}$$

Here, ν is the frequency. Then, we can combine (8.67) and (8.68) to give us the net generation rate as

$$G(\nu) = \frac{N_{ph}(\nu)}{\tau_{ph}} = \frac{8\pi\alpha\nu^2\varepsilon_\infty}{c^2}\frac{1}{e^{h\nu/k_BT} - 1}. \tag{8.69}$$

In thermal equilibrium, the recombination rate is equal to the generation rate. This fact now gives us the equilibrium recombination rate by integrating over this generation rate. This gives

$$R_0 = \frac{8\pi(k_B T)^3}{c^2 \hbar^3} \int_0^\infty \frac{x^2 \varepsilon_\infty \alpha}{e^x - 1} dx, \quad x = \frac{h\nu}{k_B T}. \tag{8.70}$$

The dielectric function has been left inside the integral, as it may have a frequency variation of its own. The major contribution to (8.70), however, comes from the frequency variation of the absorption coefficient itself. If we write $\alpha = \alpha_0 (h\nu - E_G)^r$, we can evaluate the integral as

$$R_0 = \frac{8\pi\alpha_0\varepsilon_\infty(k_B T)^{2+r}}{c^2 \hbar^3} \left(\frac{E_G}{k_B T}\right)^2 e^{-E_G/k_B T}\left[\Gamma(r + 1) + \frac{2k_B T}{E_G}\Gamma(r + 2)\right]. \tag{8.71}$$

Once the appropriate parameters are either measured or calculated, the recombination rate is easily determined. It is clear, however, that low values of absorption coefficient lead to low recombination rates.

When we shine light upon the semiconductor, we now create a non-equilibrium condition. For direct recombination, a single electron and hole recombine which eliminates an electron-hole pair. In this case, we can write the non-equilibrium recombination rate as

$$R = R_0 \frac{pn}{n_i^2}, \tag{8.72}$$

where n_i^2 is the intrinsic carrier concentration. In the following discussion, we will assume a p-type semiconductor ($p_0 > n_0$). The electrons are then the minority carrier concentration, and we can write the continuity equation (after the generation process for excess carriers has been turned off) as

$$\frac{d(\Delta n)}{dt} = -\frac{\Delta n}{\tau_n} = R_0 - R_0 \frac{pn}{n_i^2} \tag{8.73}$$

and

$$\frac{\Delta n}{\tau_n} = R_0 \frac{pn - n_i^2}{n_i^2} = R_0 \frac{(\Delta n)^2 - (n_0 + p_0)\Delta n}{n_i^2}. \tag{8.74}$$

For low densities of excess carriers and a strongly extrinsic semiconductor ($\Delta n \ll p_0$), the lifetime is relatively constant and is given by

$$\frac{1}{\tau_n} = \frac{R_0}{n_0}, \quad \Delta n, n_0 \ll p_0. \tag{8.75}$$

However, for large excess densities, which exceed the background doping concentration of holes, the recombination rate depends upon the excess density, and the decay is non-exponential in nature. In this case, the decay may be written as

$$\frac{\Delta n(t)}{\Delta n(0)} = \frac{1}{1 + \left[R_0 \Delta n(0)/n_i^2\right]t}. \tag{8.76}$$

This result holds for large density, but as the density decreases with this slow rate, the decay will eventually transition to an exponential behavior.

8.4.2 Trap recombination

As mentioned above, it is also possible for electrons and holes to recombine through a series of traps lying at intermediate levels throughout the band gap. They may be impurities or defects of various kinds. Still, recombination through these levels requires a mechanism for the electron and hole to give up their energies, basically the band gap energy plus any kinetic energy they may have. As before, this can occur through the radiation of the energy or by the emission of phonons. There exists a third approach, known as the Auger process, in the which the energy given up by the recombining electron and hole is absorbed by another electron or hole. This process is basically the inverse of impact ionization we discussed in chapter 7. In general, the trap capture time assumes a combination of these various processes. This trap recombination time is also the appropriate time for a carrier to recombine with its relevant impurity, and the band-to-band process is easily generalized from this latter approach.

For radiative processes, we can write the continuity equation exactly as was done above, which gives us

$$\frac{d(\Delta n)}{dt} = -\frac{\Delta n}{\tau_n} = g_r - r_r, \tag{8.77}$$

where g_r and $r_r = G_0$ in equilibrium, and the two rates are the generation and recombination rates, respectively. Here, the generation rate will depend upon the number of neutral donors (we assume $N_d > N_a$ here), while the recombination rate will depend upon the number of electrons as well as the number of ionized donors. Thus, these two quantities may be written as

$$r_r = R_0 \frac{nN_{di}}{n_0 N_{d0}} = R_0 \frac{n(n + N_a)}{n_0(n_0 + N_a)}$$
$$g_r = R_0 \frac{N_{dn}}{N_{dn0}} = R_0 \frac{N_d - N_a - n}{N_d - N_a - n_0}. \tag{8.78}$$

Then, the recombination may be written as

$$\frac{1}{\tau_n} = \frac{R_0}{n_0}\left[\frac{N_d - N_a}{N_d - N_a - n_0} + \frac{n_0 + \Delta n}{n_0 + N_a}\right]. \tag{8.79}$$

It is clear that only for the most unusual situations (such as excess generation across the band gap in addition to excitation from the trap level) will the lifetime deviate from its nearly constant form and affect the exponential decay. Impurity level absorption is often used for far-infrared detectors, but radiative recombination rates are typically very low and not the primary cause of the recombination.

For phonon-aided recombination, the energy lost by the carriers goes into the lattice vibrations, which is a more likely process for the small energy by the which the impurity is usually separated from the conduction band. The lifetime has the same general density dependence as that for the radiative rate, except that the equilibrium constant R is not that for radiative absorption. Rather the constant R must be evaluated for the phonon-assisted generation process.

For an n-type semiconductor, the Auger process can also occur with capture of the free carriers by the donor levels. Here, the excess energy is taken up by a second electron, which is excited to a higher energy state in the conduction band. The recombination rate will be proportional to the square of the free electron concentration, since there are two electrons involved in the overall process. The generation rate, on the other hand, is merely proportional to the number of neutral donors, as above. The inverse of Auger recombination is trap assisted impact ionization discussed in chapter 7. Thus, the forward and reverse processes are balanced to calculate proper rate constants. Thus,

$$r_a = R_a \frac{n^2 N_{di}}{n_0^2 N_{di0}} = R_a \frac{n^2(n + N_a)}{n_0^2(n_0 + N_a)} .$$
$$g_a = R_a \frac{n(N_d - N_a - n)}{n_0(N_d - N_a - n_0)}$$

(8.80)

These may now be combined to give the Auger recombination time as

$$\frac{1}{\tau_{na}} = \frac{R_a}{n_0} \frac{n_0 + \Delta n}{n_0} \left(\frac{N_d - N_a}{N_d - N_a - n_0} + \frac{n_0 + \Delta n}{n_0 + N_a} \right).$$

(8.81)

If n_0 is relatively small, such as occurs at low temperatures, Auger recombination can readily occur in a nonexponential fashion. Auger recombination is thought to occur at low temperatures in highly doped material, such as in the base and emitter regions of bipolar transistors, as well as in highly excited material. In the latter case, it can even dominate the cross gap recombination as well. This is especially the case in laser excited semiconductors where the electron-hole density is quite large.

Finally, there is the Shockley–Read–Hall theory of trap recombination [20, 21]. In this approach, there are considered to be defect levels that are electron traps and defect levels that are hole traps. Thus, electron recombination goes to the electron traps and hole recombination goes to hole traps. But, these traps can communicate so that the electron and hole eventually see each other and recombine. The recombination rate tends to show a temperature dependence that arises from the thermal excitation of carriers from the traps and this now depends upon the trap energies in the band gap. While this tends to be somewhat arbitrary in nature, this type of recombination seems to be dominant in indirect gap materials.

References

[1] Kirk A P and Fischetti M V 2012 *Phys. Rev.* B **86** 165206
[2] Esser F *et al* 2015 *Appl. Phys. Lett.* **107** 062103
[3] Weyers M, Sato M and Ando H 1992 *Jpn. J. Appl. Phys.* **31** L853
[4] Bi W G and Tu C W 1997 *Appl. Phys. Lett.* **70** 1608

[5] Naka N, Fukai K, Handa Y and Akimoto I 2014 *J. Lumines* **152** 93

[6] Castro-Neto A H *et al* 2009 *Rev. Mod. Phys.* **81** 109

[7] Beylkh V V, Yakovlev D R and Bayer M 2019 *Phys. Rev.* B **99** 161205

[8] Geltman S 1977 *J. Phys.* B **10** 831

[9] Haan S L and Geltman S J 1982 *J. Phys.* B **15** 1229

[10] Lindberg M and Koch S W 1988 *Phys. Rev.* B **38** 3342

[11] Haug H 1988 *Optical Nonlinearities and Instabilities in Semiconductors* (San Diego, CA: Academic)

[12] Kuhn T and Rossi F 1992 *Phys. Rev.* B **46** 7496

[13] Kirk A P 2015 *Photovoltaic Cells: Photons to Electricity* (London: Academic)

[14] Madelung O 2004 *Semiconductors: Data Handbook* 3rd edn (Berlin: Springer)

[15] http://ioffe.ru/SVA/NSM/Semicond/

[16] Dixon J R and Ellis J M 1961 *Phys. Rev.* **123** 1560

[17] Sturge M D 1962 *Phys. Rev.* **127** 768

[18] Muth J F *et al* 1997 *Appl. Phys. Lett.* **71** 2572

[19] Smith R A 1969 *Wave Mechanics of Crystalline Solids* (London: Chapman & Hall)

[20] Shockley W and Read W T 1952 *Phys. Rev.* **87** 835

[21] Hall R N 1952 *Phys. Rev.* **87** 387

IOP Publishing

Semiconductors (Second Edition)
Bonds and bands
David K Ferry

Chapter 9

The electron–electron interaction

In recent years, device sizes have shrunk to almost unimaginable dimensions, and carrier densities have gotten larger within these devices, if for no other reason than that device scaling requires larger densities at smaller sizes. Consequently, the interactions among the electrons have become far more important. We have encountered such interactions with brief discussions of screening and plasma effects in several earlier chapters. With the growth in carrier density within devices, it is no longer adequate simply to adopt screening lengths, and detailed handling of the direct interaction between charge carriers, and between these carriers and the impurities, has become necessary [1]. The details of the electron–electron and the electron–impurity interaction have become relevant to an overall understanding of the device behavior.

The main difficulty in dealing with the electron–electron interaction lies in the nonlinear behavior and long range of the Coulomb interaction. In the past, several approaches to the study of these interactions have occurred with various approximations. Almost all of these were based upon the assumption that the interaction potential was screened by some mechanism, usually the Debye approximation, which assumes that the distribution function is a Maxwellian. Yet, we often find today that the so-called screening length is actually smaller than the inter-electron distance. The effects are complicated by the fact that the rather high free carrier plasma frequency that accompanies the high densities in modern devices couples strongly to the polar optical phonons to produce coupled modes of the electron–plasmon–phonon interactions. To treat all of these effects properly, it is necessary to divorce ourselves from the simple screening approach and treat the full frequency- and momentum-dependent dielectric function.

To be sure, there has been considerable study of such momentum and energy dependent dielectric effects in semiconductors. But, this has usually been associated with quantum and mesoscopic effects studied at very low temperatures where the full variation of the dielectric function is not heavily damped by thermal effects. Modern

doi:10.1088/978-0-7503-2480-9ch9

devices are sufficiently small that these effects can now be observed in these devices in everyday operation at normal temperatures. Hence, it is now necessary to study how these effects should be considered in modern transport theory. In this chapter, we will first consider an approach in linear response based upon a self-consistent determination of the dielectric function. This treatment explains a great deal about the many-body properties that are normally discussed in semiconductors, such as the optical and static dielectric 'constants.' From this approach, we will be able to study many of the energy and momentum dependent properties of the dielectric function. We will then turn to a real-space consideration of the interactions among the electrons and between the electrons and the impurities. This molecular dynamics approach allows treatment of the full nonlinear interactions and is compatible with modern Monte Carlo simulations. Of course, difficulties arise in device simulation because solutions of the Poisson equation already incorporate the inter-carrier interactions up to the Hartree approximation. This would ordinarily lead to double counting of some of the interactions, but a force partitioning approach avoids this problem.

When we say 'self-consistent' simulations, we are normally talking about methods to incorporate the Poisson equation into the determination of the desired effect. Let us first discuss how this can be done classically. Consider a typical n-type semiconductor in which a local electric field or potential φ is present due e.g. to a charge fluctuation. In the classical non-degenerate limit, this potential leads to an inhomogeneous charge density given as

$$n = n_0 e^{e\varphi/k_B T}. \tag{9.1}$$

Here, n_0 is associated with the undisturbed carrier density in the semiconductor, and φ is the local potential. Let us associate the disturbance with a charge located at $\mathbf{r} = 0$ for convenience. This initial charge is generally assumed to be characterized by a Coulomb potential that falls off slowly with distance. In fact, however, the local potential must be found self-consistently by solving a Poisson's equation for it, as

$$\nabla^2 \varphi = \frac{e(n - n_0)}{\varepsilon_s} = \frac{en_0}{\varepsilon_s}(e^{e\varphi/k_B T} - 1) \approx \frac{n_0 e^2}{\varepsilon_s k_B T}\varphi, \tag{9.2}$$

in the linearized approximation. Now, one can solve this in standard manner to show that the local potential around the perturbing charge is given as

$$\varphi(r) = -\frac{e}{4\pi\varepsilon_s r}e^{-q_D r}, \quad q_D^2 = \frac{n_0 e^2}{\varepsilon_s k_B T}. \tag{9.3}$$

Hence, we find that we have classical screening in the non-degenerate limit described by Debye screening behavior.

In the next section, we will reconsider this self-consistent treatment for the full frequency- and momentum-dependent dielectric function, but still within linear response. The result we find is usually referred to as the Lindhard dielectric function. We will then explore the various ranges that can be defined for the parameters of the dielectric function to try to gain an understanding of various limits.

9.1 The dielectric function

The dielectric function tells us how the various charged species move in response to external (or internal) potentials and provides an idea about the screening of those potentials. In compound semiconductors, such as the III-V materials, these charged species include the atoms themselves as they have an effective charge arising from the polarization within the material. Since there is a small ionic contribution to the bonding from this effective charge, this contributes to the total polarization within the semiconductor, and it is this total polarization that differentiates the semiconductor from free space. As a result, we can often write the total dielectric function as

$$\varepsilon(q, \omega) = \varepsilon_0 + \delta\varepsilon_L + \delta\varepsilon_e, \tag{9.4}$$

where ε_0 is the permittivity of free space. The second and third terms of (9.4) are the lattice and electronic contributions to the total permittivity, respectively. In the last term, the electrons of interest are both the bonding electrons and the free carriers, and we will see how these two different sets contribute importantly to the dielectric function. Importantly, we will see that it is the bonding electrons which contribute to yield the so-called high frequency dielectric 'constant' ε_∞ of the semiconductor.

In the absence of free carriers, the permittivity must transition between its high frequency value and the static dielectric 'constant.' Of course, in the non-polar material, there is no distinction between these two, and we simply refer to the dielectric constant as ε_s. But, we dealt with the transition between the high- and low-frequency values in section 3.4, where we discussed the lattice polarization for the polar materials. From that section, we found that the lattice contribution to (9.4) is given by

$$\delta\varepsilon_L(\omega) = [\varepsilon(0) - \varepsilon_\infty]\frac{\omega_{TO}^2}{\omega_{TO}^2 - \omega^2}$$
$$\frac{\omega_{LO}^2}{\omega_{TO}^2} = \frac{\varepsilon(0)}{\varepsilon_\infty}, \tag{9.5}$$

where the second equation is the Lyddane–Sachs–Teller relationship.

Normally, the electronic contribution is computed in the absence of the lattice contribution, which is appropriate for linear response. Later, where it becomes important, we will reintroduce the lattice contribution to study the coupled modes that can occur in polar materials. Let us now turn to the electronic contribution.

9.1.1 The Lindhard potential

The linear response approach to determining the dielectric function means that we are only calculating to the lowest order in the various screening effects that can occur in our semiconductor. Nevertheless, this approach gives a very insightful look at a great many of the processes that affect transport within the semiconductor. Here, we will assume a perturbation in the local potential that we define as δU, and we consider that this potential is time varying in a sinusoidal manner. This potential then leads to a perturbation in the local carrier density, which in turns leads to a self-consistent potential fluctuation given by Φ. The connection between δU and Φ will give us the material response and screening of the external potential. This defines the dielectric function [2]. We do this in Fourier space in order to arrive at the momentum and energy dependent dielectric function. We express the perturbing external potential as

$$\delta U(q, \omega) = U_0 e^{i(q \cdot r - \omega t) + \alpha t}, \tag{9.6}$$

where α is a damping constant to account for the dissipative part of the carrier–carrier interaction, and allows for us to adiabatically turn on the perturbation at $t \to -\infty$.

The approach that we follow is based upon simple time-dependent perturbation theory, in which the initial wave function state at momentum k is coupled to one at $k \pm q$ by the external potential. The sinusoidal excitation by the external potential thus picks out only a few selected couplings to the initial wave function, and we may write the perturbed wave function as

$$\psi_k = |k\rangle + b_{k+q}|k + q\rangle + b_{k-q}|k - q\rangle. \tag{9.7}$$

The second and third terms here represent the deviation from the initial state that is introduced by the perturbing potential. First-order time-dependent perturbation theory then allows us to write the coefficients as

$$b_{k+q} = \frac{\langle k + q|e\delta U|k\rangle}{E(k + q) - E(k) + \hbar\omega - i\hbar\alpha} = \frac{eU_0 e^{-i(\omega t) + \alpha t}}{E(k + q) - E(k) + \hbar\omega - i\hbar\alpha}. \tag{9.8}$$

It can be noted that only a single state thus couples to the initial wave function. We can then write the change in electron density that is produced by this perturbation as

$$\delta\rho = -e\sum_k f(k)(|\psi_k|^2 - 1), \tag{9.9}$$

where the squared magnitude of the wave function yields the probability that the state actually occurs in the summation. The distribution function gives us the occupancy of that state and the distribution of states over the momentum and/or energy. That is, weighting the wave function by the distribution function actually gives us the density at the momentum. After some simple manipulation, linear response means keeping only the lowest order in the summation (9.9), and we may write the density perturbation as

$$\delta\rho = -e\sum_k f(k)\left(b_{k+q}^\dagger e^{-iq \cdot r} + b_{k+q} e^{iq \cdot r} + h.\,c.\right). \tag{9.10}$$

This linearization arises as we assume that the actual perturbation is small, and we then only need to consider the first-order terms. We can now use (9.8) to write the density perturbation as

$$\delta\rho = -e\sum_k f(k)\left[\frac{1}{E(k+q) - E(k) + \hbar\omega - i\hbar\alpha}\right.$$

$$\left. + \frac{1}{E(k-q) - E(k) + \hbar\omega + i\hbar\alpha}\right]eU_0 + c.\,c. \tag{9.11}$$

With a change of momentum $k - q \rightarrow k$, we can rewrite this as

$$\delta\rho = -e^2 U_0 \sum_k\left[\frac{f(k) - f(k+q)}{E(k+q) - E(k) + \hbar\omega - i\hbar\alpha}\right] + c.\,c. \tag{9.12}$$

This fluctuation in charge density produces, in turn, a fluctuation in the potential itself. Poisson's equation for the potential is just

$$\nabla^2\Phi = -\frac{\delta\rho}{\varepsilon_0}, \quad or \quad \Phi(\boldsymbol{q}, \omega) = \frac{\delta\rho}{q^2\varepsilon_0}e^{i\boldsymbol{q}\cdot\boldsymbol{r}} + c.\,c. \tag{9.13}$$

Hence, one obtains the resulting potential of the perturbation to be just

$$\Phi = -\frac{e^2 U_0}{q^2\varepsilon_0}\sum_k\left[\frac{f(k) - f(k+q)}{E(k+q) - E(k) + \hbar\omega - i\hbar\alpha}\right]. \tag{9.14}$$

Now, the total potential δU is the one that produces the perturbation in the charge density. But, this total potential is composed of the applied potential V plus the response term Φ. Hence, we can solve for U_0 as

$$U_0 = V + \Phi, \quad or \quad U_0 = \frac{\varepsilon_0 V}{\varepsilon(\boldsymbol{q}, \omega)}, \tag{9.15}$$

so that the *dielectric function* is just

$$\varepsilon(\boldsymbol{q}, \omega) = \varepsilon_0 + \frac{e^2}{q^2}\sum_k\left[\frac{f(k) - f(k+q)}{E(k+q) - E(k) + \hbar\omega - i\hbar\alpha}\right]. \tag{9.16}$$

We can undo the momentum shift to also give

$$\varepsilon(\boldsymbol{q}, \omega) = \varepsilon_0 + \frac{e^2}{q^2}\sum_k f(k)\left[\frac{1}{E(k+q) - E(k) + \hbar\omega - i\hbar\alpha}\right.$$

$$\left. + \frac{1}{E(k-q) - E(k) + \hbar\omega + i\hbar\alpha}\right]. \tag{9.17}$$

Both of these two last forms are referred to as the Lindhard dielectric function. It includes the free space dielectric constant as well as the linear response from the material itself. We will use both of these forms in the following developments as each has its own usefulness.

9.1.2 The optical dielectric constant

The sum over the carrier distribution in (9.16) and (9.17) is over *all* the electrons, free carriers as well as valence electrons that form the covalent bonds. At this point, we want to separate out the sum over the valence electrons. We will deal with the free carriers in several of the following sections. The denominators in the above equations indicate that we are dealing with optical excitations between at least two energy levels. With the valence electrons, one of those levels lies in the large energy spread of the valence band, while the other energy is very high in the conduction band and approaching free space levels. These excitations are at much higher energy than the band gap itself. In some sense, the excitation is between the *average* energy of the valence band and the *average* energy of the conduction band, as discussed in section 2.7. Hence, these excitations typically involve a momentum shift *k* that is close to a reciprocal lattice vector. Consequently, we will assume below that the excitation gap between the two energies is much larger than the optical frequency with which we are probing the dielectric function. To begin, we will use (9.17) and rationalize the two terms into a single expression, given by

$$
\varepsilon(\boldsymbol{q}, \omega) = \varepsilon_0 + \frac{e^2}{q^2} \sum_k f(k)
$$
$$
\times \left[\frac{2E(k) - E(k - q) - E(k + q)}{[E(k + q) - E(k) + \hbar\omega - i\hbar\alpha][E(k) - E(k - q) - \hbar\omega - i\hbar\alpha]} \right].
\tag{9.18}
$$

As we mentioned, the summation is over the valence electrons and thus runs over the entire Brillouin zone. The transitions usually involve a reciprocal lattice vector, which we will shift the momentum with. Thus, the terms in the denominator may be rewritten using

$$
E(k \pm q \pm G) - E(k) \sim E_{gap} \gg \hbar\omega,
\tag{9.19}
$$

where the gap is not the band gap, but the effective optical gap discussed above. We can also expand the numerator as

$$
2E(k) - E(k - q) - E(k + q) \approx -q^2 \frac{\partial^2 E}{\partial k^2} + \ldots \approx -\frac{\hbar^2 q^2}{m_{opt}},
\tag{9.20}
$$

Where we have used a Taylor series expansion to find the leading term, and then introduced the effective mass appropriate to the joint optical transition. With these approximations, we can rewrite (9.18) can be rewritten as

$$
\varepsilon(\boldsymbol{q}, \omega) = \varepsilon_0 + \frac{e^2}{q^2} \frac{\hbar^2 q^2}{m_{opt}} \sum_k f(k) \frac{1}{E_{gap}^2} = \frac{Ne^2\hbar^2}{m_{opt}E_{gap}^2}
$$
$$
= \varepsilon_0 \left[1 + \left(\frac{\hbar\omega_P}{E_{gap}} \right)^2 \right] \equiv \varepsilon_\infty.
\tag{9.21}
$$

Here, we have introduced the valence plasma frequency (note the capitol P here to distinguish from the free carrier plasma frequency)

$$\omega_P^2 = \frac{Ne^2}{m_{opt}\varepsilon_0},$$ (9.22)

and the high frequency dielectric permittivity. The first expression in the second line is the Penn dielectric function, that we encountered in chapter 2. Typically, the valence plasma frequency corresponds to an energy of the order of 10–20 eV. This is typically in the far ultra-violet range of optics. The quantity N is the total density of valence electrons, but in some semiconductors it will also include interactions with the d levels that overlap with the valence band energy range. This is especially true with germanium and other materials lying low in the periodic table.

With this result for the contribution of the valence electrons, we can reintroduce the contribution from the lattice polarization (9.5) to obtain the new dielectric function

$$\varepsilon(\boldsymbol{q}, \omega) = \varepsilon_\infty + [\varepsilon(0) - \varepsilon_\infty]\frac{\omega_{TO}^2}{\omega_{TO}^2 - \omega^2}$$
$$+ \frac{e^2}{q^2}\sum_k \left[\frac{f(k) - f(k+q)}{E(k+q) - E(k) + \hbar\omega - i\hbar\alpha}\right].$$ (9.23)

In this last expression, the sum now runs only over the free carriers and their distribution function. We give the valence plasma frequency and some relative dielectric permittivities for several semiconductors in table 9.1.

9.1.3 The plasmon-pole approximation

A similar approximation can be used in the free carrier term when we assume that the frequency is large compared to the energy exchange in the denominator of the

Table 9.1. Some semiconductor optical properties.

	$\hbar\omega_P(eV)$	$\varepsilon_{r,\infty}$	$\varepsilon(0)$
C	31.2		
Si	16.6	11.7	11.7
Ge	15.6	15.9	15.9
GaAs	15.6	11.1	13.1
GaP	16.5	8.5	10.2
GaSb	13.9	14.4	15.7
InAs	14.2	12.3	14.5
InP	14.8	9.6	12.4
InSb	12.7	15.7	17.9
AlAs	15.5	8.2	10.06
HgTe	12.8	6.9	14.9
CdTe	12.7	7.2	10.2

last term of (9.23). In the above section, we found a large contribution for the valence plasma effects on the dielectric function. Here, we will find a smaller, but often significant contribution due to the free carrier plasmons. While the transport of free carriers is affected by the electron–plasmon scattering, we will leave this detail until a later section where it can be treated effectively. Here, we deal only with the simple plasmon-pole approximation. While this is sufficient to deal with electron–plasmon scattering when treating the plasmon as a single coherent excitation, the simple approach will not be good enough in the case of two-dimensional systems where the plasmon frequency has a significant dependence on the momentum of the carriers. There, we will have to deal with an approach beyond this simple plasmon-pole approximation.

For this plasmon-pole approximation, we start with rewriting (9.17) exactly as (9.18). This time, however, we assume that the optical frequency is larger than excitation between the two energies, and we expand the numerator as in (9.20) using the effective mass for the free carriers. This gives us

$$\varepsilon(\mathbf{q}, \omega) = \varepsilon(0) - \frac{e^2}{q^2} \frac{\hbar^2 q^2}{m_c} \sum_k f(k) \frac{1}{\hbar^2 \omega^2} = \varepsilon_\infty \left[1 - \frac{ne^2}{m_c \varepsilon_\infty \omega^2} \right]. \qquad (9.24)$$

The free-carrier plasma frequency is defined by (note the lower case p here for the free carriers)

$$\omega_p^2 = \frac{ne^2}{m_c \varepsilon(0)}, \qquad (9.25)$$

which includes the lattice polarization in the low-frequency dielectric 'constant,' as we describe next.

The result (9.24) is the long-wavelength limit (vanishing momentum) of the free carrier contribution, and it can be combined with the lattice polarization to give

$$\varepsilon(\mathbf{q}, \omega) = \varepsilon_\infty \left[1 - \frac{\omega_p^2}{\omega^2} \right] + [\varepsilon(0) - \varepsilon_\infty] \frac{\omega_{TO}^2}{\omega_{TO}^2 - \omega^2}. \qquad (9.26)$$

We should remind ourselves that the dielectric function appears in the various scattering rates of chapter 5 in the form of $1/\varepsilon$, and this leads to singularities in the scattering that come from the zeroes of the dielectric function. In many cases, it is required to also examine the imaginary parts of the dielectric function in order to arrive at the proper scattering rates, even though (9.26) is entirely real. What this means is that we have to extend (9.26) to the momentum dependent forms, which we deal with later. Here, we want to examine just the real parts and their role in the coupled modes of the plasmons and the phonons.

The zeroes of the dielectric function are found simply by setting the right-hand side of (9.26) to zero and using the Lyddane–Sachs–Teller relation of (9.5) to yield

$$\omega^4 - \omega^2 \left(\omega_{LO}^2 + \omega_p^2 \right) + \omega_p^2 \omega_{TO}^2 = 0. \qquad (9.27)$$

This quartic equation can then be solved for a pair of quadratic roots involving the square of the frequency. But, we can also examine the various limits to gain some insight into the roots and behavior. If we set $\omega_p \ll \omega_{TO}$, we find one limiting case to yield

$$\omega_u^2 = \omega_{LO}^2 + \omega_p^2$$
$$\omega_l^2 = \frac{\varepsilon_\infty}{\varepsilon(0)}\omega_p^2 = \omega_{p'}^2, \tag{9.28}$$

for the upper and lower hybrid frequencies, where the prime on the p indicates the lower frequency dielectric permittivity. On the other hand, if we take the limit $\omega_p \gg \omega_{TO}$, we arrive at the limiting cases

$$\omega_u^2 = \omega_{LO}^2 - \omega_{TO}^2 + \omega_p^2 \approx \omega_p^2$$
$$\omega_l^2 = \omega_{TO}^2. \tag{9.29}$$

The upper hybrid mode obviously changes its character from phonon-like to plasmon like as the density (and plasma frequency) increase in magnitude. On the other hand, the lower hybrid mode changes from plasmon-like to phonon-like as the density increases. In figure 9.1, we plot the full range of the two hybrid frequencies showing how they change their nature, using the parameters for GaAs. The vertical and horizontal axes are in meV for the various quantities. The critical frequency is for $\omega_p = \omega_{TO}$, which occurs at a density of about 7×10^{17} cm^{-3} in GaAs.

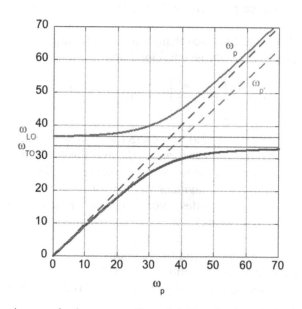

Figure 9.1. The dispersion curve for the upper- and lower-hybrid modes that arise in plasmon–polar phonon coupling. The curves are plotted for varying plasma frequency, and the units are in meV of energy for the various frequencies.

9.1.4 Static screening

So far, we have focused on the high-frequency behavior of the dielectric function. We now want to turn to the low-frequency behavior and examine how we recover the static screening, such as the Debye screening of (9.3). In this region, we can ignore both the photon energy and the damping factor in (9.16) and (9.17). So far, the contributions of the lattice polarization and the valence electrons lead us to the understanding that the first term on the right-hand side of these last two equations is replaced by $\varepsilon(0)$, which includes these contributions. We want to begin by using (9.16), and expanding the numerator and denominator as

$$
\begin{aligned}
E(\mathbf{k} + \mathbf{q}) - E(\mathbf{k}) &\sim \mathbf{q} \cdot \frac{\partial E}{\partial \mathbf{k}} + \dots \\
f(\mathbf{k}) - f(\mathbf{k} + \mathbf{q}) &\sim -\mathbf{q} \cdot \frac{\partial f}{\partial \mathbf{k}} + \dots = -\mathbf{q} \cdot \frac{\partial E}{\partial \mathbf{k}} \frac{\partial f}{\partial E} + \dots .
\end{aligned}
\tag{9.30}
$$

For most transport of interest, the distribution function is Maxwellian, especially since the inter-electronic scattering tends to drive any distribution toward this result. We will deal with the Fermi–Dirac distribution in the next section. With a Maxwellian, we can write

$$
\frac{\partial f}{\partial E} = -\frac{1}{k_\mathrm{B} T} f(E).
\tag{9.31}
$$

As previously, the sum over the various momenta just gives the integral over the distribution function and the density of states, which leads to the free carrier density n. We can now use the above equations in (9.16) to arrive at

$$
\varepsilon(\mathbf{q}, 0) = \varepsilon(0) + \frac{ne^2}{q^2 k_\mathrm{B} T} = \varepsilon(0) \left(1 + \frac{q_D^2}{q^2} \right),
\tag{9.32}
$$

where we have introduced the Debye screening wave vector from (9.3). We have used this form previously in chapter 5 for impurity and polar optical phonon scattering. Here, we see that it follows naturally from the linear response dielectric function under the appropriate approximations. It is also the classical result, as shown at the beginning of this chapter. The Debye screened potential is quite often used to describe the radial variation of the impurity potential. For example, if we have the real space Coulomb potential (energy)

$$
eV(r) = \frac{e^2}{4\pi\varepsilon r},
\tag{9.33}
$$

the Fourier transform in momentum space is

$$
eV(\mathbf{q}) = \frac{e^2}{\varepsilon(\mathbf{q}, 0) q^2} = \frac{e^2}{\varepsilon(0)\left(q^2 + q_D^2\right)}.
\tag{9.34}
$$

While it may seem that the approximations used in (9.30) are extreme, as we have assumed that $q \ll k$ in the Taylor expansions, they are actually well supported. For a free carrier concentration of e.g. 10^{17} cm^{-3}, the Debye wave vector is $q_D \sim 7.5 \times 10^5$ cm^{-1} in GaAs at 300 K. This corresponds to a screening energy of only about 3.2 meV, which is much smaller than the energy of most of the carriers. Consequently, Debye screening is almost universally used. To improve this, we need to consider what has been left out so far, and that is the dynamic momentum dependence of the dielectric response, and we now turn to that.

9.1.5 Momentum-dependent screening

At this point, we want to remove the approximation of a small value for q. We will retain the low frequency approximation, but we want to include the role of the momentum in the dielectric function. To proceed, we rewrite (9.17) as

$$\varepsilon(\boldsymbol{q}, \omega) = \varepsilon(0) + \frac{e^2}{q^2}\sum_k f(k)\left[\frac{1}{E(k+q) - E(k)} - \frac{1}{E(k) - E(k-q)}\right], \quad (9.35)$$

Where the minus sign on the second term arises from reordering the terms in the denominator. As before, we expand the two denominator terms as

$$E(k \pm q) - E(k) = \frac{\hbar^2 q^2}{2m_c} \pm \frac{\hbar^2 kq}{m_c}\cos\vartheta, \quad (9.36)$$

where the mass is the appropriate carrier effective mass and the angle is that between **k** and **q**. The summation over the momentum can be transformed into an integration as was done in chapter 5. In three dimensions, (9.35) becomes, with the above expansions,

$$\varepsilon(\boldsymbol{q}, \omega) = \varepsilon(0) + \frac{2e^2 m_c}{q^2 \hbar^2}\int_0^\infty \int_0^\pi \frac{k^2 dk \, \sin\vartheta d\vartheta}{2\pi^2}$$
$$\times \left[\frac{1}{q^2 + 2kq\cos\vartheta} + \frac{1}{q^2 - 2kq\cos\vartheta}\right] \quad (9.37)$$
$$= \varepsilon(0) + \frac{e^2 m_c}{\pi^2 q^3 \hbar^2}\int_0^\infty f(k)k\ln\left[\left|\frac{k+2q}{k-2q}\right|\right]dk.$$

The form of the argument of the logarithmic term arises from the integration over the angle and the magnitude sign is required to assure that the argument is positive definite.

To proceed further, it is convenient to introduce normalized coordinates to simplify the form of the integral. These are defined by

$$\xi^2 = \frac{\hbar^2 q^2}{8m_c k_B T}$$
$$x^2 = \frac{\hbar^2 k^2}{2m_c k_B T}. \quad (9.38)$$

It may be noted that the temperature here is that of the distribution function and represents the electron temperature. Hence, as the electrons gain energy in an electric field, the role of screening is decreased due to the rise in electron temperature. This should not be confusing, as the only temperature in the problem at this point is the electron temperature and not the lattice temperature. By incorporating the above normalizations, equation (9.37) may be rewritten as

$$\varepsilon(\boldsymbol{q},\omega) = \varepsilon(0)\left[1 + \frac{q_D^2}{q^2}F(\xi)\right],$$

$$F(\xi) = \frac{1}{\sqrt{\pi}\xi}\int_0^\infty xe^{-x^2}\ln\left|\frac{x+\xi}{x-\xi}\right|dx,$$

(9.39)

for a non-degenerate semiconductor. It should be noted that the density was introduced as a normalization of the integral, and this allowed us to introduce the Debye screening wave vector. In general, $F(\xi)$ may readily be evaluated numerically for any value of the parameter (which included the momentum wave vector). However, Hall [3] has shown that $F(\xi)$ may be rewritten as

$$F(\xi) = \frac{e^{-\xi^2}}{\xi}\int_0^\xi e^{t^2}dt,$$

(9.40)

which is related to an error function of imaginary argument. Except for the $1/\xi$ term, (9.40) is also known as Dawson's Integral [4], and is a standard tabulated function. This function is also closely related to the *plasma dispersion function*.

In the case of degenerate semiconductors, the form of the function $F(\xi)$ is somewhat more complicated due to the Fermi–Dirac function. Equation (9.37) is still correct, but the introduction of the density is more complicated due to the use of the full Fermi–Dirac integral, and (see (4.34))

$$n = \frac{2N_c}{\sqrt{\pi}}\int_0^\infty f(x)x^2dx = N_cF_{1/2}(\mu),$$

(9.41)

where $\mu = E_F/k_BT$, N_c is the effective density of states, and the other parameters have their meaning from above. In this degenerate case, the second line of (9.39) is replaced by [5]

$$F(\xi,\mu) = \frac{1}{\sqrt{\pi}\xi F_{1/2}(\mu)}\int_0^\infty \ln\left|\frac{x+\xi}{x-\xi}\right|\frac{xdx}{1+\exp(x^2-\mu)}.$$

(9.42)

If μ is very negative, the non-degenerate condition is recovered.

The behavior of $F(\xi,\mu)$ is generally not very dramatic. As $q \to 0$ ($\xi \to 0$), $F(\xi,\mu) \to 1$, for the non-degenerate case, and the usual Debye screening is recovered. On the other hand, as $q \to \infty$ ($\xi \to \infty$), $F(\xi,\mu) \to 0$, and the screening is completely broken up. Thus, for high momentum transfer in the scattering process, the scattering potential becomes almost unscreened. As the Fermi energy gets large and positive, the value of $F(\xi,\mu)$ is greatly reduced as the screening moves from the Debye version to the Fermi-Thomas screening for degenerate materials. The function $F(\xi,\mu)$ is plotted in figure 9.2 with the reduced Fermi energy as a parameter. Higher densities lead to a reduction in the classical concept of screening, but the wave vector dependence is also greatly reduced.

9.1.6 High-frequency dynamic screening

Let us now turn to the dynamic screening at a somewhat higher frequency so that we no longer ignore the photon energy. We will pursue the development separately for the non-degenerate and degenerate regimes, and will show that under certain conditions, the results are the same. In these approaches, we will find that in terms such as those of (9.30), we will have angular averages of the angle between k and q to consider, and this will lead to some complicated procedures to obtain these averages.

Non-degenerate material. We begin with the non-degenerate approach, using the expansions of (9.30), but with the expansions

$$E(k+q) - E(k) \sim q \cdot \frac{\partial E}{\partial k} + \ldots = \hbar q \cdot v + \ldots$$

$$f(k) - f(k+q) \sim -q \cdot \frac{\partial f}{\partial k} + \ldots = -\hbar q \cdot v \frac{\partial f}{\partial E} + \ldots .$$

(9.43)

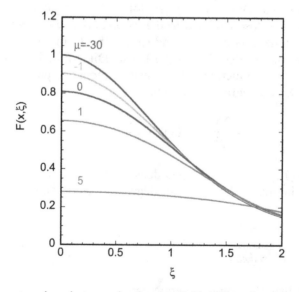

Figure 9.2. The momentum dependent screening function (9.42) is shown as a function of the reduced scattering momentum, with the reduced Fermi energy as a parameter. Adapted from [5].

This allows us to write the dielectric function from (9.16) as

$$\varepsilon(\boldsymbol{q}, \omega) = \varepsilon(0) + \frac{e^2}{q^2}\sum_k \frac{\partial f}{\partial E}\frac{-\hbar \boldsymbol{q} \cdot \boldsymbol{v}}{\hbar \boldsymbol{q} \cdot \boldsymbol{v} + \hbar\omega - i\hbar/\tau}. \tag{9.44}$$

In this last expression, we have replaced α with $1/\tau$, where the latter is the net scattering rate. For the non-degenerate distribution, we can use (9.31) and rewrite (9.44) as

$$\varepsilon(\boldsymbol{q}, \omega) = \varepsilon(0) + \frac{e^2}{q^2 k_{\mathrm{B}}T}\sum_k f(k)\frac{\boldsymbol{q} \cdot \boldsymbol{v}}{\boldsymbol{q} \cdot \boldsymbol{v} + \omega - i/\tau}. \tag{9.45}$$

The summation can be converted to an integral with the momentum (velocity) taken as the polar axis, so that the dot products introduce the polar angle. Integration over this angle is relatively complicated, so to ease this we expand the last term as

$$
\begin{aligned}
F &\equiv \frac{\boldsymbol{q} \cdot \boldsymbol{v}}{\boldsymbol{q} \cdot \boldsymbol{v} + \omega - \frac{i}{\tau}} = 1 - \frac{\omega - \frac{i}{\tau}}{\boldsymbol{q} \cdot \boldsymbol{v} + \omega - \frac{i}{\tau}} \\
&= 1 - \left[1 - \frac{\boldsymbol{q} \cdot \boldsymbol{v}}{\omega - \frac{i}{\tau}} + \frac{(\boldsymbol{q} \cdot \boldsymbol{v})^2}{\left(\omega - \frac{i}{\tau}\right)^2} + \cdots \right].
\end{aligned}
\tag{9.46}
$$

In carrying out the integration, the second term in the square brackets will always vanish. In three dimensions, the summation goes to an integral in which the differential has a term $\sin\vartheta d\vartheta$, while the dot product adds a $\cos\vartheta$. This integrates to zero, and similar effects occur in both one and two dimensions. On the other hand, the third term will involve the square of the cosine, which does not vanish in the integration, but adds a factor of $1/d$, where d is the dimensionality of the system. Then, the expansion in (9.46) can be reversed after integration, with the integration over the distribution function merely giving the density, with the result

$$\varepsilon(\boldsymbol{q}, \omega) = \varepsilon(0) + \frac{ne^2}{k_{\mathrm{B}}T}\frac{D/\tau}{(\omega - i/\tau)^2 - q^2 D/\tau}, \tag{9.47}$$

and we have introduced the diffusion coefficient as $D = v^2\tau/d$, although it must be identified with an average velocity in bringing this quantity outside the integration. We note that if we let $\omega \to 0$, we recover the Debye screening result (9.32), while if we take the limit $q \to 0$, $\omega \gg 1/\tau$, we can replace D/τ with $k_B T/m_c$ and recover the free-carrier plasma result (9.24). Hence, (9.47) satisfies the requisite limits obtained in the previous sections.

Degenerate material. A somewhat different approach must be followed for degenerate material, since we cannot relate the derivative of the distribution function merely to itself and a thermal factor. Instead, we begin with the expansion of (9.18) and use the approximations (9.20) and (9.43) to write the dielectric function as

$$\varepsilon(\boldsymbol{q},\,\omega) = \varepsilon(0) - \frac{\hbar e^2}{m_c} \sum_k f(k) \frac{1}{[\hbar \boldsymbol{q} \cdot \boldsymbol{v} + \hbar\omega - i\hbar/\tau]^2}$$

$$= \varepsilon(0) - \frac{e^2\tau^2}{m_c} \sum_k f(k) \frac{1}{[1 + i\omega\tau + i\boldsymbol{q} \cdot \boldsymbol{v}]^2}. \qquad (9.48)$$

The last factor, including the angular terms, can now be expanded and rearranged to give, after integration,

$$F = \frac{1}{[1 + i\omega\tau + i\boldsymbol{q} \cdot \boldsymbol{v}]^2} = \frac{1}{(1 + i\omega\tau)^2}$$

$$\times \left[1 - 2i\frac{\boldsymbol{q} \cdot \boldsymbol{v}\tau}{(1 + i\omega\tau)} - 3\frac{(\boldsymbol{q} \cdot \boldsymbol{v}\tau)^2}{(1 + i\omega\tau)^2} - \cdots \right] \rightarrow \frac{1}{(1 + i\omega\tau)^2 + Dq^2\tau}. \qquad (9.49)$$

We have used the same arguments from above for the angular integrations in this form, so that the dielectric function becomes

$$\varepsilon(\boldsymbol{q},\,\omega) = \varepsilon(0) - \frac{ne^2}{m_c} \frac{\tau^2}{(1 + i\omega\tau)^2 + Dq^2\tau}. \qquad (9.50)$$

It can be checked that this form gives the same limiting behavior as the non-degenerate form (9.47).

The differences in the two approaches lies in the portion of the spectrum in which the effects will occur. First, we note that for each case, the Coulomb interaction appears as a product with the Lindhard potential appropriate for a three-dimensional system. We will treat the changes necessary for reduced dimensionality in the next few sections. Here, however, we note that (9.47) may be rewritten, using the second term of (9.3) for the Debye wave vector to get

$$\varepsilon(\boldsymbol{q},\,\omega) = \varepsilon(0) \left[1 - \frac{q_D^2 D\tau}{(1 + i\omega\tau)^2 + q^2 D\tau} \right], \qquad (9.51)$$

which should be compared with (9.50), using (9.25) to give

$$\varepsilon(\boldsymbol{q},\,\omega) = \varepsilon(0) \left[1 - \frac{\omega_p^2\tau^2}{(1 + i\omega\tau)^2 + Dq^2\tau} \right]. \qquad (9.52)$$

These two are the same, if we use some definitions appropriate for non-degenerate systems as

$$q_D^2 D\tau = \frac{ne^2}{\varepsilon(0)k_{\mathrm{B}}T} \left(\frac{e\tau}{m_c} \frac{k_{\mathrm{B}}T}{e} \right) \tau = \frac{ne^2\tau^2}{\varepsilon(0)m_c} = \omega_p^2\tau^2. \qquad (9.53)$$

While the two forms have different starting places, they arrive at the same result. So, if we are not sure whether the system is degenerate or not, it is better to start with the result (9.52), and we can assure ourselves of the correct approach.

9.2 Screening in low-dimensional materials

In the above treatments, we focused mainly on three dimensional materials. In this, and some following sections, we will begin to study low-dimensional materials as well. Here, we want to consider a quasi-two-dimensional semiconductor, such as the inversion layer at the Si-SiO$_2$ interface in the conditions where several sub-bands are occupied. These sub-bands are the result of the z-directional quantization in the potential at the interface, that has been discussed several times earlier in this book. Each quantized energy level in the z-direction has free motion in the plane along the interface. And, this motion is described by the quasi-two-dimensional sub-band. This uniqueness of a discrete energy level normal to the interface (which quantized the momentum in this direction to discrete values), and free motion in the transverse directions parallel to the interface, creates new difficulties in evaluating the summations/integrals over the distribution function. While we treat the quasi-two-dimensional case, it is easily extended to a quasi-one-dimensional case for nanowires.

We start by rewriting the wave function that appears in (9.7) as a plane wave, in only two dimensions, that is coupled to a z-direction envelope function. We describe this wave function as

$$|k\rangle \rightarrow \varphi_n(z)e^{ik\cdot r} \equiv |k, n\rangle, \qquad (9.54)$$

where k and r are now two-dimensional vectors in the plane parallel to the interface. The function $\varphi_n(z)$ is the envelope function describing the variation of the wave function in the z-direction (normal to the interface). The perturbed wave function appearing in the Lindhard potential is now written as (coupling to only a single other sub-band)

$$\psi(k, z) = |k, n\rangle + b_{k+q,n'}|k + q, n'\rangle + b_{k-q,n'}|k - q, n'\rangle. \qquad (9.55)$$

The heart of the change in the Lindhard function lies in the change in the Fourier transform of the perturbing potential, and this appears as

$$V_{n,n'}(q) = \langle k, n|V(r, z)|k + q, n'\rangle$$
$$= \int d^2r \int_{-\infty}^{\infty} dz \varphi_{n'}^\dagger(z) V(r, z)\varphi_n(z)e^{iq\cdot r}e^{-qz}. \qquad (9.56)$$

In addition to the potential's variation in the z-direction, the presence of the last exponential serves to break the normal orthogonality of the envelope functions for different sub-bands. We assume that the scattering potential has only a very weak variation with z, as it is mainly a two-dimensional scattering function. Hence, we can write (9.56) in the form

$$V_{n,n'}(\boldsymbol{q}) = \frac{e^2}{2\varepsilon(0)q}\zeta_{n,n'}, \quad \zeta_{n,n'} = \int\limits_{-\infty}^{\infty} dz \varphi_{n'}^{\dagger}(z)\varphi_n(z)e^{-qz}. \tag{9.57}$$

In writing the final Lindhard dielectric function, we have to be cognizant that the envelope functions not only appear in the potential, but also in the summation over the distribution function. This greatly complicates the form, and

$$\varepsilon(\boldsymbol{q}, \omega) = \varepsilon(0)\left[\delta_{n,n'} + \frac{e^2}{2\varepsilon(0)q}\sum_{m,m'} F_{m,m'}^{n,n'}(q)L_{m,m'}(q, \omega)\right], \tag{9.58}$$

where

$$F_{m,m'}^{n,n'}(q) = \int\limits_{-\infty}^{\infty} dz' \int\limits_{-\infty}^{\infty} dz \varphi_m^{\dagger}(z)\varphi_{n'}^{\dagger}(z')\varphi_m(z)\varphi_n(z'),$$

$$L_{m,m'}(q, \omega) = \sum_k \frac{f_m(k) - f_{m'}(k + q)}{E_{m'}(k + q) - E_m(k) + \hbar\omega - i\hbar\alpha}. \tag{9.59}$$

To obtain the two-dimensional density, we have to sum over both the transverse momentum and the occupied sub-bands. This gives

$$n_s = 2\sum_{k,n} f_n(k), \quad f_n(k) = \left[1 + \exp\left(\frac{E - E_n - E_F}{k_{\mathrm{B}}T}\right)\right]^{-1}, \tag{9.60}$$

where E_n is the quantized sub-band energy level.

Quite often, at low temperatures, it is sufficient to assume the quantum limit in which only a single sub-band is occupied. In this case, the dielectric function reduces to

$$\varepsilon(\boldsymbol{q}, \omega) = \varepsilon(0)\left[1 + \frac{e^2}{2\varepsilon(0)q}F_0(q)L_{00}(q, \omega)\right]. \tag{9.61}$$

We still need to evaluate the leading function for the envelope function overlap integral, and for this we need to use a form for the overlap function. A common choice is the variational wave function [6]

$$\varphi_0(z) = \sqrt{\frac{b^3}{2}}\, z e^{-bz/2}, \tag{9.62}$$

which leads to

$$F_0 = \frac{b^6}{(b^2 + q^2)^3}, \quad \zeta_0 = \frac{b^3}{(b + q)^3}. \tag{9.63}$$

9.3 Free-particle interelectronic scattering

The study of the interacting electron gas, and the resulting electron–electron scattering process, has a long history. From the earliest days, there has been interest in computing the self-interactions that arise from the Coulomb forces between the electrons, as opposed to the interactions between the electrons and the impurities that was of interest in earlier chapters. Several times, it has been pointed out that when the electron–electron interaction was dominant, it would guide the distribution into a Maxwellian and produce an appropriate electron temperature. One of the first to examine this in semiconductors was Hearn who determined the critical concentration of electrons for the inter-carrier interactions to dominate scattering in three dimensions [7]. This was also computed for a quasi-two-dimensional electron gas [8].

In most of the treatment of this section, we assume that the Coulomb interaction is screened by the classical Debye screening. This introduces a natural cutoff which allows us to differentiate two regimes of electron–electron interactions. For momentum vectors q large than the Debye momentum vector q_D, the resulting interaction is the short range Coulomb single-particle interaction between individual electrons [9]. We will treat this in this section. However, there is another part in which the momentum is small, meaning that a long-range interaction is involved. This leads to the electron scattering from the collective plasma excitations and electron–plasmon scattering. This will be treated in a later section. It is important to understand, though, that electron–electron scattering in momentum space involves both of these processes, and treating only one of them leads to an incomplete consideration of the process.

9.3.1 Electron scattering by energetic carriers

In this section, we treat the scattering of energetic carriers by other electrons, and investigate the energy loss by such energetic carriers. We must not forget that, in the interaction between any pair of electrons, the total momentum and energy must be conserved. This does not imply that an individual electron cannot lose significant energy in the process, only that the process does not give rise to a net overall energy loss to the lattice. But, it does lead to a redistribution of the energy within the electron ensemble, such as predicated in the decay to a Maxwellian. We treat the scattering matrix element as the screened Coulomb interaction

$$M(q) = \frac{e^2}{\varepsilon_s(q^2 + q_D^2)}, \tag{9.64}$$

where ε_s is the static dielectric constant for the material of interest. Following Takanaka [10], the scattering rate can be written as

$$\Gamma_{ee}(k) = \frac{2\pi}{\hbar^2} \sum_{k_2, k', k_2'} \left[\frac{e^2}{\varepsilon_s(q^2 + q_D^2)} \right]^2 f(k_2)$$

$$\times \delta_{k+k_2, k'+k_2'} \delta\left(E_{k'} + E_{k_2'} - E_k - E_{k_2}\right). \tag{9.65}$$

Here, k and k_2 are the momentum vectors of the initial particle and the particle with which it interacts, respectively. The other two wave vectors are the resulting final momenta of the two particles as appropriate. We further assume that the distribution function is non-degenerate.

The summation over the final wave vector k_2' is carried out using the momentum conserving delta function. To accomplish this, we introduce two new wave vectors through

$$g = k_2 - k, \ g' = k_2' - k' = k + k_2 - 2k'. \tag{9.66}$$

With these definitions, the argument of the energy conserving delta function can be written as

$$E_{k'} + E_{k_2'} - E_k - E_{k_2} = \frac{1}{2}(E_{g'} - E_g). \tag{9.67}$$

In addition, we note that

$$k' = k_2 - \frac{1}{2}(g + g'), \tag{9.68}$$

so that the summation over k' can be changed into a summation over g' (there is a factor of 8 in the new denominator arising from the various factors of ½ in each of the three directions). We can then rewrite (9.65) as

$$\Gamma_{ee}(k) = \frac{\pi}{4\hbar} \sum_{k_2, g'} \left[\frac{e^2}{\varepsilon_s \left(\frac{1}{4} |g - g'|^2 + q_D^2 \right)} \right]^2 f(k_2) \delta\left(\frac{1}{2}(E_{g'} - E_g) \right). \tag{9.69}$$

To proceed further, we introduce another change of variables as

$$u = g' - g. \tag{9.70}$$

With this change, we can do the summation over g' in the following manner

$$\sum_{g'} \rightarrow \frac{1}{4\pi^2} \int u^2 du \int_0^\pi \sin\vartheta d\vartheta \frac{e^4}{\varepsilon_s^2 \left(u^2/4 + q_D^2 \right)^2} \delta\left(\frac{\hbar^2}{4m_c}(u^2 + 2ug \cos\vartheta) \right)$$

$$= \frac{e^4 m_c}{8\pi\hbar^3 q_D^2 \varepsilon_s^2} \sum_{k_2} f(k_2) \left[\frac{|k_2 - k|}{|k_2 - k|^2 + q_D^2} \right]. \tag{9.71}$$

At this point, some assumptions have to be made about the distribution function that appears in this last expression. Here, we will assume that it is a Maxwellian at an elevated electron temperature T_e. Then, we introduce the reduced quantities

$$y^2 = \frac{\hbar^2 \xi^2}{2m_c k_B T_e}, \quad \xi = k_2 - k,$$

$$x^2 = \frac{\hbar^2 k^2}{2m_c k_B T_e}, \tag{9.72}$$

$$\lambda^2 = \frac{\hbar^2 q_D^2}{2m_c k_B T_e}.$$

The final integration can now be written as

$$
\begin{aligned}
\Gamma_{ee}(k) &= \frac{e^4 m_c}{8\pi \hbar^3 q_D^2 \varepsilon_s^2} \int_0^\infty \frac{\xi^2 d\xi}{2\pi^2} \int_0^\pi \sin\vartheta d\vartheta f(\xi + k) \left(\frac{\xi}{\xi^2 + q_D^2} \right) \\
&= \frac{n e^4 m_c}{8\sqrt{\pi}\hbar^3 q_D^2 \varepsilon_s^2 k} e^{-x^2} \int_0^\infty e^{-y^2} \frac{y^2}{y^2 + \lambda^2} \sinh(2xy) dy.
\end{aligned}
\tag{9.73}
$$

This result can be rewritten as a collection of error functions and exponential integrals, but such a result provides no new information, as it still is a set of tabulated functions and not a closed form that is easily evaluated. It is best just to numerically evaluate (9.73). In figure 9.3, we plot the scattering rate (9.73) as a function of the carrier energy and for the case of Si, with the Maxwellian distribution at an elevated temperature of 2500 K and a density of 10^{17} cm^{-3}. Curves are shown for the emission and absorption of energy by the initial carrier, as well as for the total scattering rate. These values correspond to the situation in many MOSFETs today, and this scattering rate includes both emission and absorption of energy from other carriers.

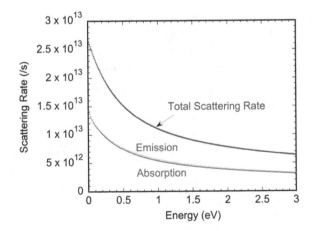

Figure 9.3. The emission, absorption, and total scattering rate for single particle electron–electron scattering. This is shown for the case of Si, doped to 10^{17} cm^{-3}, at an electron temperature of 2500 K.

9.3.2 Energy gain and loss

While the above discussion did not specifically separate the emission and absorption of energy, these were plotted separately in figure 9.3, and these have different effects on the carriers themselves. As a result, we cannot determine just how much energy is gained or lost by the initial carrier, and it may be seen from this latter figure that the two scattering rates are very close to each other, as one may expect for an electron gas that has reached a steady-state condition. It is possible, however, to reformulate the above treatment in a manner for which we can make this determination [11]. The key is to decompose the energy conserving delta function in (9.65) to a product of two such delta functions as

$$\delta(E_{k'} + E_{k_2'} - E_k - E_{k_2}) = \hbar \int d\omega \delta(E_k - E_{k'} - \hbar\omega)$$
$$\times \delta(E_{k_2} - E_{k_2'} + \hbar\omega).$$

(9.74)

In this form, integration over the frequency (the energy exchange) produces the single delta function used earlier. Hence, by leaving this frequency integration until the end of the development, the energy gained or lost by the incident electron can be studied. As previously, the integration over k_2' introduces the momentum conservation that is represented by found in the integral. The integration over k_2 now involves the frequency as well as the other variables, so that this integration can be deferred until the end. The second delta function in (9.74) involves

$$E_{k_2'} - E_{k_2} - \hbar\omega = E_{k-k'+k_2} - E_{k_2} - \hbar\omega$$
$$= E_q + \frac{\hbar^2 q k_2}{m_c} \cos\chi - \hbar\omega.$$

(9.75)

Here, χ is the angle between k_2 and q. It now makes more sense to integrate over k_2 prior to integrating over q or g. The resulting integration, as one can imagine, involves the summation over k_2, the distribution function itself, and the second delta function on the right-hand side of (9.74). This integration involves the angle shown in (9.75) which resides within the delta function, and this produces some parameters plus limits on subsequent integrations, resulting in

$$I = \sum_{k_2} f(k_2) \delta\left(E_q + \frac{\hbar^2 q k_2}{m_c} \cos\chi - \hbar\omega\right)$$
$$= \frac{1}{4\pi^2} \int f(k_2) k_2^2 dk_2 \int_0^\pi \delta\left(E_q + \frac{\hbar^2 q k_2}{m_c} \cos\chi - \hbar\omega\right) \sin\chi d\chi$$
$$= \frac{m_c}{4\pi^2 \hbar^2 q} \int_{k_0}^\infty f(k_2) k_2 dk_2.$$

(9.76)

Here, the lower limit of the integral, set by the requirement that the argument of the delta function must have a zero, is given as

$$k_0 = \frac{m_c}{\hbar^2 q}(E_q - \hbar\omega). \tag{9.77}$$

If we now assume, as in the previous section, that we have a Maxwellian distribution function at an elevated electron temperature T_e, then the last integral can be evaluated to yield

$$I = \frac{n}{\hbar q}\sqrt{\frac{m_c}{2\pi k_B T_e}}\exp\left[-\frac{\hbar^2}{8m_c k_B T_e}\left(q - \frac{2m_c\omega}{\hbar q}\right)^2\right]. \tag{9.78}$$

The result (9.78) can now be used in the integration over $\mathbf{k}'$, which involves a very similar integration to that just performed. This can now be written as

$$\Gamma_{ee}(k,\omega) = \frac{2\pi}{\hbar}\sum_{k'}\left[\frac{e^2}{\varepsilon_s(q^2 + q_D^2)}\right]^2 I\delta\left(E_q + \frac{\hbar^2 qk}{m_c}\cos\vartheta + \hbar\omega\right)$$
$$= \frac{e^4 m_c}{2\pi\hbar^3\varepsilon_s^2 k}\int_{q_-}^{q_+}\frac{Iq\,dq}{(q^2 + q_D^2)^2}. \tag{9.79}$$

The limits on the last integration arise from the delta function and are

$$k - \sqrt{k^2 - \frac{2m_c}{\hbar q}} = q_- < q < q_+ = k + \sqrt{k^2 - \frac{2m_c}{\hbar q}}. \tag{9.80}$$

The reader will note that limits such as these, arising from the delta function, were already encountered in chapter 5 for optical phonon scattering by the electrons.

We can now re-introduce (9.78) and perform the final integration over the frequency to give the result

$$\Gamma_{ee}(k) = \frac{ne^4 m_c}{4\pi\hbar^3\varepsilon_s^2 k}\sqrt{\frac{m_c}{2\pi k_B T_e}}\int_0^{E(k)/\hbar}d\omega\int_{q_-}^{q_+}\frac{q\,dq}{(q^2 + q_D^2)^2}$$
$$\times\exp\left[-\frac{\hbar^2}{8m_c k_B T_e}\left(q - \frac{2m_c\omega}{\hbar q}\right)^2\right]. \tag{9.81}$$

This result is for the case of energy loss by the incident electron. For the case of energy gain, the result is slightly different in the various limits and the result is given by

$$\Gamma_{ee}(k) = \frac{ne^4 m_c}{4\pi\hbar^3\varepsilon_s^2 k}\sqrt{\frac{m_c}{2\pi k_B T_e}}\int_0^{\infty}d\omega\int_{q_-}^{q_+}\frac{q\,dq}{(q^2 + q_D^2)^2}$$
$$\times\exp\left[-\frac{\hbar^2}{8m_c k_B T_e}\left(q + \frac{2m_c\omega}{\hbar q}\right)^2\right], \tag{9.82}$$

with the limits now given by

$$\sqrt{k^2 + \frac{2m_c}{\hbar q}} - k = q_- < q < q_+ = \sqrt{k^2 + \frac{2m_c}{\hbar q}} + k. \tag{9.83}$$

In figure 9.4, we plot the relative scattering rates for the loss of energy for a given energy of the initial electron (taken to be 0.5, 1.0, and 2.0 eV) for the same distribution function and electron temperature as used in figure 9.3. The integration over these curves in figure 9.4 gives the same total scattering rates as shown in figure 9.3.

Goodnick and Lugli [12] studied non-equilibrium transport in a quantum well excited by a short pulse laser and incorporated the electron–electron scattering explicitly in their simulation for a two-dimensional electron gas. They found that, indeed, the energy exchange was quite small, but non-zero. The interaction was still effective in transferring energy between different sub-bands, and would lead to thermalization of the excited electrons into a steady-state distribution function within a few picoseconds.

9.4 Electron–plasmon scattering

As we discussed earlier, the longer range electron–electron scattering takes into account the scattering of an incident electron by the long-range collective excitations of the electron gas—the plasmons. Here, as in the single particle case, the strength of the scattering is determined by the dielectric function. The zeroes of the dielectric function at the plasmon frequency (the plasmon-pole approximation) correspond to the excited modes of the two hybrid frequencies in the case of the plasmon–phonon interaction. But, we cannot use just the simple plasmon-pole approximation, as it gives a real dielectric function, and the scattering and dissipation entail determining the imaginary part. Moreover, the plasmon frequency is a set quantity only in three

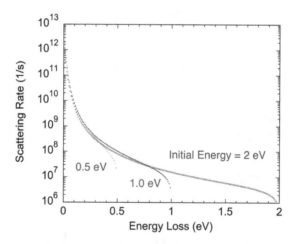

Figure 9.4. The scattering rate for the amount of energy loss from the initial electron energy. These are plotted for initial energies of 0.5, 1.0, and 2.0 eV for the primary electron.

dimensions. In lower dimensions, the plasmon frequency depends upon the momentum in the scattering event. As a result, we have to use a more complete, momentum dependent form for the dielectric function. Indeed, the importance of the plasmons on transport in semiconductors has been known for quite some time [13–15]. The full dielectric function has zeroes at both zero frequency and at the plasmon frequency, which leads to the two parts of the interaction [16, 17]. Hence, in order to capture the details around the plasmon pole, especially in low-dimensional systems, we need to work with the proper incorporation of the imaginary part of the inverse of the dielectric function. It is the poles of this latter function which give rise to the important scattering processes involving the plasmons.

9.4.1 Plasmon scattering in a three-dimensional system

To proceed, we will calculate the inverse dielectric function, but will actually ignore the photon frequency, assuming that its energy is small compared to that of the plasmon. For this, we take the dielectric function itself in the form (9.17) but with the expansion (9.20). We can then write the momentum dependent dielectric function, in three dimensions as

$$
\begin{aligned}
\varepsilon(\boldsymbol{q}) = \varepsilon(0) &- \frac{e^2}{q^2} \frac{h^2 q^2}{m_c} \sum_k f(k) \\
&\times \frac{1}{[E(k+q) - E(k) - i\hbar\alpha][E(k) - E(k-q) - i\hbar\alpha]} \\
= \varepsilon(0) &- \frac{e^2}{m_c} \sum_k f(k) \frac{1}{(q \cdot v - i\alpha)^2}.
\end{aligned}
\tag{9.84}
$$

The angle in the denominator will be ignored for now, but will be evaluated later after we have restored the full form of the energy functions. The dielectric function may now be written as

$$
\varepsilon(\boldsymbol{q}) = \varepsilon(0)\left[1 - \frac{\omega_p^2}{(q \cdot v - i\alpha)^2}\right].
\tag{9.85}
$$

The scattering strength is now given by

$$
\begin{aligned}
\frac{1}{\pi}\mathrm{Im}\left(\frac{1}{\varepsilon(\boldsymbol{q})}\right) &= \frac{1}{\pi\varepsilon(0)} \frac{(q \cdot v - i\alpha)^2}{(q \cdot v - i\alpha)^2 - \omega_p^2} \\
&= \frac{1}{\pi\varepsilon(0)} \frac{(q \cdot v)^2}{(q \cdot v - i\alpha - \omega_p)(q \cdot v - i\alpha + \omega_p)} \\
&= \frac{1}{2\varepsilon(0)}[\delta(q \cdot v + \omega_p) + \delta(q \cdot v - \omega_p)].
\end{aligned}
\tag{9.86}
$$

In this last expression, we have used the relationship

$$\lim_{\alpha \to 0} \frac{1}{x - i\alpha} = P\frac{1}{x} + i\pi\delta(x),$$ (9.87)

where P denotes the principal part of the integral (over the distribution function that will be used). The two delta functions correspond to the absorption and the emission of a plasmon by the electron in the interaction. Now, restoring the full arguments of the delta functions, we can write the scattering rate as

$$\Gamma_{e-pl}(k) = \frac{2\pi}{\hbar} \frac{1}{4\pi^2} \int q^2 dq \frac{\hbar\omega_p}{2\varepsilon(0)q^2} \int_0^{\pi} \sin\vartheta d\vartheta$$

$$\times [N_{pl}\delta(E(k + q) - E(k) + \hbar\omega_p)$$

$$+ (N_{pl} + 1)\delta(E(k + q) - E(k) - \hbar\omega_p)].$$ (9.88)

As usual, the arguments of the delta functions include the angles in the polar angle integration and lead to limits on the remaining integrations and some numerical factors. In addition, here the Bose–Einstein distribution is for the plasmons themselves, as

$$N_{pl} = \left[1 + \exp\left(\frac{\hbar\omega_p}{k_BT_e}\right)\right]^{-1}.$$ (9.89)

We note that this distribution is a function of the *electron* temperature and not the lattice temperature. This follows as the plasmons are collective modes of the electron gas, and should evolve to have the same electron temperature as found in the single particle scattering. With this in mind, we can carry out the integrations (exactly similar in scope as for the optical phonons in chapter 5) to yield

$$\Gamma_{e-pl}(k) = \frac{m_ce^2\omega_p}{4\pi\hbar^2k}\left[N_{pl}\int_{q_-^a}^{q_D}\frac{dq}{q} + (N_{pl} + 1)\int_{q_-^e}^{q_D}\frac{dq}{q}\right]$$

$$= \frac{m_ce^2\omega_p}{4\pi\hbar^2k}\left[N_{pl}\ln\left(\frac{q_D/k}{\sqrt{1 + \hbar\omega_p/E} - 1}\right)\right.$$ (9.90)

$$\left. + (N_{pl} + 1)\ln\left(\frac{q_D/k}{1 - \sqrt{1 - \hbar\omega_p/E}}\right)\right].$$

One important difference for this scattering as opposed to phonon scattering is that the Bose–Einstein distribution does not stay in equilibrium, but evolves as the electron temperature evolves. In this case, it is much more like the hot phonons that were discussed in chapter 7. Fortunately, studies have shown that the plasmon temperature comes into a quasi-equilibrium with the free carrier temperature, at a given high electric field [15]. This provides a reasonable assurance that the approach is consistent with the single-particle scattering discussed previously.

9.4.2 Scattering in a quasi-two-dimensional system

As we have discussed many times, electrons that are in excited states with energies well above the mean energy of the electron gas will, on average, lose energy to the overall electron gas. This will occur either through single-particle scattering or through the net emission of plasmons, as well as by phonon emission. This is true regardless of the dimensionality, but new concepts and problems arise in low-dimensional systems. We have discussed the plasmon scattering in the previous paragraphs, where the three dimensional system has a well-defined plasmon mode with a frequency that depends mainly on the electron density. And, this frequency remains well defined even when $q = 0$. In low dimensional systems, this is no longer the case. When we move to two-dimensional systems, the density becomes a sheet density, and formulas such as (9.25) are no longer dimensionally correct. In fact, in two dimensions, the square of the plasma frequency is given by

$$\omega_p^2 = \frac{n_s e^2}{2m_c \varepsilon(0)} q \equiv \chi_2 q. \tag{9.91}$$

Hence, as the momentum vector goes to zero, the plasma mode disappears. Even for non-zero momentum vector, the plasma frequency is a dispersive quantity and the plasmon-pole approach becomes a nebulous concept, and certainly not one for use in scattering theory. Consequently, we have to result to using a more complete dielectric function as was done in the preceding section. But, we have to also maintain both frequency and momentum variables in the computation, not the least because the so-called plasmon frequency entering into the emission and absorption calculation will be momentum dependent.

A second factor that will enter this problem is that the lower limit for the momentum integration cannot be set to zero, as this integral will then diverge. Hence, a lower limit will need to be utilized, and this is taken from a natural cutoff of the interaction. In two dimensions, this cutoff arises from the inverse of the mean free path, the longest allowed real space quantity in our region of interest. This follows from the assumption that other scattering would breakup any process that would lead to plasmon scattering with a smaller wave number. Thus, we set the minimum momentum wave number to be $q_{min} \sim 1/v\tau = 1/\sqrt{2D\tau}$.

We can now formulate the process of energy loss by an energetic electron in a quasi-two-dimensional electron gas. We treat a nearly free electron in a fairly high mobility semiconductor such as the quasi-two-dimensional electron gas at a heterostructure interface; e.g., Si-SiO$_2$ or GaAs/AlGaAs for example. Our approach will be quite similar to that of the previous paragraph, except that we take the fact that the plasmon energy will likely be small, so that we do not need to differentiate between emission and absorption. Then, we can write the Bose–Einstein factors as

$$N_{pl} + (1 + N_{pl}) = \frac{1}{\exp\left(\frac{\hbar\omega}{k_B T_e}\right) - 1} + \frac{\exp\left(\frac{\hbar\omega}{k_B T_e}\right)}{\exp\left(\frac{\hbar\omega}{k_B T_e}\right) - 1}$$

$$= \coth\left(\frac{\hbar\omega}{2k_B T_e}\right). \tag{9.92}$$

We do not use the subscript on this frequency, as the actual plasmon frequency is dispersive and depends upon the actual scattering wave momentum. We can now write the scattering rate in analogy with (9.88) as

$$\Gamma_{e-pl}(k) = \frac{2\pi}{\hbar} \sum_q \int_{-\infty}^{\infty} \frac{d\omega}{2\pi} \coth\left(\frac{\hbar\omega}{2k_B T_e}\right) \frac{e^2}{2q}$$

$$\times \mathrm{Im}\left(\frac{1}{\varepsilon(\boldsymbol{q}, \omega)}\right) \delta\left(\omega - \frac{E(k+q) - E(k)}{\hbar}\right). \tag{9.93}$$

For the dielectric function, we use (9.52), adapted to two dimensions, as

$$\varepsilon(\boldsymbol{q}, \omega) = \varepsilon(0)\left[1 - \frac{n_s e^2 q}{2m_c \varepsilon(0)} \frac{\tau^2}{(1 + i\omega\tau)^2 + Dq^2\tau}\right]$$

$$= \varepsilon(0)\left[1 - \frac{\chi_2 q\tau^2}{(1 + i\omega\tau)^2 + Dq^2\tau}\right]. \tag{9.94}$$

In the low frequency limit, we can rewrite the imaginary part of the inverse dielectric function, along with the bare Coulomb potential, as

$$V(q)\mathrm{Im}\left(\frac{1}{\varepsilon(\boldsymbol{q}, \omega)}\right) \sim -\frac{e^2}{2\varepsilon(0)q} \frac{2\omega}{Dq^2 + \chi_2 q\tau}. \tag{9.95}$$

The integral over the frequency can now be written as, after taking the small argument limit of $\coth\left(\frac{x}{2}\right) \sim 2/x$,

$$\int_{-\infty}^{\infty} \frac{d\omega}{2\pi} \coth\left(\frac{\hbar\omega}{2k_B T_e}\right) \frac{e^2}{2\varepsilon(0)q} \frac{2\omega}{Dq^2 + \chi_2 q\tau} \delta\left(\omega - \frac{E(k+q) - E(k)}{\hbar}\right)$$

$$= \frac{e^2 k_B T_e}{\pi\hbar\varepsilon(0)q(Dq^2 + \chi_2 q\tau)}. \tag{9.96}$$

With this result, we can now evaluate the scattering rate, using the lower cutoff momentum value, as

$$\Gamma_{e-pl}(k) = \frac{2\pi}{\hbar} \sum_q \frac{e^2 k_B T_e}{\pi\hbar\varepsilon(0)q(Dq^2 + \chi_2 q\tau)}$$

$$= \frac{e^2 k_B T_e}{\pi\hbar^2\varepsilon(0)} \int_{q_{min}}^{\infty} \frac{dq}{Dq^2 + \chi_2 q\tau} \sim \frac{e^2 k_B T_e}{\pi\hbar^2\varepsilon(0)\chi_2\tau} \ln\left(1 + \chi_2\tau\sqrt{\frac{2\tau}{D}}\right). \tag{9.97}$$

We notice that this scattering rate has a characteristic role-off with temperature as $\tau_{e-pl} \sim 1/T$, which is quite typical, and seen experimentally, for the dephasing rate of the electrons in a quasi-two-dimensional semiconductor at low temperature.

In figure 9.5, we plot the measured phase-breaking time in a pair of open quantum dots (non-tunneling transport) in a GaAs/AlGaAs heterostructure at low temperature [18]. To get the effective electron temperature, the resistance of the structure was measured as a function of lattice temperature and then as a function of the applied electric field, measured at the lowest temperature. This gives an effective electron temperature at each electric field, and thus provides the thermometer for the system. The electron density in the array was $3.6 \pm 0.1 \times 10^{11}$ cm^{-2}. In the figure, it is clearly seen that the phase breaking time falls with $1/T$ at higher temperatures. Now, it may be noted that in the above derivation, the electron temperature enters the problem through the hyperbolic cotangent, particularly in the small argument limit. At very low temperatures, one may take the large argument limit in which $\coth(x/2) \sim 1$. Then, the result will be independent of the electron temperature, and that is seen very clearly at the lowest temperatures.

9.4.3 Plasmon energy relaxation in graphene

We have found through-out this book that carriers can be driven out of equilibrium by the application of a high electric field. When this happens, we are interested in how these electrons are driven back towards equilibrium by the scattering processes. We have also seen how plasmons can be an important part of this process. In chapter 7, we clearly began to understand that the distribution function becomes quite

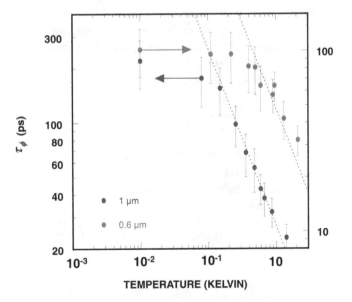

Figure 9.5. The measured phase-breaking time in two different open quantum dots of different dimension. The dotted lines are guides to the eye for a $1/T$ variation in the phase-breaking time. Reprinted with permission from [18]. Copyright 2005 IOP Publishing.

elongated in the direction of the electric field. In the work of Baraff [19], this was taken to be a pure delta function for consideration of the impact ionization, but the general nature of this form was known earlier [20, 21]. Electrical transport measurements probe only the non-symmetric part of the distribution, and hence probe this elongated portion of the actual distribution. As we have discussed above, plasmon emission can contribute to the decay of this elongated portion of the distribution. In graphene, the nature of these processes are somewhat more complicated due to the unusual band structure of this material (see section 2.3.2 for a discussion of the graphene band structure). Generally, the optical phonons in graphene have a very high energy, and are not particularly important for transport except at room temperature and above [22]. At lower temperatures, some have thought that the major cooling process of carriers in the high energy elongation of the distribution function was via the acoustic phonons. Due to momentum conservation constraints, however, this process is quasi-elastic, and so is very inefficient at mediating energy transfer. One process that can explain the energy relaxation is through the emission of plasmons. In comparing results with experiments, it is convenient to express the power input per carrier, which gives also the power loss in steady state, in the form [23]

$$P_e(T_e, T_L) = A\left(T_e^p - T_L^p\right). \tag{9.98}$$

Indeed, the aforementioned acoustic phonon process gives $p = 4$ [24, 25], while electrical transport experiments have suggested that $p = 3$ [26, 27]. In this section, we will explore the energy loss that is possible in graphene due to interactions with the plasmons [28, 29], and show that it indeed has the exponent 3.

Basically, graphene is a two-dimensional semiconductor, so our starting place is the full momentum and frequency dependent dielectric function given in (9.94) with the two-dimensional plasma frequency given by (9.91). Here, however, we will ignore the diffusion term in the denominator, as we take the leading order in q only. Hence, we use the potential and inverse dielectric function as

$$V(q)\mathrm{Im}\left(\frac{1}{\varepsilon(\boldsymbol{q}, \omega)}\right) \sim -\frac{e^2}{2\varepsilon(0)q}\frac{2\omega}{\chi_2 q\tau}. \tag{9.99}$$

Because we are interested in the energy loss, the development will depend upon the *difference* between the emission and absorption terms (we used the sum in the previous section) and we will have to add the energy $\hbar\omega$ that is exchanged in each interaction with the plasmons.

The general approach we use to describe this process is quite straight-forward, and builds upon the results given throughout this chapter. We begin with a standard equation for the energy loss rate, which can be written as [23, 30, 31]

$$\left\langle\frac{dE}{dt}\right\rangle = \sum_{k,q,\omega} \{f(E_k)[1 - f(E_k - \hbar\omega)](N_{pl} + 1)W_{k\to k-q} \\ - f(E_k)[1 - f(E_k + \hbar\omega)]N_{pl}W_{k\to k+q}\}, \tag{9.100}$$

where W represents the scattering function

$$W_\pm = \frac{2\pi}{\hbar}V(q)\text{Im}\left(\frac{1}{\varepsilon(\boldsymbol{q},\omega)}\right)\delta\left(\omega - \frac{E(k\pm q) - E(k)}{\hbar}\right). \tag{9.101}$$

Here, we have kept the Fermi–Dirac distributions because of the peculiarity of the energy dispersion relations in graphene (we ignored them in the previous section). Using these preliminaries, we can rewrite (9.100) in terms of the power loss as

$$P = -2\pi\sum_q \int \frac{\omega d\omega}{2\pi}\frac{e^2\omega}{\varepsilon(0)\chi_2 q^2\tau}\delta_\omega\Phi(\hbar\omega), \tag{9.102}$$

where δ_ω represents the delta function in (9.101) and the various distribution functions are pulled into the expression

$$\Phi(x) = \sum_k \{(N(x,T)+1)f(E_k)[1-f(E_k-x)] \\ - N(x,T)f(E_k)[1-f(E_k+x)]\}, \tag{9.103}$$

and the carrier distribution functions are at the electron temperature. Price [32] has shown how the integral in (9.103) can be carried out and how the electron temperature then appears within the Bose–Einstein distributions as well. Using the properties of the linear bands in graphene, his results lead to

$$\Phi(x) = \frac{xE_F}{\pi(\hbar v_F)^2}[N(x,T_e) - N(x,T_L)]. \tag{9.104}$$

We are now ready to evaluate the remaining integrals in (9.102). Using the delta function, these may be carried out in a straight-forward manner. We begin by writing the energy conservation for the emission process in terms of the magnitudes of the various vectors, using the linear dispersion for graphene, as

$$\hbar v_F(k - |\boldsymbol{k}-\boldsymbol{q}|) = \hbar\omega = \hbar\sqrt{\chi_2 q}, \tag{9.105}$$

where we have used the plasma frequency from (9.91). The proper dynamic mass at the Fermi energy has been used in the last term (the susceptibility). Expanding the terms in the first equation, we can solve for the angle variation as

$$\cos\vartheta = -\left[\frac{q}{2k} - \frac{\chi_2}{2kv_F^2} + \frac{1}{v_F}\sqrt{\frac{\chi_2}{q}}\right]. \tag{9.106}$$

Thus, we require the terms in the square brackets to be positive. For the absorption process, we proceed in the same manner, with the result

$$\cos\vartheta = \left[-\frac{q}{2k} + \frac{\chi_2}{2kv_F^2} + \frac{1}{v_F}\sqrt{\frac{\chi_2}{q}}\right]. \tag{9.107}$$

It is clear that there are solutions to both of the above equations which allow for the emission and absorption of plasmons. In the emission case, the scattering lies in a very small range of angles, of the order of 6–10° from back-scattering, although the direct back-scattering is forbidden by chiral symmetry in equilibrium in graphene. For absorption, there is a very large range of angles that are allowed, but the momentum exchange is quite high. There is almost no density dependence for the scattering when the density is in the range 10^{11}–10^{12} cm^{-2}, but it gets stronger as one approaches the Dirac point, although the plasmons may be heavily damped in this region by the single-particle scattering.

Using these above results, we may now write the integrals in (9.102) in the reduced form, using the fact that there are two terms in the result, each of which is at a different temperature. Each of these two terms is thus in the form

$$F_i(T_i) = \frac{e^2 E_F (k_B T_i)^3}{2\pi\hbar^4 v_F^2 \chi_2 \tau} \int_0^{u_{max}} \frac{u^2 du}{e^u - 1}, \tag{9.108}$$

where

$$u = \frac{\hbar q v_F}{k_B T_i}$$

$$u_{max} = \frac{2\hbar v_F k_F}{k_B T_i}. \tag{9.109}$$

In general, the upper limit of the integral is both density and temperature dependent, but the integral approaches a value of about 2.4 as the upper limit gets large. Hence, we may write the power, or average of the energy loss as

$$\left\langle \frac{dE}{dt} \right\rangle = F_e(T_e) - F_L(T_L) = \frac{2.4 k_B^3 \left(T_e^3 - T_L^3 \right)}{2\pi\hbar^2 v_F^2 \tau}. \tag{9.110}$$

An interesting point is that, with the unique band structure of graphene, the explicit density dependence has dropped out of this final expression for the power loss, but will return when we normalize this expression to get the power loss per electron. The second important point is that the loss rate is characterized by the exponent $p = 3$, consistent with experiments. The only parameter in (9.110) is the scattering time τ, and we can determine this by measuring the mobility as a function of temperature and density. However, the mobility is related to the momentum relaxation time and not just the scattering time. In general for graphene at low temperature, it has been found that the reduction of the scattering time to get the momentum relaxation time varies from about 3.95 at a density of 10^{10} cm^{-2} up to about 4.85 at a density of 10^{13} cm^{-2} [28].

In figure 9.6, we compare the computed power loss per electron to the experimental values of [26]. Here, we take the measured mobility at each electron density and temperature in order to compute the relaxation time τ. This is then used to evaluate (9.110) at each data point from the experiments. Then, the computed power loss is plotted with the actual data points, and the exponent is shown to compare very favorably. Using this approach to determine the actual scattering time

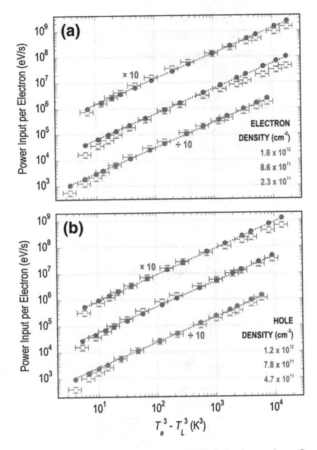

Figure 9.6. The energy loss per carrier for (a) electrons and (b) holes in graphene. In each case, the open symbols are experimental data from [26], and the solid symbols are the theory from (9.110). Each color refers to a different density of carriers, and error bars are shown for the experimental data only. Dotted lines through the theory points are guides to the eye that indicate a linear variation for (9.98) with $p = 3$. Reproduced from [28].

means that there are no adjustable parameters in the fits to the data shown in figure 9.6.

9.4.4 Scattering in a quasi-one-dimensional system

We now turn to the case of a nanowire, which is a quasi-one-dimensional system. While we like to think of these as one dimensional, they are usually created by a narrow quasi-two-dimensional system by the use of, for example, lateral gates to create a narrow constriction for transport in the two-dimensional system. In such a case, the one-dimensional density is created from the two-dimensional density by $n_l = n_s W$, where W is the width of the constriction. The Fermi level is usually then set by that of the 2D regions away from the constriction, but the need to satisfy (4.27), relating the one-dimensional density and the Fermi energy in the nanowire, may lead to a potential barrier between the two-dimensional and the one-dimensional regions.

With these caveats, we will continue to discuss the nanowire on its own without further comment about the external regions. Within a quasi-one-dimensional nanowire, the Coulomb interaction is further modified from the normal form to be

$$V(q) = \frac{e^2}{4\pi\varepsilon(0)\ln\left(1 + q_0^2/q^2\right)}.$$

(9.111)

In the following, we will ignore the multi-sub-band possibilities and consider only the transport in the lowest sub-band, hence ignoring the possible lateral wave function overlap integrals, such as described in section 9.2. The logarithmic factor in (9.111) will eventually cancel with an equivalent factor, and so we will ignore it in the following development. In one dimension, the plasmon dispersion curve becomes

$$\omega_p^2 = \frac{n_1 e^2}{4\pi m_c \varepsilon(0)} q^2 \equiv \chi_1 q^2,$$

(9.112)

where χ_1 is the one-dimensional susceptibility. Then, the momentum dependent dielectric function of (9.52) becomes

$$\varepsilon(q, \omega) = \varepsilon(0)\left[1 - \frac{\chi_1 q^2 \tau^2}{(1 + i\omega\tau)^2 + Dq^2\tau}\right].$$

(9.113)

From this, we can find the imaginary part of the inverse dielectric function and the potential as

$$V(q)\mathrm{Im}\left(\frac{1}{\varepsilon(q, \omega)}\right) \sim \frac{e^2}{2\pi\varepsilon(0)} \frac{\omega}{q^2(D + \chi_1\tau)}.$$

(9.114)

This result can now be used in (9.93).

In evaluating (9.93) for the one-dimensional case, it will be more convenient to carry out the integration over q prior to the integration over the frequency. The first integration still involves the delta function and we will need to invoke some cutoffs for the limits of integration on the subsequent frequency integration. In fact, we are interested in frequencies that lie between $1/\tau_{e\text{-}pl}$ and $1/\tau$ (going from the lower frequency to the higher one). We can now write the momentum integration as

$$\sum_q \rightarrow \sum_q \left| V(q)\mathrm{Im}\left(\frac{1}{\varepsilon(q, \omega)}\right) \right| \delta\left(\omega - \frac{E(k + q) - E(k)}{\hbar}\right)$$
$$= \sum_q \frac{e^2\omega}{2\pi\varepsilon(0)q^2(D + \chi_1\tau)} \delta\left(\omega - \frac{E(k + q) - E(k)}{\hbar}\right).$$

(9.115)

Now, there still must be a summation over the forward-scattering and the back-scattering. However, the dominant contribution comes from the latter term through plasmon emission, so that $q > k$, and the integration yields

$$\sum_q \frac{e^2\omega}{2\pi\varepsilon(0)q^2(D+\chi_1\tau)}\delta\left(\omega - \frac{E(k+q)-E(k)}{\hbar}\right)$$

$$= \frac{m_c e^2}{2\pi\hbar\varepsilon(0)(D+\chi_1\tau)} \times \sqrt{\frac{\hbar}{2m_c\omega}}. \tag{9.116}$$

We can now use this in the remainder of (9.93), the frequency integral, to give

$$\frac{1}{\tau_{e-pl}} = \frac{e^2}{2\hbar\varepsilon(0)(D+\chi_1\tau)}\int_{1/\tau_{e-pl}}^{1/\tau}\frac{d\omega}{2\pi}\frac{2k_BT}{\hbar\omega}\frac{1}{\sqrt{\omega}}$$

$$= \frac{e^2 k_B T}{\hbar^{3/2}\sqrt{2m_c}\,\varepsilon(0)(D+\chi_1\tau)}\sqrt{\tau_{e-pl}}. \tag{9.117}$$

In the first line, we have introduced the two cutoff frequencies, although only the lower one was used to avoid the divergence. The second term in the integral is the expansion of the hyperbolic cotangent term already used in the quasi-two-dimensional expression. We can now solve for the electron–plasmon scattering time to give

$$\Gamma_{e-pl} = \frac{1}{\tau_{e-pl}} = \left[\frac{e^2 k_B T}{\hbar^{3/2}\sqrt{2m_c}\,\varepsilon(0)(D+\chi_1\tau)}\right]^{2/3}. \tag{9.118}$$

This shows us that the plasmon induced dephasing time varies as $T^{-2/3}$, which is different from the rate in other dimensionalities. This resulting temperature dependence has been found in other approaches as well [33, 34].

At lower temperatures, the hyperbolic cotangent function will saturate at unity, and the integral in (9.117) now becomes

$$\Gamma_{e-pl} = \frac{1}{\tau_{e-pl}} = \frac{e^2}{2\hbar\varepsilon(0)(D+\chi_1\tau)}\int_{1/\tau_{e-pl}}^{1/\tau}\frac{d\omega}{2\pi}\frac{1}{\sqrt{\omega}}$$

$$= \frac{e^2}{2\pi\sqrt{2\hbar m_c\tau}\,\varepsilon(0)(D+\chi_1\tau)}, \tag{9.119}$$

where the upper cutoff frequency is important. Just as in the quasi-two-dimensional case, the electron–plasmon scattering time becomes independent of the temperature at very low temperatures, as found in experiment [35, 36]. This functional form has also been found by [34].

9.5 Molecular dynamics

In the preceding sections of this chapter, we have examined the electron–electron interaction entirely in momentum and frequency space. Just as in chapter 2, where we discussed band structure in both real space and momentum space, it is possible to treat the electron–electron interaction in real space and time. In this latter approach, we do not have to make approximations as have appeared in each section of this chapter, but we may make other adaptations. Of course, other problems arise which have a somewhat different nature and cause these adaptations. We will discuss these as they come along. The real space approach is a variant of molecular dynamics, in which the forces between the particles are updated each time step. That is, the force on each carrier, due to the other carriers and the charge impurities is computed anew at each time step in order to maintain the temporal evolution of these forces. Then this force is applied to the carrier during the next time step. It is important to note that this is not a perturbative approach.

Consider an electron distribution (an n-type semiconductor, and we ignore the charged donors for the moment) in which normal transport and scattering is treated through an ensemble Monte Carlo process, as described in chapter 7. Now, the inter-electronic Coulomb interaction is retained as a real-space potential. We will consider later the situation in which the local potential is determined self-consistently. In this approach, the local force on each electron due to the electric field and the field due to the other electronics is computed at each time step of the Monte Carlo simulation [37, 38]. This approach has the advantage that no approximation to the range of the Coulomb force has to be made, and there is no need to separate the interaction into single particle and plasmon contributions. The down side is that only a finite number of particles can be included due to the large computational load of recomputing the inter-particle forces at each time step. The potential between the electrons is given by the standard form

$$V(r) = \frac{1}{2} \sum_{i,j} \frac{e^2}{4\pi\varepsilon(0)|r_i - r_j|},$$ (9.120)

and the force on a given particle is

$$F(r_i) = -\nabla V(r) = \sum_j \frac{e^2}{4\pi\varepsilon(0)|r_i - r_j|^2}.$$ (9.121)

In (9.120), the factor of ½ arises from a double counting of the potential when summing over all values of i and j. Now, one may note that the potential (energy) is in fact a solution to the Lienard–Wiechert potential for the static wave equation for the scalar potential—the normal solution to the Poisson equation. This is an important point to which we will return in a later section.

9.5.1 Homogeneous semiconductors

First, we note that when charged impurities are included, there are two terms in (9.120) and (9.121). The first term represents the force on the particle from other free particles, and the second is the force on the particle from the charged impurities. In a homogeneous semiconductor, we can ignore the second term if we treat the impurity scattering by the normal momentum space interaction. Or, we can also include the impurities in the molecular dynamics simulation [39]. In general, however, we can use only a limited number of impurities due to the computational load molecular dynamics imposes on the simulation (we will see how this load is eased in the successive sections). Using a finite number of particles introduces a three-dimensional box, whose size is determined by the number of particles and the doping level considered in the simulation. The volume of this box is given as $\Omega = N/n$, where n is the assumed doping level and N is the number of particles in the simulation. The appearance of this induced box can affect the simulation by generating artefacts, and this must be avoided as much as possible. The artefacts arise because the box introduces an artificial periodicity into the simulation.

The Coulomb force is considered only within the shortest interconnecting length between any two particles, and this means that we need two boxes in the simulation. This is illustrated in figure 9.7. First is the box introduced by the size of the volume given in the previous paragraph (this box is the dark black box in figure 9.7). The second box has to be centered on the particle at which the force is being computed, so that artificial non-centrosymmetric forces are not generated in the force calculation (this is the green box in figure 9.7). Thus, when the inter-particle force is computed at each time step, each of the other simulated particles interact with the particle of interest only through the shortest possible vector, which may be shifted by a lattice vector of the artificial periodic structure. Replicas of the original box and the particles in this box are shown in figure 9.7 as well. Since the simulated volume may not correspond to each electron interacting through the shortest distance with the carrier of interest, replicas of the electrons that lie within the second neighbor boxes must be used in computing the forces through the well-known form for the electric field at the particle of interest, determined from (9.121), as

$$\mathbf{E}_i(\mathbf{r}_i) = \mathbf{E}_{\text{ext}}(\mathbf{r}_i) - \frac{e}{4\pi\varepsilon(0)} \sum_{j\neq i} \frac{1}{r_{ij}^2} \mathbf{a}_{ij}, \qquad (9.122)$$

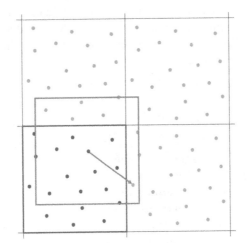

Figure 9.7. Various boxes for determining the inter-particle forces in molecular dynamics. The proper simulation box is outlined in dark blue with red particles. The centered box for computing the forces on a particle is the green box that yields a centro-symmetric potential. Replicas of the simulation box have the magenta particles.

where r_{ij} and $\boldsymbol{a}_{ij}$ are the length and unit vector direction of the vector between the two particles. The unit vector points *from* the particle of interest *toward* the interacting particle.

The two boxes are explained best by reference to figure 9.7. The replicated rectangular cells are the basic computational volume repeated over an assumed lattice of grid points. Four of these basic cells are shown in the figure and these cells form a 'superlattice' defined by the vectors (the volume is $\Omega = L_x L_y L_z$)

$$\boldsymbol{L} = L_x \boldsymbol{a}_x + L_y \boldsymbol{a}_y + L_z \boldsymbol{a}_z. \tag{9.123}$$

Each of these cells contains the basic number of particles N used in the Monte Carlo simulation. Only a small number is shown in the figure for clarity. The on-site particle, where the force is being computed is illustrated as well in the figure. But, in order to have the force be centro-symmetric, we need the second box, shown in green, which includes replicas of some of the particles in the second cell of the lattice (the replica particles are shown in magenta rather than red). The total potential arising from the sum over the nearest particle is an approximation to (9.120).

If we were to sum over all the particles *and* their replicas, we would be introducing superlattice effects. We can see this from

$$\sum_{\text{cells}} \frac{1}{r} = \frac{1}{r} + \sum_{L \neq 0} \frac{1}{|\boldsymbol{r} + \boldsymbol{L}|} \sim \frac{1}{r} + \sum_{L \neq 0} \frac{1}{L}$$
$$- r \sum_{L \neq 0} \frac{1}{L^2} + r^2 \sum_{L \neq 0} \frac{1}{L^3} + \ldots. \tag{9.124}$$

The denominator has been expanded, since in general we have $L > \sqrt{3}\, r$ when the centered cell is used for the direct field calculation. The summations over the lattice vectors are sometimes called Ewald sums [40], since they were developed for x-ray scattering in periodic structures. The method of evaluating these sums have been studied for quite some time in connection with molecular dynamics in other fields [41–43], and have well-developed formulas. This approach has been shown to be very effective in determining the inter-carrier scattering in femtosecond laser excitation of semiconductors [44, 45].

9.5.2 Incorporating Poisson's equation

In today's world, the most significant use of transport and simulation is in modeling the real semiconductor devices that are being produced by industry, whether they are the ubiquitous MOSFET, or other devices such as microwave HEMTs or solar cells for energy conversion. In these simulations, the internal self-consistent potential, that arises as charges (free carriers) move (or not when the impurities are included). This self-consistent potential is found by the solution of Poisson's equation, as discussed above below (9.121). Generally, this means that a grid is superimposed on the device, and the potential is determined by Poisson's equation on these grid points (the solution is often referred to as the mesh potential). This grid is much different than the grid we used in chapter 2 for the band structure. In chapter 2, the grid is part of a periodic structure, and the periodicity led to a number of effects. Here, the grid is *not* part of a periodic structure, but must replicate the actual device structure. We will discuss this only in terms of the grid used to develop finite-difference equations, although commercial software more often uses finite elements to solve Poisson's equation; the differences in these approaches is not important to the points we want to discuss in this section.

It is important to note that the potential that comes from solving Poisson's equation in the confined dimensions of the actual device leads to additional problems. First, one must incorporate the complicated set of boundary conditions that are imposed upon the solution. These boundary conditions may be actual potentials at some boundaries and forced derivatives of the potential at other boundaries. The potentials (and their derivatives) are usually different at different boundaries. In some cases, boundaries may be specified by boundary and/or surface charges, and these charges must be incorporated within the general Poisson equation scheme. Second, the internal charges, as well as any boundary/surface charge, seldom sits exactly upon the grid point at which Poisson's equation will be saved. Hence, one must use a method of projecting the charge onto the grid point in an equitable method that does not introduce artificial forces. There are many such approaches, and we will not discuss those here [46].

Then, one discovers that the existence of the mesh introduces a cutoff at short distance for the actual potential effects. That is, the grid spacing provides an upper limit to the momentum wave vectors that can be simulated, just as it does in band structure. The periodicity introduces an artificial band structure that is superposed upon the real one and leads to an upper limit to the energy at the zone edge. That is, the grid spacing introduces some phase shifts that limit the effective energy to a cosinusoidal behavior limiting the upper value of the momentum wave vector. Hence, within a size of the order of the grid box, the potential will not be accurate [47]. The importance of this limitation is that the mesh potential is no longer the complete inter-particle potential that would be found by detailed molecular dynamics. We can see this by a simple simulation. We place a fixed particle in a position that is not one of the grid points. Then, we let a test particle move to, and past, the fixed particle, solving Poisson's equation for each position of the test particle. In figure 9.8, we depict the results in terms of the magnitude of the electric field acting upon the test particle [47]. The true Coulomb field has a singularity at the position of the fixed particle because the distance in (9.122) goes to zero. However, the mesh field deviates from the true Coulomb field near the fixed particle, primarily because of the cutoff in allowed momentum states. The difference in the two potentials is a very short range potential, and it is this potential that must be used in the molecular dynamics calculation. In essence, we are splitting the actual molecular dynamics force into a short range part and a long range part, the latter of which is handled within the Poisson equation solution.

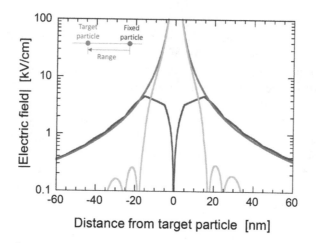

Figure 9.8. Forces acting upon a 'target particle' from a 'fixed particle'. The red curve is the exact Coulomb force, while the blue curve is the mesh force from a Poisson solution to the potential with a grid size of 10 nm. The difference is the green curve which is the short-range potential to be used in the molecular dynamics calculation in situations in which Poisson's equation is also solved. Adapted from [47].

9.5.3 Splitting the Coulomb potential

The basic approach followed in the coupling of the Monte Carlo and molecular dynamics approaches gives us the ability to follow all the particles in real space and time. The problem that arises, as discussed above, is that the electron–electron and electron–ion interactions are already included in the Poisson solution found for the device. This leads to double counting of the particle interactions. However, as noted above, the mesh potential is not accurate on length scales smaller than the grid spacing. So, one brute force method to overcome this double counting is to separate the total Coulomb force and the mesh force to yield a short-range molecular dynamics force.

The separation of the inter-particle Coulomb force from the Poisson forces has been discussed for quite some time. Such a split was used by Kohn for electronic structure calculations, where inter-electronic forces have to be added to the potential solutions from Poisson's equation for the atomic potentials in the density-functional approach [48]. The split was also used earlier in order to treat molecular dynamics problems in plasma physics where the short-range interaction also appears between ions and electrons [49]. And, this general splitting of the Coulomb force has been shown to be useful in general many-body problems [50]. The general approach is similar to dividing a delta function, where we can write

$$\delta(r) \rightarrow \delta(r) + f(r) - f(r), \tag{9.125}$$

from which the Coulomb potential goes into something like

$$\frac{1}{r} \rightarrow [1 - erf(r)]\frac{1}{r} + \frac{1}{r}erf(r). \tag{9.126}$$

The first term on the right-hand side is a short-range function that vanishes as the magnitude of r increases. On the other hand, the last term on the right-hand side is a long-range function that vanishes at short distances. This is, of course, the entire principle of the potential splitting in which the long-range function is found from Poisson's equation. Using a splitting of the form of (9.126) makes the Poisson's equation more difficult as the actual charge density has the more cumbersome multiplicative function.

When we actually compute the mesh force and the proper Coulomb force, using the short-range difference actually accomplishes the split (9.125) more effectively. Moreover, the new short-range force has several advantages. First, it is a relatively universal force curve. Secondly, because it is relative universal, it can be tabulated for use in the actual molecular dynamics program. Third, we do not have to sum over all electrons, only those who lie within a short range of the particle of interest, because the short-range force dies off so quickly. There is also the advantage that, at high densities where the grid size usually has to be small, a larger grid size can be used with the corrections coming from the short-force. This makes both the Poisson solution and the molecular dynamics more efficient in computer resources [51]. One can go further, by also incorporating into the molecular dynamics the degeneracy of the semiconductor through the rejection for filled states discussed in section 7.2.5 [52]. In extremely high densities, we can also correct the inter-particle force for the quantum mechanical exchange interaction energy [53].

9.5.4 Problems with ionized impurities

In the source and drain regions of a device, one has to be careful about the implementation of the molecular dynamics treatment of the short-range force. In these regions, the charged impurities are of the opposite sign to the particles being considered. This means that the force between the impurities and the particles is positive and can lead to an unphysical attractive force. This will inevitably lead to unphysical trapping of the particles by the charged impurities. This occurs, when the particles can no longer leave the vicinity of the impurity atom, and this leads to very few particles actually being injected (from the source) into the channel region. To avoid this problem, one needs to modify the short-range force in the proximity of the charged impurity, usually when the range is below 2 nm, which is approximately the Bohr radius of the charged impurity.

Effectively, one needs to truncate the short-force in the region where the range is below 2 nm, and there are various ways to do this. In figure 9.9(a), we illustrate three methods to accomplish this truncation [54]. These include: (a) a sharp cutoff of the potential at the Bohr radius, shown in green, (b) a linear decrease of the force, shown in black, and (c) setting the force to a constant for the small values of the range, shown in red. We have evaluated all of these approaches [54]. To select the most reasonable approach, we need a measure, and we have taken the average energy of the particles at very low values of the applied electric field. A free particle traveling through the semiconductor should have an average energy of $3k_BT/2 \sim 38.8$ meV. In figure 9.9(b), we plot the average energy per particle as a function of the

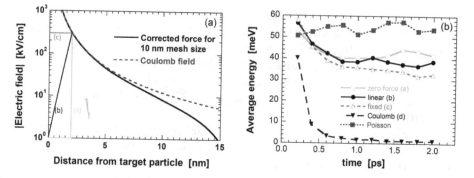

Figure 9.9. (a) Cutoff methods for the short-range molecular dynamics field in the region of a donor (for electron transport). (b) Average energy of the carriers with various types of cutoff. The individual curves are discussed in the text. Adapted from [54].

simulation time. First, it may be noted that merely using the mesh force from the Poisson solver, gives an average energy well above the physical value, which is often a problem with ensemble Monte Carlo techniques which do not include the interparticle forces. The three approaches to the truncated inter-particle forces all come reasonably close to the desired average energy per particle, but only the linear force (b) gives a truly small noise about the proper energy. On the other hand, if we do not correct for this attractive force, the average energy drops down to only about 2 meV, which clearly indicates a trapped particle. From studies such as this, it may be concluded that the best way to truncate the attractive inter-particle force is to use the linear reduction method for ranges below the Bohr radius.

While the inter-particle force was first used in homogeneous semiconductors [39], its use to treat the impurities as atomistic quantities with molecular dynamics in real devices seems to have been from IBM [55]. We have subsequently used the approach in HEMTs [56] as well as MOSFETs. We have used this simulation approach for a short channel (not so short for the typical state of the art devices in the 21st century) of 80 nm, with an oxide thickness of 3 nm. The source and drain regions are about 50 nm extent from the channel, and the channel doping is 3×10^{18} cm^{-3} [54]. In figure 9.10, the drift velocity of the electrons throughout the device is plotted. Without the inter-particle force, it may be seen that the velocity is too high throughout the device, but especially so at the channel-drain interface. With the short-range force added to the calculation, the drift velocity is not as high and has a smoother variation throughout the device. We should note also that the average energy drops rapidly as soon as the particles enter the drain region due to the higher doping in this region.

9.5.5 Accounting for finite size of the charges

In modern nano-scale semiconductor devices, physical dimensions in some cases are of the order of 10–20 nm or less. Quantum effects have begun to appear both in experiment and in theoretical simulations [57]. But, one must begin to ask just how large the electron (or hole) is quantum mechanically [58]. Certainly, in classical

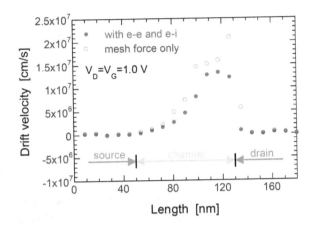

Figure 9.10. Average velocity of the carriers along the device length, with and without the inclusion of the electron–electron and electron–ion molecular dynamics forces. Adapted from [54].

physics, the electron is incredibly small. But, this is not the case in quantum mechanical systems, particularly in semiconductor devices. In the small devices mentioned, the key question is how to fit the electrons (or holes) into these small structures. Now, the size of the electron has been argued for some time, and becomes important in setting an effective potential [59]. This discussion *is not separate* from the preceeding sections, as we will see below.

Effective potentials have appeared in the literature for many decades, particularly in regard to the question of how quantum effects modify equilibrium thermodynamics. As early as 1926, Madelung developed a set of hydrodynamic equations from the Schrödinger equation and showed that an additional quantum potential was present in the formulation [60] (later made famous by Bohm [61]). Shortly after this, Kennard demonstrated that quantum particles would exactly follow classical potential forces plus those from any additional quantum potentials [62]. Wigner followed by developing the Wigner function and showing that the first order correction to the classical thermodynamic potential was a term that depended upon the spatial second derivative of this classical potential [63]. Somewhat later, Feynman and Hibbs showed that the normal partition function would have a quantum correction that involved the second derivative of the classical potential, but could be much more complicated in nature [64]. Finally, it was shown that the effective potential could be found from an effective size of the particle [58].

The connection between the size of the particle and the thermodynamic partition function corrections arises easily when it is recognized that the effective quantum size of the particle depends upon the statistical properties [59]. If the particle is part of a dense high density Fermi gas, then the effective size of the particle is approximately

$$\delta r = 2\Delta r \sim \frac{\lambda_F}{\pi}, \tag{9.127}$$

where λ_F is the Fermi wavelength and $\Delta r \sim 1/k_F$ is a mean radius of the wave function. On the other hand, if the carrier density is non-degenerate, then the classical size of the particle is approximately

$$\delta r = 2\Delta r \sim \sqrt{\frac{3}{8}}\lambda_D \sim 0.61\lambda_D, \qquad (9.128)$$

where

$$\lambda_D = \sqrt{\frac{2\pi\hbar^2}{m_c k_B T}} \qquad (9.129)$$

is the thermal de Broglie wavelength. Some typical numbers can be considered for a HEMT or a MOSFET with a sheet density of 10^{12} cm^{-2}. This may give a size of about 8 nm for the HEMT, and the thermal de Broglie wavelength for the Si device is about 4.3 nm. Hence, the effective size is not negligible.

When we write the total energy via the Hamiltonian in an general inhomogeneous system, the scalar potential enters through the term

$$H_V = \int dr\, V(r)n(r). \qquad (9.130)$$

Let us now make the assumption that each particle can be described as a Gaussian with a standard deviation (one half the particle size) α. Then, the density can be written as

$$n(r) = \sum_i n_i(r) = \sum_i \int dr' \exp\left(-\frac{|r - r'|^2}{\alpha^2}\right)\delta(r' - r_i), \qquad (9.131)$$

and the potential can then be written as

$$\begin{aligned}
H_V &= \int dr\, V(r)n(r) \\
&= \sum_i \int dr\, \delta(r - r_i) \int dr'\, V(r')\exp\left(-\frac{|r - r'|^2}{\alpha^2}\right).
\end{aligned} \qquad (9.132)$$

The last form has been achieved by interchanging the two integration variables, as both are essentially dummy variables. The first integral and the summation treats the particles as true point charges and the Gaussian smoothing has been transferred to the potential to give an *effective potential*. In the case of the MOSFET, the effective potential is a very smoothed version of the sharp high interface barrier between the oxide and the channel and the almost linear field in the Si itself due to the depletion charge. The result of the smoothing moves the classical charge away from the interface and raises the bottom of the conduction band just as the quantization in z-momentum would do. Both of these are observable effects in the characteristics of the MOSFET [65, 66].

As we discussed in the last section, there is a problem when treating ionized impurities that have the opposite sign of the particle, due to the strong attractive

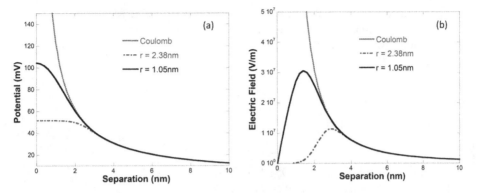

Figure 9.11. (a) The modified effective potential for the donor and (b) the electric field arising from that donor. Two different cutoff distances are used, and the smaller distance compares very well with the linear cutoff shown in figure 9.9(a), although it is certainly smoothed by the effective potential. This smoothing is described in the text and in [67, 68].

force and trapping that can occur. In that section, we discussed how to truncate the short-range force at the Bohr radius of the impurity atom. This truncation gives very good results, but it is ad hoc in nature. It has been found that the effective potential gives a better way of truncating the short-range force [67]. This new approach has two steps in the process. First, the charge of the impurity atom is considered to be uniformly distributed within a sphere whose radius is the Bohr radius (the approximately 2 nm of the earlier work). When the charge is uniformly distributed, the field inside the sphere increases almost linearly with distance. Then, the total potential of this charged sphered is smoothed with the effective potential according to (9.132). The results of this new quantum Coulomb potential is compared with the classical Coulomb potential in figure 9.11. Here, the smoothing is given by $\alpha_y = 0.52$ nm and $\alpha_x = 1.14$ nm. One can discern only a slight difference in the two quantum curves, but both peak at 50.7 nm, close to the ionization energy of the donor. In Si, the phosphorous donor ionization energy is 45 meV and that of arsenic is 54 meV. This approach is found to give excellent results for simulation of MOSFETs using Monte Carlo techniques with the molecular dynamics simulations [67, 68].

References

[1] Ramey S M and Ferry D K 2003 *IEEE Nanotechnol.* **2** 193
[2] Ziman J 1964 *Principles of the Theory of Solids* (Cambridge: Cambridge University Press)
[3] Hall G L 1962 *J. Chem. Phys. Sol.* **23** 1147
[4] Abramowitz M and Stegun I A 1964 *Handbook of Mathmatical Functions* (Washington D C: Government Printing Office)
[5] Chung W-Y and Ferry D K 1988 *Sol. State Electron.* **31** 1369
[6] Ando T, Stern F and Fowler A B 1982 *Rev. Mod. Phys.* **54** 437
[7] Hearn C J 1965 *Proc. Phys. Soc.* **86** 881
[8] Ferry D K 1977 *Phys. Lett.* **60A** 243
[9] Madelung O 1978 *Introduction to Solid State Theory* (Berlin: Springer)

[10] Takanaka N, Inoue M and Inuishi Y 1979 *J. Phys. Soc. Jpn.* **47** 861
[11] Ferry D K, Goodnick S M and Hess K 1999 *Physica* B **272** 538
[12] Goodnick S M and Lugli P 1988 *Phys. Rev.* B **37** 2578
[13] Varga B B 1965 *Phys. Rev.* **137** 1896
[14] Mooradian A and Wright G B 1966 *Phys. Rev. Lett.* **16** 999
[15] Kim M E, Das A and Senturia S D 1978 *Phys. Rev.* B **18** 6890
[16] Lugli P and Ferry D K 1985 *Appl. Phys. Lett.* **46** 594
[17] Lugli P and Ferry D K 1985 *IEEE Electron. Dev. Lett.* **6** 25
[18] Ferry D K, Akis R and Bird J P 2005 *J. Phys. Cond. Matt.* **17** S1017
[19] Baraff G A 1962 *Phys. Rev.* **128** 2507
[20] Yamashita J and Watanabe M 1954 *Prog. Theor. Phys.* **12** 443
[21] Reik H G and Risken H 1961 *Phys. Rev.* **124** 777
[22] Fischetti M V *et al* 2013 *J. Phys. Cond. Matter.* **25** 473202
[23] Gantmahker V F and Levinson Y B 1987 *Carrier Scattering in Metals and Semiconductors* (Amsterdam: North Holland) sec 6.4
[24] Bistritzer R and MacDonald A H 2009 *Phys. Rev. Lett.* **102** 206410
[25] Kubakaddi S 2009 *Phys. Rev.* B **79** 075417
[26] Somphonsane R *et al* 2013 *Nano Lett.* **13** 4305
[27] Viljas J K and Heikkilä T T 2010 *Phys. Rev.* B **81** 245404
[28] Ferry D K, Somphonsane R, Ramamoorthy H and Bird J P 2016 *J. Comp. Electron.* **15** 144
[29] Ferry D K, Somphonsane R, Ramamoorthy H and Bird J P 2015 *Appl. Phys. Lett.* **107** 262103
[30] Conwell E M 1967 *High Field Transport in Semiconductors* (New York: Academic)
[31] Ferry D K 1991 *Semiconductors* (New York: Macmillan) sec 10.3.1
[32] Price P J 1982 *J. Appl. Phys.* **53** 6863
[33] Altshuler B L, Aronov A G and Khmelnitskii D E 1982 *J. Phys.* C **15** 7367
[34] Golobev D S and Zaikin A D 1998 *Phys. Rev. Lett.* **37** 2578
[35] Ikoma T, Odagiri T and Hirakawa K 1992 *Quantum Effects Physics, Electronics, and Applications* ed K Ismail, T Ikoma and H I Smith (Bristol: IOP Conference Series), 127 346
[36] Mohanty P, Jariwala E M Q and Webb R A 1997 *Phys. Rev. Lett.* **78** 3366
[37] Jacoboni C 1976 *Proceedings of the 13th International Conference on the Physics of Semiconductors* (Rome: Marves) p 1195
[38] Lugli P and Ferry D K 1986 *Phys. Rev. Lett.* **56** 1295
[39] Joshi R P and Ferry D K 1991 *Phys. Rev.* B **43** 9734
[40] Ewald P P 1921 *Ann. Phys.* **64** 253
[41] Brush S C, Salikin H L and Teller F 1964 *J. Chem. Phys.* **45** 2101
[42] Potter D 1973 *Computational Physics* (London: Wiley)
[43] Adams D J and Dubey G S 1987 *J. Comp. Phys.* **72** 156
[44] Kann M J, Kriman A M and Ferry D K 1989 *Sol.-State Electron.* **32** 1831
[45] Ferry D K, Kann M J, Kriman A M and Joshi R P 1991 *Comp. Phys. Commun.* **67** 119
[46] Hockney R W and Eastwood J W 1981 *Computer Simulation Using Particles* (Maidenhead: McGraw-Hill)
[47] Gross W J, Vasileska D and Ferry D K 1999 *IEEE Electron. Dev. Lett.* **20** 463
[48] Kohn W, Meir Y and Makarov D E 1998 *Phys. Rev. Lett.* **80** 4153
[49] Kelbg G 1964 *Ann. Phys.* **468** 385
[50] Morawetz K 2002 *Phys. Rev.* E **66** 022103

[51] Gross W J, Vasileska D and Ferry D K 2000 *VLSI Des.* **10** 437
[52] Lugli P and Ferry D K 1986 *IEEE Trans. Electron. Dev.* **32** 2431
[53] Kriman A M, Kann M J, Ferry D K and Joshi R 1990 *Phys. Rev. Lett.* **65** 1619
[54] Gross W J, Vasileska D and Ferry D K 2000 *IEEE Trans. Electron. Dev.* **47** 2000
[55] Wong H S and Taur Y 1993 *Proc. IEEE Intern. Electron. Dev. Mtg.* (New York: IEEE Press) 29.2.1
[56] Zhou J-R and Ferry D K 1995 *IEEE Comp. Sci. Engr.* **2** 30
[57] Ferry D K and Grubin H L 1996 *Solid State Physics* (New York: Academic), 49 283
[58] Ferry D K 2000 *Superlatt. Microstruc.* **28** 419
[59] Ferry D K and Nedjalkov M 2018 *The Wigner Function in Science and Technology* (Bristol: Institute of Physics Publishing) ch 4
[60] Madelung E 1926 *Z. Phys.* **31** 876
[61] Bohm D 1952 *Phys. Rev.* **85** 166
[62] Kennard E H 1928 *Phys. Rev.* **31** 876
[63] Wigner E 1932 *Phys. Rev.* **40** 749
[64] Feynman R P and Hibbs A R 1965 *Quantum Mechanics and Path Integrals* (New York: McGraw-Hill) 10.3
[65] Vasileska D and Ferry D K 1999 *Nanotechnology* **10** 192
[66] Vasileska D *et al* 2001 *Microelectron. Eng.* **63** 233
[67] Ramey S M and Ferry D K 2003 *IEEE Trans. Nanotechnol.* **2** 193
[68] Ramey S M and Ferry D K 2004 *Semicond. Sci. Technol.* **19** S238

CPSIA information can be obtained
at www.ICGtesting.com
Printed in the USA
LVHW062106130922
728269LV00006B/280